Periodic table

s-block

1A	2A
1 H	
3 Li	4 Be
11 Na	12 Mg
19 K	20 Ca
37 Rb	38 Sr
55 Cs	56 Ba
87 Fr	88 Ra

d-block — transition elements

3A	4A	5A	6A	7A		(8)		1B	2B
21 Sc	22 Ti	23 V	24 Cr	25 Mn	26 Fe	27 Co	28 Ni	29 Cu	30 Zn
39 Y	40 Zr	41 Nb	42 Mo	43 Tc	44 Ru	45 Rh	46 Pd	47 Ag	48 Cd
57* La	72 Hf	73 Ta	74 W	75 Re	76 Os	77 Ir	78 Pt	79 Au	80 Hg
89† Ac									

p-block

3B	4B	5B	6B	7B	(0)
					2 He
5 B	6 C	7 N	8 O	9 F	10 Ne
13 Al	14 Si	15 P	16 S	17 Cl	18 Ar
31 Ga	32 Ge	33 As	34 Se	35 Br	36 Kr
49 In	50 Sn	51 Sb	52 Te	53 I	54 Xe
81 Tl	82 Pb	83 Bi	84 Po	85 At	86 Rn

f-block

* the lanthanoids

58 Ce	59 Pr	60 Nd	61 Pm	62 Sm	63 Eu	64 Gd	65 Tb	66 Dy	67 Ho	68 Er	69 Tm	70 Yb	71 Lu

† the actinoids

90 Th	91 Pa	92 U	93 Np	94 Pu	95 Am	96 Cm	97 Bk	98 Cf	99 Es	100 Fm	101 Md	102 No	103 Lr

Introduction to
Inorganic Chemistry

By the same author:
Introduction to Physical Chemistry
A New Introduction to Organic Chemistry

Introduction to Inorganic Chemistry

Second Edition

G I Brown BA, BSc

Formerly Assistant Master, Eton College

Longman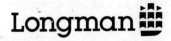

LONGMAN GROUP UK LIMITED
*Longman House, Burnt Mill, Harlow, Essex CM20 2JE, England
and Associated Companies throughout the world*

First published 1974
Second edition 1985
Sixth impression 1990

Set in Monophoto Times New Roman, Series 569-B

*Produced by Longman Group (FE) Ltd
Printed in Hong Kong*

ISBN 0-582-35459-5

Acknowledgements

We are grateful to the following Examining Bodies for permission to reproduce questions from past examination papers:

The Associated Examining Board (AEB); Joint Matriculation Board (JMB); Managing Committee for Entrance Scholarships of the Cambridge Colleges (Group III) (CS); Oxford and Cambridge Schools Examination Board (OC); Oxford Colleges Admissions Office (OS); Southern Universities Joint Board for School Examinations (SUJB); University of Cambridge Local Examinations Syndicate (C); University of Oxford Delegacy of Local Examinations (O); University of London University Entrance and School Examinations Council (L); Welsh Joint Education Committee (W).

Preface

The second edition of this book has been somewhat simplified, as compared with the first edition, in order to reflect changes in syllabus contents and examination requirements, but it is still intended to provide a thorough introduction to modern inorganic chemistry and to arouse a student's interest in it. All the basic requirements of the A level syllabuses are met, but some important topics are treated more extensively in order to make them as complete and relevant as possible at this level.

The first chapters deal with the general chemical principles that are applied, in later chapters, in considering the chemistry of the main elements. The emphasis has been laid on general aspects of periodicity; on group comparisons; on size, shape and structure; on energy changes; and on the uses of chemicals.

The presentation is in numbered sections so that topics may be omitted or rearranged to suit individual tastes. In particular, the more difficult ideas have been concentrated in Chapter 6 and, to a lesser extent, in Chapters 8 and 12. There is a wide range of questions; those marked with an asterisk requiring the use of the data given in the appendices, or of a separate data book. Summarising charts are given for the most important elements.

SI units have been used, and the nomenclature is based on the recommendations of the ASE report, *Chemical Nomenclature, Symbols and Terminology* (1979).

Contents

Charts

1
Electrons in atoms

1 Introduction

Inorganic chemistry is the study of substances other than carbon compounds, though a few of the latter are conventionally included in inorganic chemistry. The field is absolutely vast for there are about 100 elements concerned and it is possible to make endless numbers of compounds from them.

Early studies in inorganic chemistry were directed towards solving analytical and mineralogical problems. There is, too, a mass of relatively factual information concerning the preparation, manufacture, properties and uses of inorganic chemicals.

More recently, however, inorganic chemistry has developed into a comparative study, the properties of one element or compound being compared with those of another. Such comparisons are based on the positions of the elements concerned in the periodic table; on the nature of the bonding between atoms; on the size, structure and shape of individual molecules or crystals; and on energy changes.

2 The periodic table

The original form of the periodic table was put forward by Mendeleev, in 1869, when he arranged the known elements in the order of their relative atomic masses. Only about 60 elements had been discovered at that time (see front end-paper), so Mendeleev's periodic table was incomplete. It was clear, however, that the elements fell into vertical groups and horizontal rows (known as periods or series), and that they exhibited periodicity of both physical and chemical properties.

A modern form of the periodic table is given on the front end-paper. The elements in any one vertical group have similar chemical and physical properties, the similarities being greater in some groups than in others. Those groups with marked similarity are often given trivial names, e.g. the noble gases, the halogens (Group 7B), the alkali metals (Group 1A) and the alkaline earths (Group 2A). The A and B nomenclature arises because Mendeleev's original, numbered groups were each divided into two subgroups, labelled A and B.

The lanthanoids (formerly known as the rare earths) and the actinoids each occupy one single position in Group 3A. Mendeleev originally called the nine elements, Fe, Co, Ni, Ru, Rh, Pd, Os, Ir and Pt the transition elements and placed them all together in what he called Group 8, but the term transition element is now used in a wider sense (p. 341).

Once in order, the elements are numbered (from 1 to 103) by what is called their *atomic number*. This is equal to the number of protons or the number of electrons in an atom of the element concerned.

3 Electrons in atoms

The detailed arrangement of electrons within an atom is of fundamental importance because it is the interaction between the outer electrons of two or more atoms that leads to their possible chemical combination.

As a first approximation, the electrons can be regarded as very small, negatively charged, particles. The electrons in an atom occupy *orbitals*, each representing a certain energy level, and electrons move from lower to higher energy levels (they are said to be 'promoted' or 'excited') when atoms absorb the requisite amounts of energy. Alternatively, electrons move from higher to lower energy levels when atoms emit energy; when this happens the various quantities of energy emitted show up as lines in the line spectrum of the element concerned. Study of such spectra provides information about the various orbitals available for occupation by electrons. These orbitals are labelled using quantum numbers as follows:

a Principal quantum number   The innermost orbital, i.e. the one nearest to the nucleus, has a principal quantum number of 1, and this is the orbital of lowest energy. Other orbitals, with principal quantum numbers of 2, 3, 4 . . . , are further from the nucleus and represent higher energy levels.

The number of electrons in an atom that can occupy orbitals with the same principal quantum number is limited and is given by $2n^2$, where n is the principal quantum number concerned. Thus

principal quantum number (n)	1	2	3	4	5
maximum number of electrons	2	8	18	32	50

A group of orbitals with the same principal quantum number is referred to as a *shell*.

b Subsidiary quantum number   For each value of the principal quantum number greater than 1, there are several closely associated orbitals. So a principal quantum number of 2, 3, 4 . . . represents a group or shell of orbitals, the orbitals within each group being labelled by a subsidiary quantum number, l, which may have values of 0, 1, 2 or 3. This quantum number is, however, usually denoted by using an s, p, d or f symbol instead of 0, 1, 2 or 3.

These orbitals, e.g. 1s, 2s, 2p, 3d, 4f, represent different energy levels within an atom which can be occupied by electrons. The electrons are referred to as 1s electrons, 2p electrons, and so on.

c Magnetic quantum number   p orbitals are subdivided into three, but in a free atom the three p orbitals all represent the same energy level. They are said to be *degenerate* orbitals, and because they have equal energies it is not always necessary to distinguish between them. They differ so far as their directional nature is concerned and are envisaged as being in different planes (Fig. 1). They are labelled p_x, p_y and p_z to distinguish them.

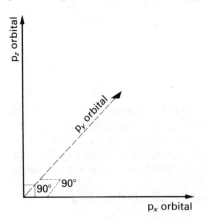

Fig. 1 *The directional arrangement of p orbitals.*

When an external magnetic field is applied to an atom, the directional nature of the p_x, p_y and p_z orbitals is fixed in relation to the direction of the magnetic field, and this causes slight differences between the energies of the three orbitals. It is because of this magnetic effect that the three p orbitals are labelled by allotting a magnetic quantum number, m, which can have values of -1, 0 and $+1$.

d orbitals are subdivided into five with m having values of -2, -1, 0, $+1$ and $+2$. f orbitals are subdivided into seven, with $m = -3, -2, -1, 0, +1, +2$ and $+3$, but these subdivisions are not significant in simple chemical problems.

d Spin quantum number Each orbital having the same principal, subsidiary and magnetic quantum numbers can hold only two electrons. These are envisaged as having different spins and are denoted by a spin quantum number, s, which can have a value of $+\frac{1}{2}$ or $-\frac{1}{2}$.

e Summary Four quantum numbers are required to characterise completely any particular electron, and no two electrons in an atom can have all their four quantum numbers the same (this is the *Pauli principle*). The quantum numbers that an electron may have can, therefore, be indicated as follows.

n	1	2			3								
l	0(s)	0(s)	1(p)			0(s)	1(p)			2(d)				
m	0	0	-1	0	$+1$	0	-1	0	$+1$	-2	-1	0	$+1$	$+2$
s	$\pm\frac{1}{2}$	$\pm\frac{1}{2}$	$\pm\frac{1}{2}$	$\pm\frac{1}{2}$	$\pm\frac{1}{2}$	$\pm\frac{1}{2}$	$\pm\frac{1}{2}$	$\pm\frac{1}{2}$	$\pm\frac{1}{2}$	$\pm\frac{1}{2}$	$\pm\frac{1}{2}$	$\pm\frac{1}{2}$	$\pm\frac{1}{2}$	$\pm\frac{1}{2}$
number of electrons	2	2	6			2	6			10				
	2	8				18								

When n, l and m are all alike, only two electrons can occupy the particular orbital, and these two must differ in spin. Any s orbital can hold two electrons; p orbitals can hold six electrons in three pairs; d orbitals can hold ten electrons in five pairs; f orbitals can hold 14 electrons in seven pairs.

4 Energy levels of orbitals

The energy levels of the orbitals in an atom of an element can be discovered from a study of the spectrum of the element. The arrangement of electrons in the atom can then be built up because, in an atom in its normal or ground state, the arrangement is that which minimises the energy. In other words, the electrons occupy the orbitals in energy order, the lowest energy ones being filled first.

The actual energy order is conveniently summarised in Fig. 2.

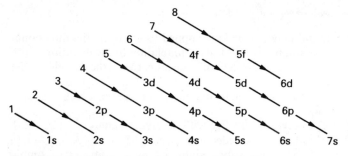

Fig. 2 *The order of filling of orbitals in an atom (simplified).*

The one electron in a hydrogen atom will occupy the most stable orbital, i.e. the 1s, and the second electron in the helium atom will occupy the same orbital but will have a different spin. The 1s orbital is now full. A third electron will occupy the next most stable orbital, i.e. the 2s. When both the 1s and the 2s orbitals are full, as in beryllium, the next electron goes into the 2p orbital, and so on. The best 'seats' are occupied first.

The electronic arrangements of all the elements are shown on the back end-papers. These arrangements are derived from spectroscopic data; they do not differentiate between the three p, five d and seven f levels, and, as will be seen, there are a few minor irregularities, e.g. Cr and Cu.

A common way of writing the arrangement of electrons in a helium atom is $1s^2$; the arrangement in argon would be written as $1s^2$, $2s^2$, $2p^6$, $3s^2$, $3p^6$.

5 The rule of maximum multiplicity

This empirical rule, suggested by Hund, states that the distribution of electrons in a free atom between the three p, five d and seven f orbitals is

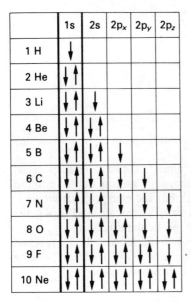

	1s	2s	2p$_x$	2p$_y$	2p$_z$
1 H	↓				
2 He	↓↑				
3 Li	↓↑	↓			
4 Be	↓↑	↓↑			
5 B	↓↑	↓↑	↓		
6 C	↓↑	↓↑	↓	↓	
7 N	↓↑	↓↑	↓	↓	↓
8 O	↓↑	↓↑	↓↑	↓	↓
9 F	↓↑	↓↑	↓↑	↓↑	↓
10 Ne	↓↑	↓↑	↓↑	↓↑	↓↑

Fig. 3 *The distribution of electrons in the elements from He to Ne.*

such that as many orbitals as possible are occupied by single electrons before any pairing of electrons with opposed spins takes place. Thus, if three electrons are to occupy the three p orbitals in any one shell, one will go into each of the three available orbitals. This can be interpreted as meaning that electrons repel each other and keep as far away from each other as is possible.

These more detailed arrangements, as shown in Fig. 3, are important, but are not always necessary. There are, in general, three ways of expressing an electronic arrangement within an atom. Fluorine, for example, has nine electrons. The arrangement of the electrons can be written as 2.7, showing, simply, the numbers of electrons with the different principal quantum numbers. In more detail, the arrangement can be written as $1s^2$, $2s^2$, $2p^5$, and the arrangement given in Fig. 3, showing the spin and the subdivision between the p orbitals, is still more detailed. The method most appropriate to the problem being considered must be chosen.

For d-block elements, up to ten electrons are distributed amongst five d orbitals. The rule of maximum multiplicity applies in the same way as for p orbitals (p. 341).

6 The shapes of atomic orbitals

The electron has, so far, been regarded as a tiny, negatively charged, particle but the fact that a beam of electrons can be diffracted shows that moving electrons have wave-like properties. It is not easy to obtain any

Fig. 4 *The spherical shape of an s atomic orbital.*

pictorial idea of these wave-like properties, but they can be treated mathematically.

On this basis, a 1s electron, or a 1s orbital, is depicted as a spherical charge cloud, a distribution of electrical charge of varying density around the nucleus. A 2s orbital is also spherical, but the charge distribution is different. s orbitals are said to be spherically symmetrical (Fig. 4); the electrical charge is not concentrated in any particular direction.

p orbitals, on the other hand, are dumb-bell shaped and have a marked directional character depending on whether a p_x, p_y or p_z orbital is being considered. The shapes of the p orbitals are shown in Fig. 5.

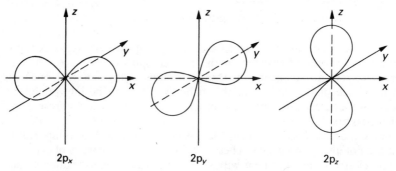

Fig. 5 *The dumb-bell shapes of $2p_x$, $2p_y$ and $2p_z$ atomic orbitals.*

The five d orbitals and seven f orbitals are also directional in space.

Questions on Chapter 1

1 The arrangement of electrons in Cr, Cu, Mo, Pd, Ag, Pt and Au is slightly anomalous. Compare the actual arrangement with that which you might have expected. Can you draw any conclusions?
2 Write down the detailed arrangements of the electrons (including spin) for all the elements from calcium to zinc inclusive. What conclusions can you draw?
3 'The position of an element in a group in the periodic table and its chemical properties are related to the number of outer electrons which the atom of the element contains.' Discuss, with illustrative examples, this

statement with respect to the following elements: (a) hydrogen, (b) oxygen, (c) sodium, (d) neon, (e) chlorine. [Atomic numbers: H = 1; O = 8; Na = 11; Ne = 10; Cl = 17.] (L)

4 The elements silicon, phosphorus, sulphur and chlorine occur in the second short period of the periodic table. With reference to the properties of the elements and their simple compounds, discuss the gradation in properties which they exhibit, and show how far these properties are consistent with the positions of the elements in the periodic table. (L)

5 What part did a study of X-ray spectra play in the elucidation of the periodic table arrangement?

6 'The original arrangement of elements in the periodic table was based on relative atomic mass values, but it is now known that atomic number values are of much greater significance.' Comment on this statement.

7 Do you think it likely, or not, that the number of elements is limited to 103? Give reasons.

8 Describe the important features in the electronic structure of atoms, taking examples up to atomic number 36 (Kr). Describe briefly the supporting experimental evidence. (C)

9 What evidence is there for the existence of electronic energy levels in atoms? Discuss concisely the importance of the existence of electronic energy levels in chemistry. The lanthanoids (rare earths) are the elements of atomic numbers 58 to 71 in the periodic system; outline how it can be shown experimentally that there are fourteen such elements without actually isolating them. (CS)

2
Ions and ionic bonding

1 Introduction

In the formation of an ionic bond, atoms of an electropositive element lose electrons to form positive ions and the electrons are transferred to the atoms of an electronegative element which form negative ions. Thus

$$Na\bullet + {}^{\times}_{\times}\overset{\times\ \times}{Cl}{}^{\times}_{\times} \longrightarrow [Na]^{+}\left[\ {}^{\bullet}_{\times}\overset{\times\ \times}{Cl}{}^{\times}_{\times}\ \right]^{-}$$

$$[Ca]^{2+}\left[\ {}^{\bullet}_{\times}\overset{\times\ \times}{Br}{}^{\times}_{\times}\ \right]_{2}^{-} \qquad [K]_{2}^{+}\left[\ {}^{\bullet}_{\bullet}\overset{\times\ \times}{S}{}^{\times}_{\times}\ \right]^{2-}$$

The formulae for ionic compounds, as above, show only the relative numbers of each ion present in the compound. There are no individual molecules, for the ions are held together in an ionic crystal (p. 14) by electrostatic forces.

It requires a lot of thermal energy to break down the forces holding the crystal together, so ionic solids have high melting and boiling temperatures (p. 26). The strength of the forces also means that ionic crystals are hard and brittle.

When the forces holding the crystal together are broken down, either by melting or by solution in an ionising solvent such as water, the 'free' ions can move under the influence of an applied potential difference, so ionic compounds are invariably electrolytes.

Ionic compounds are also commonly soluble in water and other ionising solvents, and insoluble in benzene and other organic solvents.

The formation of a free gaseous ion from an element involves atomisation followed by ionisation. This is followed by hydration when a hydrated ion is concerned, i.e.

$$X(s)\xrightarrow{\text{atomisation}}X(g)\xrightarrow{\text{ionisation}}X^{n\pm}(g)\xrightarrow{\text{hydration}}X^{n\pm}(aq)$$

| standard | free | free | hydrated |
| state | atom | ion | ion |

It is the energy, or enthalpy, changes in the various processes that mainly determine how readily an ion will be formed.

2 Enthalpy of atomisation

The first step in the formation of an ion involves the atomisation of the element, i.e. the breaking down of a molecule or a crystal to give free atoms.

This involves an input of energy which is, generally, expressed as the enthalpy of atomisation. *It is the enthalpy change when 1 mol of gaseous atoms is formed from an element in its standard state.* For example

$$Na(s) \longrightarrow Na(g) \qquad \Delta H_{a,m}^{\ominus}(298\,K) = +108\,kJ\,mol^{-1}$$

$$\tfrac{1}{2}Cl_2(g) \longrightarrow Cl(g) \qquad \Delta H_{a,m}^{\ominus}(298\,K) = +121\,kJ\,mol^{-1}$$

For diatomic molecules, such as Cl_2, the enthalpy of atomisation is half the bond enthalpy (p. 61).

The way in which the enthalpy of atomisation of the elements varies with atomic number is shown in Fig. 62 on p. 122 and values for individual elements are given in Appendix V.

3 Ionisation energy and ionisation enthalpy

Energy is required to remove an electron from a free, gaseous atom to form a cation. For a mole of atoms, the value is called the *first molar ionisation energy* or simply the first ionisation energy. The energy change is expressed as a ΔU_m value at $0\,K$, e.g.

$$Na(g) \longrightarrow Na^+(g) + e^- \qquad \Delta U_m(0\,K) = +496\,kJ\,mol^{-1}$$

Corresponding first ionisation *enthalpy* values, at 298 K, which are also quoted and used, are about $6\,kJ\,mol^{-1}$ higher,* e.g.

$$Na(g) \longrightarrow Na^+(g) + e^- \qquad \Delta H_m(298\,K) = +502\,kJ\,mol^{-1}$$

Second, third . . . *n*th ionisation energies refer to the loss of a second, third . . . *n*th electron. Thus the second ionisation energy of X is the first ionisation energy of X^+. It is the *sum* of the first *n* ionisation energies that gives the total energy change for the formation of the X^{n+} ion:

$$X \xrightarrow{-e^-} X^+ \xrightarrow{-e^-} X^{2+} \qquad X \xrightarrow{-ne^-} X^{n+}$$

The second ionisation energy is always higher than the first, and the third higher than the second. This is because each succeeding electron has to be withdrawn against the attractive force of a more strongly charged positive ion. It is, then, more difficult to form ions with higher positive charges. That is why simple cations are limited to a charge of $+4$, and such highly charged ions are only formed by large atoms such as tin and lead.

Ionisation energies are expressed in $kJ\,mol^{-1}$, but older figures in eV are

---

*The difference between ΔU and ΔH values is explained in books of physical chemistry. The numerical difference between the two values is relatively small so that the values, and the terms ionisation energy and ionisation enthalpy, are often used interchangeably.

sometimes quoted ($1\,\text{eV} = 96.487\,\text{kJ}\,\text{mol}^{-1}$) and may be called ionisation potentials.

The variation of first ionisation energy with atomic number is explained on p. 123.

4 Electron affinity

This relates to the gain of electrons by an atom in the formation of an anion. The first electron affinity of an element is concerned with the gain of one electron, i.e.

atom $(X) + e^- \longrightarrow X^-(g)$ $\Delta U_m(0\,\text{K})$ = first electron affinity*

$Cl(g) + e^- \longrightarrow Cl^-(g)$ $\Delta H_m(298\,\text{K}) = -354\,\text{kJ}\,\text{mol}^{-1}$

$O(g) + e^- \longrightarrow O^-(g)$ $\Delta H_m(298\,\text{K}) = -148\,\text{kJ}\,\text{mol}^{-1}$

The second electron affinity refers to the gain of a second electron, e.g.

$X^-(g) + e^- \longrightarrow X^{2-}(g)$ $\Delta U_m(0\,\text{K})$ = second electron affinity*

$O^-(g) + e^- \longrightarrow O^{2-}(g)$ $\Delta H_m(298\,\text{K}) = +838\,\text{kJ}\,\text{mol}^{-1}$

The second electron affinity is positive because a negatively charged ion 'opposes' the addition of a second electron. Some typical values of electron affinities, in $\text{kJ}\,\text{mol}^{-1}$, are given below.

H -72 O -142 O$^-$ $+844$ S -200 S$^-$ $+532$

F -333 Cl -348 Br -328 I -295

It is easier to form an A^- anion than an A^{2-} one, e.g.

$Cl(g) \longrightarrow Cl^-(g)$ $\Delta H_m = -354\,\text{kJ}\,\text{mol}^{-1}$

$O(g) \longrightarrow O^{2-}(g)$ $\Delta H_m = +690\,\text{kJ}\,\text{mol}^{-1}$

and, as a generalisation, it is the smallest atom in any periodic table group that most readily forms the anion. The electrons are closer to the nucleous

The simple anions are limited to H^-, F^-, Cl^-, Br^-, I^-, O^{2-}, S^{2-}, Te^{2-}, N^{3-}, P^{3-} and C^{4-}.

---

*As for ionisation energy (p. 9) electron affinities are commonly quoted as ΔU_m values at 0 K, and these are only approximately equal to the corresponding enthalpy changes, generally quoted as ΔH_m values at 298 K. To convert the ΔU_m value to the ΔH_m value it is necessary to add approximately $-6\,\text{kJ}\,\text{mol}^{-1}$. Nowadays, the ΔH_m value may be referred to as the *enthalpy of electron attachment*.

5 Lattice enthalpy (starting from gaseous ions.)

The ions in an ionic crystal are strongly bound together by electrostatic forces between the ions, and there is a considerable release of energy when such a crystal is formed from the free gaseous ions, e.g.

$$Na^+(g) + Cl^-(g) \longrightarrow NaCl(s) \qquad \Delta H_m(298\,K) = -788\,kJ\,mol^{-1}$$

The value quoted for the enthalpy change is known as the *lattice enthalpy.**
Care must be taken, however, for the term is sometimes used for the reverse process and this changes the sign, e.g.

$$NaCl(s) \longrightarrow Na^+(g) + Cl^-(g) \qquad \Delta H_m(298\,K) = +788\,kJ\,mol^{-1}$$

Some typical values are quoted below, in $kJ\,mol^{-1}$.

NaF -924	NaCl -788	$MgCl_2 -2525$	MgO -3890
NaCl -788	KCl -718	$CaCl_2 -2254$	CaO -3499
NaBr -748	RbCl -695	$SrCl_2 -2152$	SrO -3319
NaI -704	CsCl -675	$BaCl_2 -2054$	BaO -3157

Tin(IV) oxide has a value of $-11\,779\,kJ\,mol^{-1}$.

The values show, as would be expected, that the energy required to 'break up' a crystal increases with the charge on the ions, and as the distance between the ions gets smaller.

Lattice enthalpy values can be obtained from a Born–Haber cycle, or by considering the energy changes as the ions in a crystal come together (p. 13).

6 The Born–Haber cycle

A Born–Haber cycle (Fig. 6) summarises the relationships between the various enthalpy changes involved in forming an ionic crystal from its elements in their standard states. The overall enthalpy change in the formation of sodium chloride from its elements is given by the standard enthalpy of formation, i.e.

standard enthalpy

$$Na(s) + \tfrac{1}{2}Cl_2(g) \longrightarrow NaCl(s) \qquad \Delta H_m^{\ominus}(298\,K) = -411\,kJ\,mol^{-1}$$

The process can, however, be broken down into the five stages summarised as follows.

---

*The lattice enthalpy is only approximately equal to the lattice energy, though the difference between the two is often ignored, and the two terms are often used as though they were synonymous. To get the lattice enthalpy for an M^+X^- salt, $\Delta H_m(298\,K)$, it is necessary to add $-5\,kJ\,mol^{-1}$ to the lattice energy, $\Delta U_m(298\,K)$.

			$\Delta H_m(298\,\mathrm{K})$
i	atomisation of sodium	$Na(s) \longrightarrow Na(g)$	$+108\,\mathrm{kJ\,mol^{-1}}$
ii	formation of sodium ion	$Na(g) \longrightarrow Na^+(g)+e^-$	$+502\,\mathrm{kJ\,mol^{-1}}$
iii	atomisation of chlorine	$\frac{1}{2}Cl_2(g) \longrightarrow Cl(g)$	$+121\,\mathrm{kJ\,mol^{-1}}$
iv	formation of chloride ion	$Cl(g)+e^- \longrightarrow Cl^-(g)$	$-354\,\mathrm{kJ\,mol^{-1}}$
v	formation of crystal	$Na^+(g)+Cl^-(g) \longrightarrow NaCl(s)$	$-788\,\mathrm{kJ\,mol^{-1}}$

In accordance with Hess's law, the algebraic sum of the five enthalpy changes must be equal to the standard enthalpy of formation of NaCl(s), i.e. $-411\,\mathrm{kJ\,mol^{-1}}$, and it is these figures that are summarised in the Born–Haber cycle in Fig. 6.

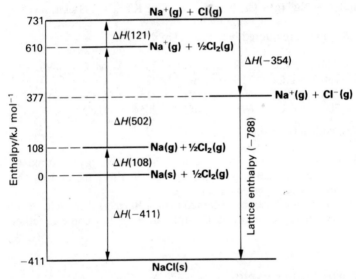

Fig. 6 *The Born–Haber cycle used to find the lattice enthalpy of sodium chloride.*

Lattice enthalpy values are, generally, obtained from such cycles by using measured values of enthalpies of formation and *i, ii, iii* and *iv*. Alternatively, any of those quantities can be obtained by using a calculated value for a lattice enthalpy. This can be very valuable for finding electron affinity values, which are not easy to measure directly, particularly for anions which are polyatomic, e.g. CN^-, or multiply charged, e.g. O^{2-} and SO_4^{2-}.

7 The stoicheiometry of ionic compounds

It is the high negative value of the lattice enthalpy which makes the highest contribution to the negative enthalpy of formation of sodium chloride, and this is so for other ionic compounds. It is not just a matter of how easily ions may be formed; it is also a matter of how strongly they bind together in the ionic crystal.

Born–Haber cycles can be used to obtain estimates for the enthalpy of formation of a hypothetical compound. The lattice enthalpy for $NaCl_2$, for example, can be calculated or taken as approximately equal to that of the very similar $MgCl_2$. By fitting this value into a Born–Haber cycle together with the other necessary values, the enthalpy of formation of the hypothetical $NaCl_2$ can be estimated. The figures, summarised below at 298 K,

$$Na(s) \longrightarrow Na(g) \qquad \Delta H_m = +108 \, kJ \, mol^{-1}$$
$$Na(g) \longrightarrow Na^{2+}(g) + 2e^- \qquad \Delta H_m = +5069 \, kJ \, mol^{-1}$$
$$Cl_2(g) \longrightarrow 2Cl(g) \qquad \Delta H_m = +242 \, kJ \, mol^{-1}$$
$$2Cl(g) + 2e^- \longrightarrow 2Cl^-(g) \qquad \Delta H_m = -708 \, kJ \, mol^{-1}$$
$$Na^{2+}(g) + 2Cl^-(g) \longrightarrow NaCl_2(s) \qquad \Delta H_m = -2525 \, kJ \, mol^{-1}$$
$$Na(s) + Cl_2(g) \longrightarrow NaCl_2(s) \qquad \Delta H_m = +2186 \, kJ \, mol^{-1}$$

give a value of $+2186 \, kJ \, mol^{-1}$ as an estimate of the enthalpy of formation of $NaCl_2$. The value estimated for $NaCl_3$ is $+5919 \, kJ \, mol^{-1}$, so NaCl, which actually exists, is much stabler than either $NaCl_2$ or $NaCl_3$.

Similarly, the enthalpies of formation for the hypothetical MgCl and $MgCl_3$ can be shown to be -130 and $+3909 \, kJ \, mol^{-1}$, respectively. $MgCl_2$, with a value of $-642 \, kJ \, mol^{-1}$, is clearly the most stable.

8 Calculated lattice energy values

Approximate lattice energy values for crystals can be calculated by regarding the ions concerned as charged spheres and by considering the attractive and repulsive forces between them. The Kapustinskii equation, which is commonly used, relates the lattice energy to n (the number of ions, i.e. 2 for NaCl, 3 for CaF_2), z^+ and z^- (the charge numbers of the ions, i.e. 1 for Na^+, 2 for Ca^{2+}) and r^+ and r^- (the ionic radii, in nm, for octahedral coordination):

$$\text{lattice energy} = \frac{-121.4nz^+z^-}{r^+ + r^-}\left(1 - \frac{0.0345}{r^+ + r^-}\right)$$

The calculations can be done in a number of ways depending on the degree of refinement required. The agreement between the calculated and the measured values is within 2 per cent for the alkali metal halides, but for other substances, e.g. AgI and CuI, there is a discrepancy of the order of 12 per cent. Such discrepancies may be caused by polarisation, i.e. the effect of the electrical field of one ion on that of its neighbours (p. 16). This introduces some degree of covalent bonding so that the resulting crystal is not fully ionic. Indeed, the best way of defining an ionic compound may well be to state that it is one in which the measured and calculated lattice energies are equal.

9 Ionic crystals

An ionic crystal is made up of ions held together by ionic bonds. In a crystal containing A^+ and B^- ions the number of B^- ions surrounding an A^+ ion is called the coordination number of A^+, and vice versa. In a binary compound, AB, the coordination numbers of A^+ and B^- must be equal so that the crystal as a whole will be electrically neutral. In AB_2 compounds, the coordination number of A^{2+} will be twice that of B^-.

a The radius ratio   The value of the coordination number will depend on the relative sizes of A^+ and B^-, i.e. on the radius ratio, r^+/r^-. Cations are generally smaller than anions (p. 132), and it is a matter of how many anions can pack round one cation. The anions will repel each other and take up positions as far apart as possible.

For a very small cation and a very large anion, i.e. for a small radius ratio, it may only be possible to pack two anions around the cation; the AX_2 grouping will be linear and A has a coordination number of 2. For a larger cation, three anions might pack around it with their centres at the corners of an equilateral triangle; the coordination number of A will be 3. For still larger cations the coordination number will rise to 4 (tetrahedral arrangement), 6 (octahedral arrangement) or 8 (cubical arrangement).

Simple geometrical calculations show that a radius ratio of less than 0.155 would lead to a linear arrangement with other limiting values as follows: triangular (0.155–0.225), tetrahedral (0.225–0.414), octahedral (0.414–0.732) and cubical (0.732–1). These theoretical figures generally, but not invariably, apply in practice.

b The sodium chloride structure   The radius ratio for Na^+ and Cl^- ions is 0.95/1.81 or 0.525, so an octahedral structure with coordination number 6 results (Fig. 7).

c The caesium chloride structure   The radius ratio for Cs^+ and Cl^- is 1.69/1.81 or 0.93, and the crystal structure is cubic with a coordination number of 8 (Fig. 8).

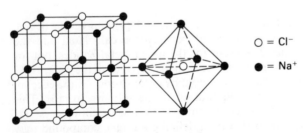

○ = Cl^-

● = Na^+

Fig. 7 *Crystal structure of sodium chloride, showing (right) the octahedral arrangement of six Na^+ ions around one Cl^- ion.*

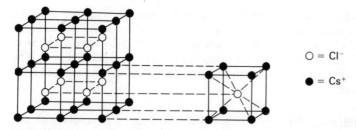

$O = Cl^-$

$\bullet = Cs^+$

Fig. 8 *Crystal structure of caesium chloride, showing (right) the cubic arrangement of eight Cs^+ ions around one Cl^- ion.*

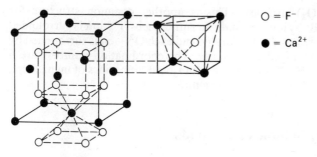

$O = F^-$

$\bullet = Ca^{2+}$

Fig. 9 *The fluorite structure, showing (right) the tetrahedral arrangement of four Ca^{2+} ions around one F^- ion, and (below) the cubic arrangement of eight F^- ions around one Ca^{2+} ion.*

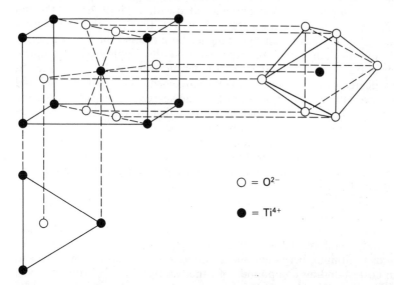

$O = O^{2-}$

$\bullet = Ti^{4+}$

Fig. 10 *The rutile structure, showing (right) the octahedral arrangement of six O^{2-} ions around one Ti^{4+} ion, and (below) the triangular arrangement of three Ti^{4+} ions around one O^{2-} ion.*

d The fluorite (CaF_2) structure This is a common structure amongst $A^{2+}B^-_2$ compounds. Each Ca^{2+} ion is surrounded by eight F^- ions at the corners of a cube, and each F^- ion is surrounded by four Ca^{2+} ions arranged tetrahedrally. The coordination numbers are 8 and 4 (Fig. 9). The difluorides of calcium, strontium, barium, cadmium and mercury all have this structure.

e The antifluorite structure This is like the fluorite structure with the anions and cations interchanged, i.e. $A^+_2B^{2-}$. It occurs in oxides and sulphides of lithium, sodium and potassium.

f The rutile (TiO_2) structure This is another common structure for $A^{2+}B^-_2$ or $A^{4+}B^{2-}_2$ ionic crystals. In rutile, each Ti^{4+} ion is surrounded by six O^{2-} ions arranged octahedrally, and each O^{2-} ion is surrounded by three Ti^{4+} ions arranged triangularly. The coordination numbers are 6 and 3 (Fig. 10). GeO_2, SnO_2, PbO_2, MnO_2 and difluorides of manganese, magnesium and zinc all have this structure.

10 Covalency in ionic compounds

A fully ionic crystal would consist entirely of ions, and there would be no sharing of electrons, i.e. no covalency. But such completely ionic crystals do not exist. In sodium chloride, for example, it might be expected that the Na^+ ion would have 10 electrons associated with it, whilst the Cl^- ion had 18. X-ray diffraction studies show, however, that the electron density around the Na^+ ion is somewhat higher (and around the Cl^- ion somewhat lower) than would be expected, and this implies an incomplete electron transfer. Some of the electron density is being shared, i.e. there is some degree of covalency. The good agreement between measured and calculated lattice energies suggests that the degree of covalency in sodium chloride can only be very small, but it is higher in compounds where the agreement between the two values is less good.

This is thought to arise because of *polarisation*, which is the electrical effect of one ion on its neighbours. A positive cation, for example, will attract the electron cloud of a neighbouring anion and the consequent build-up of charge between the two ions will represent some measure of covalent bonding. It is difficult to define polarisation very precisely or to quantify it, but the general idea is useful.

A small positive ion with a high charge will have the highest electrical field and, therefore, the highest polarising power. That is why the compounds of lithium, beryllium and aluminium are so much more covalent than corresponding compounds of other elements in the same periodic table groups.

A small negative ion will be least easily polarised (have a low *polarisability*) because its electrons are less diffusely spread and are closer to the

(handwritten annotations: "less polanti ⇒ mostionic." "the more the polarisity ⇒ the more covalent")

nucleus than those of a larger ion. That is why fluorides are more ionic than other halides, and why oxides are more ionic than sulphides.

11 Hydration of ions

Isolated ions, such as those that exist in the gaseous state, cannot exist in aqueous solutions because they become hydrated by the polar water molecules that are attracted to them. The number of water molecules that may be said to be attached to any ion in an aqueous solution is variable because there are sheaths of molecules around both anions and cations (Fig. 11). But in solid hydrates, definite numbers of water molecules are linked to ions and there are definite geometrical arrangements, as in the octahedral $[Fe(H_2O)_6]^{3+}$ ion.

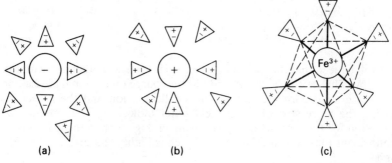

(a)   (b)   (c)

Fig. 11 *Hydration of ions.* (*a*) *Random distribution of water molecules around an anion in aqueous solution.* (*b*) *Random distribution about a cation.* (*c*) *Octahedral arrangement of six water molecules around an Fe^{3+} ion in solid hydrates.*

a Enthalpy of hydration This refers to the enthalpy change when 1 mol of a gaseous ion is hydrated, i.e.

$$M^{n\pm}(g) + aq \longrightarrow M^{n\pm}(aq) \qquad \Delta H_{hyd,m}^{\ominus}(298\,K) = \text{enthalpy of hydration}$$

Values may be quoted as *relative* values, on the basis that the value for $H^+(aq)$ is zero, e.g.

$$Na^+(g) + aq \longrightarrow Na^+(aq) \qquad \Delta H_{hyd,m}^{\ominus}(298\,K) = +685\,kJ\,mol^{-1}$$
$$H^+(g) + aq \longrightarrow H^+(aq) \qquad \Delta H_{hyd,m}^{\ominus}(298\,K) = 0$$
$$Cl^-(g) + aq \longrightarrow Cl^-(aq) \qquad \Delta H_{hyd,m}^{\ominus}(298\,K) = -1469\,kJ\,mol^{-1}$$

Alternatively, an *absolute* value for $H^+(aq)$ can be decided ($-1091\,kJ\,mol^{-1}$ is a common choice, but other figures may be used) and other values adjusted accordingly, e.g.

$$Na^+(g)+aq \longrightarrow Na^+(aq) \quad \Delta H_{hyd,m}^{\ominus}(298\,K) = -406\,kJ\,mol^{-1}$$
$$H^+(g)+aq \longrightarrow H^+(aq) \quad \Delta H_{hyd,m}^{\ominus}(298\,K) = -1091\,kJ\,mol^{-1}$$
$$Cl^-(g)+aq \longrightarrow Cl^-(aq) \quad \Delta H_{hyd,m}^{\ominus}(298\,K) = -378\,kJ\,mol^{-1}$$

The absolute values are obtained from the relative ones by subtracting $1091n\,kJ\,mol^{-1}$, where n is the charge number on the ion (including its positive or negative sign). Some absolute values, on this basis, are given below.

Absolute enthalpies of hydration of ions/$kJ\,mol^{-1}$

H^+ -1091			
Li^+ -520	Be^{2+} -2484	Al^{3+} -4680	F^- -524
Na^+ -406	Mg^{2+} -1926		Cl^- -378
K^+ -320	Ca^{2+} -1579		Br^- -348
Rb^+ -296	Sr^{2+} -1446		I^- -308
Cs^+ -264	Ba^{2+} -1309		

It will be seen that the hydration of an ion is always exothermic, with the enthalpy value getting more negative as the charge on the ion increases, or as the size of the ion decreases. The smaller an ion and the higher its charge, the more strongly it attracts water molecules. The enthalpy of hydration for the very small H^+ ion is much higher than those for other monovalent ions, and that for Al^{3+}, which is highly charged, is also very high.

To get the sum of the enthalpies of hydration of an anion and a cation, either the relative or the absolute values can be added together.

12 Solubility of ionic solids in water

When an ionic solid dissolves in water, the crystal lattice must be broken down and this requires a large input of energy. This input is, however, offset by a release of energy caused by hydration of the ions. Whether or not a solid will dissolve is, broadly, a matter of the balance between these two energy terms.

If the lattice enthalpy is much higher than the hydration enthalpy, the process of solution would be endothermic and, therefore, not likely to take place. If the hydration enthalpy outweighs the lattice enthalpy, solution is highly likely (Fig. 12).

These generalisations must, however, be treated with caution. The factors that favour high lattice enthalpies (small ionic sizes and high ionic charges) also favour high enthalpies of hydration. The enthalpy change on solution is, then, the difference between two large numbers, and its value may be quite small. Moreover, the entropy change in the process of

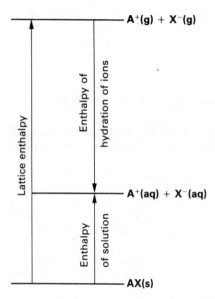

Fig. 12 *A high lattice enthalpy and a low enthalpy of hydration give a high positive value for the enthalpy of solution, and favour insolubility.*

solution (p. 70) may be more important than the enthalpy change, particularly if the latter is small. Conclusions drawn about solubility from enthalpy values can, therefore, be very misleading. It is the free energy values that must be considered (p. 72).

For sodium chloride, for example, $788 \, \text{kJ mol}^{-1}$ (the lattice enthalpy with the sign reversed) is required to form free gaseous ions from the solid crystal. But the hydration of the $Na^+(g)$ and $Cl^-(g)$ ions releases $784 \, \text{kJ mol}^{-1}$ ($406 \, \text{kJ mol}^{-1}$ from the $Na^+(g)$ ion and $378 \, \text{kJ mol}^{-1}$ from the $Cl^-(g)$ ion). The overall enthalpy of solution is, therefore, $4 \, \text{kJ mol}^{-1}$, i.e. the process is endothermic:

$$NaCl(s) \longrightarrow Na^+(aq) + Cl^-(aq) \qquad \Delta H_{s,m}^{\ominus}(298 \, \text{K}) = +4 \, \text{kJ mol}^{-1}$$

Sodium chloride is, nevertheless, soluble in water because the free energy change is negative ($-8.9 \, \text{kJ mol}^{-1}$).

Silver chloride, AgCl, has an enthalpy of solution of $+66 \, \text{kJ mol}^{-1}$ (and a positive free energy change) so that it is insoluble. For silver fluoride, which is soluble, both the enthalpy and the free energy of solution are negative. entropy = randomness

There are, then, some very odd (and not always explainable) solubilities amongst ionic substances which might have been expected to be similar to each other. For example, CsF is about 1400 times more soluble than LiF; CsOH is 25 times more soluble than LiOH; BaF_2 is about 15 times more soluble than MgF_2; $Ba(OH)_2$ is about 4000 times more soluble than $Mg(OH)_2$; and $MgSO_4$ is about 140 000 times more soluble than $BaSO_4$.

The carbonates, sulphates and hydroxides of Group 1A metals are, in general, much more soluble than those of Group 2A metals.

It is important to know the solubility of a chemical, for its preparation, properties and uses depend heavily on it.

13 Standard enthalpy of formation of hydrated ions

This refers to the enthalpy change when 1 mol of a hydrated ion is formed from its element in its standard state. Once again, relative or absolute values can be obtained.

For *relative* values, the value for $H^+(aq)$ is taken as zero, i.e.

$$\tfrac{1}{2}H_2(g) + aq \longrightarrow H^+(aq) + e^- \qquad \Delta H_{f,m}^{\ominus}(298\,K) = \quad 0$$
$$Na(s) + aq \longrightarrow Na^+(aq) + e^- \qquad \Delta H_{f,m}^{\ominus}(298\,K) = -240\,kJ\,mol^{-1}$$
$$\tfrac{1}{2}Cl_2(g) + e^- + aq \longrightarrow Cl^-(aq) \qquad \Delta H_{f,m}^{\ominus}(298\,K) = -167\,kJ\,mol^{-1}$$

Absolute values can be obtained from the enthalpies of atomisation and ionisation, together with the absolute enthalpies of hydration (p. 18). Thus

$$X \xrightarrow{\text{atomisation}} X(g) \xrightarrow{\text{ionisation}} X^{\pm}(g) \xrightarrow{\text{hydration}} X^{\pm}(aq)$$

standard   single   free   hydrated
state   atom   ion   ion

$$\tfrac{1}{2}H_2(g) \xrightarrow{+218} H(g) \xrightarrow{+1317} H^+(g) \xrightarrow{-1091} H^+(aq)$$
$$Na(s) \xrightarrow{+108} Na(g) \xrightarrow{+502} Na^+(g) \xrightarrow{-406} Na^+(aq)$$
$$\tfrac{1}{2}Cl_2(g) \xrightarrow{+121} Cl(g) \xrightarrow{-354} Cl^-(g) \xrightarrow{-378} Cl^-(aq)$$

which, taken algebraically, give absolute enthalpies of formation for $H^+(aq)$, $Na^+(aq)$ and $Cl^-(aq)$ of $+444$, $+204$ and $-611\,kJ\,mol^{-1}$, respectively, i.e.

$$\tfrac{1}{2}H_2(g) + aq \longrightarrow H^+(aq) + e^- \qquad \Delta H_{f,m}^{\ominus}(298\,K) = +444\,kJ\,mol^{-1}$$
$$Na(s) + aq \longrightarrow Na^+(aq) + e^- \qquad \Delta H_{f,m}^{\ominus}(298\,K) = +204\,kJ\,mol^{-1}$$
$$\tfrac{1}{2}Cl_2(g) + e^- + aq \longrightarrow Cl^-(aq) \qquad \Delta H_{f,m}^{\ominus}(298\,K) = -611\,kJ\,mol^{-1}$$

These absolute values can be obtained from the relative values by adding $444n\,kJ\,mol^{-1}$, where n is the charge number on the ion concerned (including its positive or negative sign).

The enthalpy change involved in the formation of a *pair* of ions can be obtained by adding either the relative or the absolute values for each ion, e.g.

$$\Delta H_m^{\ominus}(298\,K)$$

$Na(s) + \tfrac{1}{2}Cl_2(g) \longrightarrow Na^+(aq) + Cl^-(aq)$	$-407\,kJ\,mol^{-1}$
$\tfrac{1}{2}H_2(g) + \tfrac{1}{2}Cl_2(g) \longrightarrow H^+(aq) + Cl^-(aq)$	$-167\,kJ\,mol^{-1}$

14 Standard electrode potential

The standard enthalpies of formation of ions give a measure of their stability as compared with the $H^+(aq)$ ion. A similar electrical measure can be obtained by using standard electrode potentials (p. 400).

A standard hydrogen electrode, consisting of an $H^+(aq)/\frac{1}{2}H_2(g)$ couple under standard conditions, is given an arbitrary standard electrode potential, $E^\ominus$, of 0 V. The $E^\ominus$ values for other couples, e.g. $Zn^{2+}(aq)/Zn(s)$ and $\frac{1}{2}Cl_2(g)/Cl^-(aq)$, can then be measured, in a cell, against that of the standard hydrogen electrode. Some $E^\ominus$ values are summarised below.

	$E^\ominus/V$
$Li^+(aq)+e^- \longrightarrow Li(s)$	-3.04
$Na^+(aq)+e^- \longrightarrow Na(s)$	-2.71
$Al^{3+}(aq)+3e^- \longrightarrow Al(s)$	-1.67
$Zn^{2+}(aq)+2e^- \longrightarrow Zn(s)$	-0.54 — 0.74
$H^+(aq)+e^- \longrightarrow \frac{1}{2}H_2(g)$	0
$Ag^+(aq)+e^- \longrightarrow Ag(s)$	$+0.80$
$\frac{1}{2}Cl_2(g)+e^- \longrightarrow Cl^-(aq)$	$+1.36$
$\frac{1}{2}F_2(g)+e^- \longrightarrow F^-(aq)$	$+2.87$

An element with a high *negative* $E^\ominus$ value is one which readily loses electrons and forms positive ions; such elements, are, however, said to be *electropositive*. A large *positive* $E^\ominus$ value indicates a readiness to form negative ions by gaining electrons; such elements are said to be *electronegative*. The $E^\ominus$ value for an element depends on its position in the periodic table (p. 128). It is the Group 1A elements (particularly lithium) that are most electropositive, whilst it is the halogens (particularly fluorine) that are most electronegative.

There is an approximate relationship between the $E^\ominus$ value and the enthalpy of formation* of the ion concerned per unit charge on the ion. That is, ions with high negative standard enthalpies of formation per unit charge have high negative electrode potentials, and vice versa. For example

	Na^+	Al^{3+}	H^+	Ag^+	F^-
$E^\ominus$	-2.71	-1.67	0	0.80	2.87
$\Delta H_{f,m}{}^\ominus/n$	-240	-175	0	$+106$	$+333$

---

*The accurate relationship is between $E^\ominus$ and the standard free energy of formation of the ion concerned (p. 76), i.e.

$$\Delta G_m{}^\ominus = -nFE^\ominus$$

where n is the charge number on the ion concerned, $\Delta G_m{}^\ominus$ is expressed in J mol^{-1} and F is the Faraday constant (96.487×10^3 C mol^{-1}).

Questions on Chapter 2

1 What are ions? Explain in detail the evidence for their existence in solutions, gases and solids. What are the most important differences in the nature and behaviour of ions in these three media? Give as many equations, or statements of the quantitative physico-chemical behaviour of ions as you can.

2 How do the atoms and ions of fluorine and neon (a) resemble, and (b) differ from, each other?

3 List some exceptions to the general rules given on p 8 about the characteristics of ionic compounds. Can you account for any of the exceptions you mention?

4 'Ionic bonds are formed by a transfer of electrons from one atom to another.' Illustrate the statement. To what extent is it true?

5 Why is it that the first electron affinity of an atom is negative whilst the second is positive?

6 Calculate the lattice enthalpies of the five halides from the following data and comment on any noticeable trends or discrepancies:

	LiCl	NaCl	KCl	RbCl	AgCl
enthalpy of solution/ $kJ\,mol^{-1}$	-37	4	17	19	66
enthalpy of hydration/ $kJ\,mol^{-1}$	-882	-765	-690	-674	-845

7 How far is it true to say that substances with high lattice enthalpies will be insoluble?

8 What is meant by the terms polarisation and polarisability? Why are they useful ideas?

9 'A rough guide to the polarising power of a cation is provided by the charge/radius ratio.' Comment.

10 Using the data on pp. 12 and 404, and taking a hypothetical lattice enthalpy of $-787\,kJ\,mol^{-1}$ for the hypothetical MgCl crystal, calculate the enthalpy of formation of MgCl. Why is it not formed?

11 Explain carefully why CaCl and KCl_2 do not exist whilst KCl and $CaCl_2$ do.

12 Explain how the value of the lattice energy of an ionic compound may be determined by means of a Born–Haber cycle. Indicate as far as you can how the energy data required for a Born–Haber cycle may be determined experimentally.

Lattice energies may also be determined from theoretical considerations. For substances like sodium chloride the results agree well with experimental data, but the agreement is not good for the silver halides. Comment. (L)

13 Sulphur dioxide, sodium chloride, anhydrous copper(II) sulphate, ammonia and ethanal are readily soluble in water. On the other hand, mercury(II) iodide, methane, butan-1-ol, mercury and sulphur are not readily soluble in water. Account, as far as you can, for the difference in behaviour pattern between the two groups. (L)

14* Record, and comment on, the enthalpies of atomisation of (a) the alkali metals, (b) carbon and (c) tungsten.

15* Plot the values of the first, second . . . nth ionisation energies against n for boron, carbon, nitrogen and oxygen. What can you conclude?

16* Use the Kapustinskii equation to calculate the lattice energy of sodium chloride.

17* Summarise the enthalpy changes in the atomisation, ionisation and hydration processes involved in the formation of hydrated ions from (a) fluorine, (b) lithium and (c) magnesium. What values do you obtain for the enthalpies of formation of the hydrated ions?

18* What kind of relationship would you expect between the melting temperatures of ionic solids and their lattice enthalpies? Take some examples to test your suggestion.

19* Use values for the lattice enthalpies and the enthalpies of hydration to calculate the enthalpies of solution of the halides of sodium and silver. Comment on any points of significance.

20* Use a Born–Haber cycle to show why it is that magnesium oxide exists as $Mg^{2+}O^{2-}$ and not $Mg^{+}O^{-}$.

21* Give ten examples of pairs of chemicals that have similar formulae but widely different solubilities.

---

*Questions marked with an asterisk require the use of a book of data.

3
The covalent bond

1 Introduction

Ionic bonds are formed by a transfer of electrons from one atom to another. In covalent bonding, electrons are shared, in pairs, between two bonded atoms.

In a single covalent bond between two atoms, one electron from each is held in common by both. The shared pair of electrons constitutes a single covalent bond, which can be represented as in the following typical examples.

$$H \overset{\times}{\cdot} H \qquad \overset{\cdot\cdot}{\underset{\cdot\cdot}{:}}\overset{\times\times}{\underset{\times\times}{Cl}} \overset{\times}{\cdot} \overset{\times\times}{\underset{\times\times}{Cl}}\overset{\times}{\times} \qquad H \overset{\times\times}{\cdot} \overset{\times\times}{\underset{\times\times}{Cl}}\overset{\times}{\times} \qquad H \overset{\times\times}{\cdot} \overset{\times\times}{\underset{\times\times}{O}}\overset{\times}{\times} \overset{\circ\circ}{\underset{\circ\circ}{O}}\overset{\circ}{\cdot} H$$

$$H-H \qquad\qquad Cl-Cl \qquad\qquad H-Cl \qquad\qquad H-O-O-H$$

$$\overset{\times\times}{\underset{\times\times}{\underset{\times\times}{\times F \times}}} \qquad\qquad H \qquad\qquad H \qquad\qquad \overset{\times\times}{\underset{\cdot\cdot}{\times O \times}}$$

$$\underset{\times F \times}{\overset{\cdot\times}{B}} \qquad H \overset{\times}{\cdot}\overset{\times\,\cdot}{\underset{\cdot}{C}}\overset{\times}{\cdot} H \qquad H \overset{\times}{\cdot}\overset{\times\cdot}{\underset{\times\times}{N}}\overset{\times}{\cdot} H \qquad H \qquad\qquad H$$

$$\qquad\qquad\qquad\quad H$$

$$BF_3 \qquad\qquad CH_4 \qquad\qquad NH_3 \qquad\qquad H_2O$$

Although they are useful, such dot-and-cross diagrams are over-simplifications of the actual arrangements of the electrons in molecules, which is discussed on p. 29.

Double and triple covalent bonds involving two or three pairs of shared electrons are also common, e.g.

$$\overset{\times\times}{\underset{\times}{\times O \times}}\overset{\circ\circ}{\underset{\circ\circ}{O}} \qquad \overset{\times}{\underset{\times}{\times N}}\overset{\circ}{\underset{\circ}{N}}\overset{\circ}{} \qquad \underset{H}{\overset{H}{\underset{\cdot}{\overset{\times}{\times}}C}}\overset{\times}{\underset{\cdot}{\times}}\underset{H}{\overset{H}{\overset{\circ}{C}\overset{\cdot}{\cdot}}} \qquad H \overset{\times}{\cdot} C \overset{\circ}{\underset{\circ}{\times}} C \overset{\circ}{\cdot} H$$

$$O{=}O \qquad\quad N{\equiv}N \qquad\quad \text{ethene} \qquad\qquad \text{ethyne}$$

$$H \overset{\times}{\cdot} C \overset{\circ}{\underset{\circ}{\times}} N \overset{\circ}{\circ} \qquad \overset{\circ\circ}{\underset{\circ\circ}{\circ}} O \overset{\times}{\underset{\times}{\circ}} C \overset{\times}{\underset{\times}{\circ}} O \overset{\circ\circ}{\underset{\circ\circ}{\circ}} \qquad \underset{H}{\overset{H}{H \overset{\times}{\cdot} C \overset{\times}{\underset{\times}{\times}} O \overset{\circ\circ}{\underset{\circ\circ}{\circ}}}}$$

$$H{-}C{\equiv}N \qquad\quad O{=}C{=}O \qquad\quad \text{methanal}$$

The majority of covalent compounds are formed by combination between non-metallic elements, with hydrogen and carbon forming a particularly large number.

a Dative bonding   The shared pair of electrons in a covalent bond may also be formed by one of the two bonded atoms providing both electrons. If so, the bond is sometimes called a dative bond but, as it is similar to a covalent bond once formed, the two are not always distinguished. The atom providing the two electrons to make up the dative bond is known as the donor. It must, of course, have an 'unused' pair of electrons available and such a pair is known as a lone pair. The atom sharing the pair of electrons from the donor is known as the acceptor.

When it is not necessary to distinguish between a dative and a covalent bond, the — symbol is used for both. A specific dative bond between atoms A and B is written as A→B, A being the donor and B the acceptor.

Typical molecules involving dative bonding are as follows.

aluminium chloride   isocyanomethane   ammonium ion

                                    nitromethane   oxonium ion

nitric acid   carbon monoxide   ozone (trioxygen)

b Typical properties of covalent substances   Covalent substances are made up of individual covalent molecules, and the solid forms of the substances are either amorphous or exist as molecular crystals (p. 56). The forces between the molecules in the crystals are weak and the substances are soft if they are solid. At room temperature they are generally gases or liquids.

Covalent compounds do not contain ions; they have low melting and boiling temperatures because little energy is needed to break down the weak intermolecular forces of attraction (p. 56); they generally dissolve in benzene and other organic solvents, but not in water (unless they contain polar bonds (p. 39) such as O—H).

A comparison of the properties of sodium chloride and tetrachloromethane illustrates the general differences between a typically ionic and a typically covalent compound.

sodium chloride	tetrachloromethane
composed of Na^+ and Cl^- ions in an ionic crystal	composed of individual CCl_4 molecules with weak inter-molecular forces; forms a molecular crystal when solid
electrolyte	non-electrolyte
hard solid at room temperature	liquid at room temperature; soft when solid
melts at 803 °C	melts at −28 °C
boils at 1430 °C	boils at 77 °C
soluble in water; insoluble in benzene	soluble in benzene; insoluble in water

c Types of covalent bond   It is useful to think of single, double, triple and dative covalent bonds as distinct types of bonding, and to represent them by using dot-and-cross diagrams. Such ideas are, however, over-simplified and must be treated with some caution; they cannot account for many of the known properties of many chemicals.

It is, for example, inaccurate to regard all chemicals as being either ionic or covalent, for many of them have properties that are intermediate between those of NaCl and those of CCl_4. Aluminium chloride, for example, is a white, fibrous solid, which sublimes at about 180 °C. It both dissolves in, and reacts with, water; and it is soluble in benzene and other organic solvents. (non-polar.)

To account for such intermediate properties, it is necessary to think in terms of intermediate types of covalent bonding. Ionic bonding (with a complete transfer of electrons) and covalent bonding (with a complete, equal sharing) are two extreme types of bonding, and all the in-between possibilities of partial transfer and partial sharing occur. Such bonds are called polar covalent bonds (p. 39).

There are, moreover, two sorts of covalent bonds, known as σ bonds and π bonds (p. 30). And, when covalent bonding occurs between more than two atoms, it is possible for the electron sharing to take place amongst all the atoms concerned, instead of being localised between two adjacent atoms. This leads to so-called delocalised bonds, which are neither single, double nor triple, but something in between (p. 41). A still more detailed study involves the consideration of molecular orbitals (p. 29).

In considering any chemical problem, it is a matter of choosing the right degree of sophistication. Simple dot-and-cross models give very satisfactory answers to many issues, whilst even the most mathematical treatment of molecular orbitals does not resolve others; for, indeed, the theories of chemical bonding are incomplete and still under active development. It will be a long time before all the properties of all chemicals can be accounted for simply in terms of any theory of the nature of the bonding within the chemical.

2 Electron pair repulsion

The shapes of many simple molecules and ions of non-transitional elements can be predicted using the idea that electron pairs distributed round a central atom will repel each other and take up positions as far apart as possible. The electron pairs concerned may be bonded or shared pairs constituting covalent bonds, or unbonded lone pairs. As the latter are closer to the central atom they cause greater repulsion than shared pairs and the repulsions are in the order

lone pair/lone pair > lone pair/shared pair > shared pair/shared pair

When multiple bonds occur, they act as a single shared pair of electrons.

a Two electron pairs Beryllium chloride, $BeCl_2$, is linear because the two bonded pairs round the beryllium atom repel each other as fully as possible. The situation is summarised in Fig. 13. Other examples include $HgCl_2$, $O{=}C{=}O$, $S{=}C{=}S$, $H{-}C{\equiv}N$ and the $-C{\equiv}$ bond as in ethyne.

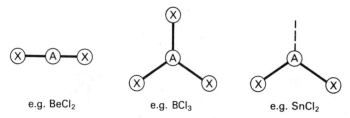

e.g. $BeCl_2$ e.g. BCl_3 e.g. $SnCl_2$

Fig. 13 *Shapes of molecules with two or three electron pairs.*

b Three electron pairs Boron has three bonded electron pairs in all its BX_3 compounds. As these pairs repel each other equally, the molecules, e.g. BCl_3 and $B(CH_3)_3$, are planar with bond angles of 120° (Fig. 13). Other examples of similar groupings include SO_3, $COCl_2$, NO_3^-, CO_3^{2-} and the $\diagdown C{=}$ bond as in ethene.

In tin(II) chloride, $SnCl_2$, in the vapour state, the tin atom is surrounded by two bonded pairs and one lone pair. The molecule is V-shaped (Fig. 13), with a bond angle smaller than 120° because of the increased repulsion caused by the lone pair. Other similar examples are SO_2, $NOCl$ and O_3.

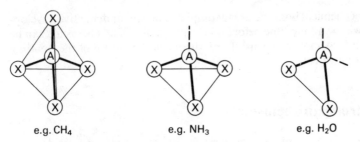

e.g. CH₄   e.g. NH₃   e.g. H₂O

Fig. 14 Shapes of molecules with four electron pairs.

c Four electron pairs, all shared Four electron pairs will be distributed round a central atom tetrahedrally (Fig. 14), and this accounts for the tetrahedral shape of a methane molecule, CH_4, with a bond angle of 109° 28′. Other similar species include $CHCl_3$, SO_2Cl_2, $POCl_3$, $Pb(C_2H_5)_4$, SO_4^{2-}, CrO_4^{2-}, PO_4^{3-}, ClO_4^-, MnO_4^- and NH_4^+.

d Four electron pairs, including one lone pair With three bonded pairs and one lone pair the resulting AX_3 molecule will be trigonal pyramidal in shape (Fig. 14). The increased repulsion caused by the lone pair will lead to a bond angle smaller than the 109° 28′ found in methane. Ammonia, NH_3, for example, has a bond angle of 107°.

The bond angle in phosphine, PH_3, is still lower (93° 20′). This is because phosphorus has a lower electronegativity than nitrogen, so the shared pairs are closer to the nitrogen atom in NH_3 than they are to the

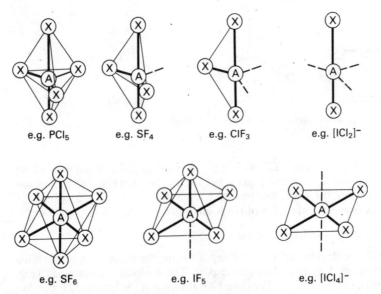

e.g. PCl₅   e.g. SF₄   e.g. ClF₃   e.g. [ICl₂]⁻

e.g. SF₆   e.g. IF₅   e.g. [ICl₄]⁻

Fig. 15 Shapes of molecules with five or six electron pairs.

phosphorus atom in PH_3. This causes an increased repulsion between the shared pairs in NH_3 compared with PH_3. The bond angles in AsH_3 ($91° 51'$) and SbH_3 ($91° 20'$) are lower still.

PCl_3, SO_3^{2-} and ClO_3^- have similar shapes.

e Four electron pairs, including two lone pairs With two lone pairs the resulting AX_2 molecule will be V-shaped (Fig. 14). The increased repulsive effect of one lone pair lowers the bond angle from $109° 28'$ in methane to $107°$ in ammonia. In the H_2O molecule the effect of two lone pairs is to lower the angle to $104° 31'$. The bond angles for H_2S ($93°$), H_2Se ($91°$) and H_2Te ($89° 30'$) are lower still because of the decrease in electronegativity in passing from sulphur to tellurium.

Cl_2O and ClO_2^- have similar shapes.

f Five, six and seven electron pairs. Summary The possible arrangements of higher numbers of electron pairs which are actually found in some molecules are shown in Fig. 15 and details are included in the following summary.

number of shared pairs	number of lone pairs	shape of molecule	example
2	0	linear	$BeCl_2$
3	0	trigonal planar	BCl_3
2	1	V-shaped	$SnCl_2$
4	0	tetrahedral	CH_4
3	1	trigonal pyramidal	NH_3
2	2	V-shaped	H_2O
5	0	trigonal bipyramidal	PCl_5
4	1	irregular tetrahedral	SF_4
3	2	T-shaped	ClF_3
2	3	linear	$[ICl_2]^-$
6	0	octahedral	SF_6
5	1	square pyramidal	IF_5
4	2	square planar	$[ICl_4]^-$
7	0	pentagonal bipyramid	IF_7

3 Molecular orbitals

A single covalent bond can be written as ' — ' or ' ⁚ ' and this can be useful when, for example, considering the shapes of simple molecules. But the shared electron pair idea gives little or no indication as to why a covalent bond should hold two atoms together. A better understanding comes from a consideration of molecular orbitals (MO for short) using the charge-cloud representation of atomic orbitals (AO for short) instead of the idea of electrons as single point charges (p. 6).

A full treatment of molecular orbitals is mathematical, but they can be considered as being formed by overlap of atomic orbitals from two adjacent atoms. As will be seen, it is the shapes of molecular orbitals that are important, as for atomic orbitals. Likewise, the Pauli principle (p. 3) applies to both molecular and atomic orbitals, so that no molecular orbital can contain two precisely similar electrons. This means that a molecular orbital can only contain two electrons, and that the two must have different spins. The nomenclature, s, p and d, used for atomic orbitals is replaced by σ, π and δ for molecular orbitals. σ molecular orbitals lead to the formation of what are known as σ bonds and π molecular orbitals give what are known as π bonds.

a σ bonds   In a hydrogen molecule, H_2, the two 1s atomic orbitals, one from each hydrogen atom, are thought of as overlapping to form a plum-shaped molecular orbital (Fig. 16) which can hold two electrons of different spins. It is the accumulation of negative charge between the two positively charged nuclei of the two hydrogen atoms that holds the molecule together. An electron density map for the molecule can be calculated, and this shows that the greatest electron density does occur between the two atoms and along the axis.

The molecular orbital in the H_2 molecule is called a σ1s orbital: σ because it is symmetrical about the molecular axis, and 1s because it is formed by the overlap of 1s atomic orbitals. The bond that is formed is called a σ bond.

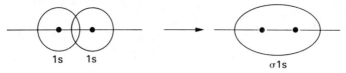

Fig. 16 *The overlap of two 1s atomic orbitals to form a plum-shaped σ1s molecular orbital, as in H_2.*

Fig. 17 *The overlap of a 1s atomic orbital in hydrogen with a $3p_x$ atomic orbital in chlorine to give a σ bond in HCl.*

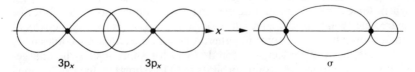

Fig. 18 *The overlap of two $3p_x$ atomic orbitals from chlorine atoms to form a σ bond in Cl_2.*

Similar σ bonds can be formed by the overlap of s and p atomic orbitals, as in the HCl molecule (Fig. 17), and by *end-on* overlap of two p_x orbitals, as in the Cl_2 molecule (Fig. 18).

In any σ bond, the charge build-up is between the two bonded atoms and it is symmetrical about the bond axis.

b π bonds   Two p atomic orbitals can overlap *end-on* to form a σ bond, but they can also overlap *broadside-on*, as shown in Fig. 19. The accumulation of negative charge alongside the molecular axis constitutes what is known as a π bond. The molecular orbital made up from 2p atomic orbitals is labelled π2p.

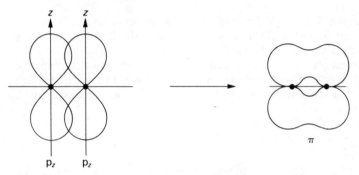

Fig. 19 *The formation of a π bond by broadside-on overlap of two p_z atomic orbitals.*

There are, then, really two kinds of covalent bond. σ bonds hold two atoms together by an accumulation of charge *between* their nuclei; π bonds hold them together by an accumulation of charge *alongside* the line joining the two nuclei.

4 Covalent bonding in simple molecules

The formation of a molecular orbital, and the corresponding covalent bond, by the overlap of two atomic orbitals requires that each of the two orbitals must contain one unpaired electron. The number of bonds that an atom will form depends very largely, therefore, on the number of unpaired electrons in the atom.

A summary of this number for the elements hydrogen to neon, using the table on p. 5, shows that the number of unpaired electrons is, in fact, equal to the common covalency of the element concerned, except for beryllium, boron and carbon, i.e.

	H	He	Li	Be	B	C	N	O	F	Ne
number of unpaired electrons	1	0	1	0	1	2	3	2	1	0
common covalency	1	0	1	2	3	4	3	2	1	0

The noble gases, helium and neon, have no unpaired electrons and form no bonds. Fluorine, oxygen and nitrogen, with one, two and three unpaired electrons respectively, commonly form one, two and three bonds. The nature of the bonding in beryllium, boron and carbon compounds is, however, rather different, and is explained below.

a Bonding in beryllium compounds   The arrangement of electrons in the beryllium atom,

$$\begin{array}{cccccc} & 1s & 2s & 2p_x & 2p_y & 2p_z \\ \text{Be} & \uparrow\downarrow & \uparrow\downarrow & & & \end{array}$$

would suggest that beryllium might form no compounds at all, since its atom has no unpaired electron, whereas it actually forms two covalent bonds, as in $BeCl_2$, which is a collinear molecule.

To account for this, it is assumed that, in the process of combination, one of the 2s electrons is promoted into a 2p orbital. The energy required for this is very small and an excited atom, denoted by an asterisk, results:

$$\begin{array}{cccccc} & 1s & 2s & 2p_x & 2p_y & 2p_z \\ \text{Be*} & \uparrow\downarrow & \uparrow & \downarrow & \downarrow & \end{array}$$

This excited atom has two unpaired electrons and would be expected to form two bonds as it does, for example, in $BeCl_2$. The two bonds are collinear and they are both alike, but it is not easy to describe the bonds in terms of the overlap of simple atomic orbitals. For if one bond involved overlap of the Be 2s orbital with a $3p_x$ orbital of Cl, and the other involved overlap of the $2p_x$ orbital of Be with a $3p_x$ orbital of Cl, it would be expected that the two bonds would be different.

The simplest way of 'picturing' the situation is to use the concept of hybridisation.* The 2s and $2p_x$ atomic orbitals of Be can be combined, mathematically, to form two collinear orbitals, with shapes different from those of s or p orbitals, as shown in Fig. 20. These *hybrid orbitals* are referred to as sp orbitals to indicate their formation from one s and one p orbital taken together. On this basis, the bonding in the $BeCl_2$ molecule is

Fig. 20 *Two collinear sp orbitals.*

---

*Whenever hybridisation is used to describe the nature of the bonds in a molecule it is possible, in a more sophisticated treatment, to consider the molecular orbitals that are available within the molecule, and to 'feed' the available electrons into those orbitals, but a discussion of the precise nature of such orbitals is too advanced for a book at this level.

regarded as being made up of two σ bonds formed by overlap of the two sp orbitals of Be with $3p_x$ orbitals from the Cl atoms (Fig. 21). This accounts satisfactorily for the collinear shape of the molecule.

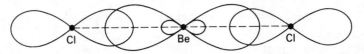

Fig. 21 *The formation of two σ bonds in $BeCl_2$ by overlap of $3p_x$ orbitals from two chlorine atoms with two collinear sp orbitals from a beryllium atom.*

b Bonding in boron compounds The B atom has only one unpaired electron,

$$\text{1s} \quad \text{2s} \quad 2p_x \quad 2p_y \quad 2p_z$$
$$\text{B} \quad \uparrow\downarrow \quad \uparrow\downarrow \quad \downarrow$$

but it can form three bonds, as in BCl_3, which is a trigonal planar molecule. This is accounted for by promoting one of the 2s electrons so that the excited atom,

$$\text{1s} \quad \text{2s} \quad 2p_x \quad 2p_y \quad 2p_z$$
$$\text{B*} \quad \uparrow\downarrow \quad \downarrow \quad \downarrow \quad \downarrow$$

has three unpaired electrons available.

The individual molecular orbitals that the 2s, $2p_x$ and $2p_y$ atomic orbitals can form can be considered but, as for beryllium, the situation can be simplified by thinking in terms of three coplanar orbitals directed at angles of 120° to each other (Fig. 22). As these hybrid orbitals originate from one s and two p atomic orbitals, they are called sp^2 hybrid orbitals. Their use in bonding accounts for the trigonal planar shape of molecules such as BCl_3 and $AlCl_3$. The boron and aluminium atoms in these molecules only share six electrons. As a result, they can act as acceptors very readily in the formation of dative bonds (p. 190).

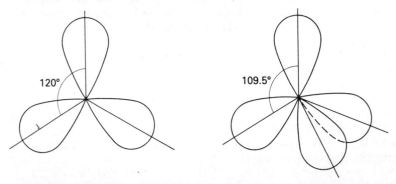

Fig. 22 *The shapes of the three sp^2 orbitals formed by boron (left) and the four sp^3 orbitals formed by carbon (right).*

c Bonding in carbon compounds Carbon forms a number of different types of bond, which are summarised below, together with typical bond angles and examples.

$\overset{\displaystyle\mid}{\underset{\diagup\ \vdots\ \diagdown}{C}}$	$\overset{\diagdown}{\underset{\diagup}{C}}=$	$=C=$	$-C\equiv$
109° 28′	120°	180°	180°
diamond	graphite	CO_2	C_2H_2
CH_4	C_2H_4	CS_2	HCN
CCl_4	$COCl_2$		

The carbon atom contains only two unpaired electrons,

$$
\begin{array}{cccccc}
 & 1s & 2s & 2p_x & 2p_y & 2p_z \\
C & \uparrow\downarrow & \uparrow\downarrow & \downarrow & \downarrow & \\
\end{array}
$$

so that it is necessary to think in terms of the activated atom,

$$
\begin{array}{cccccc}
 & 1s & 2s & 2p_x & 2p_y & 2p_z \\
C^* & \uparrow\downarrow & \downarrow & \downarrow & \downarrow & \downarrow \\
\end{array}
$$

to account for the various bonds, and to combine the four available atomic orbitals in different ways.

The 2s and three 2p orbitals can form four sp^3 hybrid orbitals; they are directed towards the corners of a tetrahedron (Fig. 22). It is the use of such orbitals, overlapping with orbitals of other atoms, that results in the formation of four σ bonds, as in CH_4 and CCl_4. The bond angles are always close to the expected value of 109° 28′.

When double bonding occurs, the carbon atom uses sp^2 hybrid orbitals to form three coplanar σ bonds at 120° to each other. The carbon atom still has an unused 2p orbital in a plane at right angles to the plane of the σ bonding, and this can overlap with p orbitals on adjacent atoms, such as C or O, to form a π bond. The double bond consists, therefore, of a σ bond and a π bond.

In triple bonding, the carbon atom uses sp hybrid orbitals to form two collinear σ bonds. Two 2p atomic orbitals remain and they can form π bonds by overlapping with two orbitals from adjacent atoms, such as C or N. The triple bond consists, therefore, of a σ bond and two π bonds.

In some molecules and ions, the carbon atom still has an unused lone pair that can be used for dative bonding. That is why, for example, CO forms many carbonyls, e.g. $[Ni(CO)_4]$ (p. 206), and why there are very many complex cyanides, e.g. $[Fe(CN)_6]^{4-}$.

d Bonding in N_2 and nitrogen compounds A nitrogen atom has three unpaired 2p electrons, and the bond in the N_2 molecule is a very strong triple bond. It consists of a σ bond formed by overlap of two $2p_x$ atomic orbitals, and two π bonds formed by overlap of two $2p_y$, and two $2p_z$, atomic orbitals. A simplified representation of the bonding is given in Fig. 23.

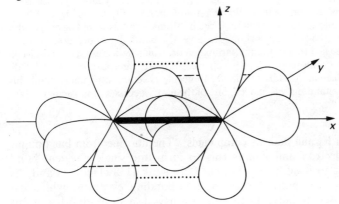

Fig. 23 *Diagrammatic representation of the triple bonding in a nitrogen molecule, N_2. The bold line shows the σ bond formed by the two $2p_x$ orbitals. The dotted and dashed lines show the π bonds formed by the $2p_z$ and $2p_y$ orbitals respectively.*

The nitrogen atom can use its three 2p electrons to form three bonds in different ways, e.g.

$$\diagdown \!\!\!\!{\underset{\diagup}{N}}\!\!- \qquad -N= \qquad N\equiv$$

$$NH_3 \qquad H-O-N=O \qquad N\equiv N$$
$$NCl_3 \qquad\qquad\qquad\qquad H-C\equiv N$$

The remaining lone pair of electrons enables it to form dative bonds as, for example, in NH_4^+ and HNO_3 (p. 25), $H_3N \rightarrow BF_3$ (p. 192) and many ammines (p. 234).

e Bonding in O_2 and oxygen compounds   The oxygen atom has two un-paired 2p electrons and it generally uses these to form two bonds in its covalent compounds, e.g.

$$-O- \qquad O=$$

$$\begin{array}{ll} H_2O & CO_2 \\ H_2O_2 & \text{aldehydes} \\ OF_2 & \text{ketones} \\ \text{alcohols} \end{array}$$

Two lone pairs remain on the oxygen atom. One is quite commonly used to form dative bonds, as in H_3O^+ (p. 25) and aqua-complexes (p. 268). But bonding by both lone pairs is rare; it occurs in basic beryllium ethanoate, $Be_4O(COOCH_3)_6$ (p. 267).

The bonding in the O_2 molecule is, however, unusual. By analogy with N_2, it would be expected that there would be a double bond in O_2, formed by overlap of the two available 2p atomic orbitals. The $2p_x$ atomic orbitals would be expected to give a σ bond, and the $2p_y$ orbitals to give a π bond. This does not, however, account for the fact that oxygen is paramagnetic, which indicates that the molecule contains one or more unpaired electrons. A more rigorous treatment of the available molecular orbitals in the O_2 molecule is required to understand this. Such an understanding was, indeed, one of the early successes of molecular orbital theory.

f Bonding in F_2 and fluorine compounds   The fluorine atom has one un-paired 2p electron and it uses this to form one single covalent bond. The bonding to itself, in F_2, is unusually weak ($158\,kJ\,mol^{-1}$) and this is attributed to repulsion between non-bonding electrons, which can approach closely because the fluorine atoms are small. With other atoms, fluorine forms strong single covalent bonds.

A fluorine atom forming one covalent bond has three lone pairs of electrons, but these are not generally used to form dative bonds. The number of bonds that fluorine will form is, therefore, lower than the four which can be formed by carbon, nitrogen and oxygen.

However, the lone pairs in a fluoride ion, F^-, are used very readily in the bonding in fluoro-complexes, e.g. $[SiF_6]^{2-}$.

g The use of d orbitals in bonding and maximum covalency   There is no way in which the elements lithium to fluorine can form more than four bonds; they are said to have a maximum covalency of 4.

However, elements from sodium to chlorine, and beyond, can form more than four covalent bonds and, therefore, many more compounds. This is because they have d orbitals available within their atoms, even though they are not occupied in the ground state of the atom. These d orbitals can participate in bonding in two different ways.

In SiF_4, for example, the s and p orbitals of the silicon atom are fully used in the formation of four covalent bonds. The d orbitals, however, enable bonding with some ion that can act as a donor, such as F^-. That is why the $[SiF_6]^{2-}$ ion is formed; the coordination limit of 6 is due to steric reasons for it is not possible to pack more than six fluorine atoms around a silicon atom. CF_4 cannot form $[CF_6]^{2-}$ because the carbon atom has no d orbitals.

Alternatively, the existence of d orbitals in an atom enables the formation of excited atoms, because s, p and d orbitals within the same shell have similar energies, so that it is easy to promote s or p electrons into the d level. The chlorine atom, for example, can have three excited forms,

	3s	$3p_x$	$3p_y$	$3p_z$	3d	3d	3d	3d	3d
Cl	↓↑	↓↑	↓↑	↓					
Cl*	↓↑	↓↑	↓	↓	↓				
Cl*	↓↑	↓	↓	↓	↓	↓			
Cl*	↓	↓	↓	↓	↓	↓	↓		

with three, five or seven unpaired electrons. These atoms can, at least theoretically, form three, five or seven covalent bonds, as, for example, in the oxoacid ions (p. 319),

$$[O\!-\!Cl\!=\!O]^- \qquad \left[\begin{array}{c} O \\ \| \\ O\!-\!Cl\!=\!O \end{array}\right]^- \qquad \left[\begin{array}{c} O \\ \| \\ O\!-\!Cl\!=\!O \\ \| \\ O \end{array}\right]^-$$

and in the fluorides (p. 325):

$$\begin{array}{c} F \\ | \\ \underset{F}{\overset{}{Cl}} \\ \diagup\quad\diagdown \\ F\qquad F \end{array} \qquad\qquad \begin{array}{c} F\quad F \\ \diagdown\;\diagup \\ F\!-\!\underset{F}{\overset{}{Cl}}\!-\!F \\ | \\ F \end{array}$$

ClF_7 does not exist because it is not possible to pack seven fluorine atoms around a chlorine atom; but IF_7 is known, together with IF_3 and IF_5. It is fluorine and, to a lesser extent, oxygen that 'bring out' the highest covalency in other elements.

5 Bond enthalpies and energies*

a Meaning of terms   The splitting of a single covalent bond in a diatomic molecule involves a standard molar enthalpy change which is known as the *bond enthalpy* of the bond concerned, e.g.

$$H_2(g) \longrightarrow 2H(g) \qquad \Delta H_m^{\ominus}(298\,K) = +436\,kJ\,mol^{-1}$$
$$HCl(g) \longrightarrow H(g) + Cl(g) \qquad \Delta H_m^{\ominus}(298\,K) = +431\,kJ\,mol^{-1}$$

For an X_2 molecule, the value is twice the enthalpy of atomisation per mole of X atoms. In polyatomic molecules, with more than one bond, the bond enthalpy (or energy) can be regarded in two ways. The N—H bond

---

*Bond enthalpies are related to $\Delta H_m^{\ominus}(298\,K)$ values. The corresponding $\Delta U_m^{\ominus}(0\,K)$ values are known as bond energies. There is a small difference between the values of the two terms, but it is commonly disregarded. The average bond enthalpy for the N—H bond in ammonia, for example, is $391\,kJ\,mol^{-1}$; the average bond energy is $386\,kJ\,mol^{-1}$.

enthalpy in ammonia, for example, might be taken as one-third of the enthalpy change for the reaction

$$NH_3(g) \longrightarrow N(g) + 3H(g) \qquad \Delta H_m^{\ominus}(298\,K) = +1172\,kJ\,mol^{-1}$$

i.e. $391\,kJ\,mol^{-1}$. Such a bond enthalpy is, really, an *average bond enthalpy*. Alternatively, the N—H bond enthalpy in ammonia might be taken as the enthalpy change when the NH_3 molecule splits into NH_2 and H. This bond enthalpy is called the *bond dissociation enthalpy*, and the particular bond to which it refers must be specified. There are, for instance, three bond dissociation values for N—H bonds in ammonia:

$$NH_3(g) \longrightarrow NH_2(g) + H(g) \qquad \Delta H_m^{\ominus}(298\,K) = +448\,kJ\,mol^{-1}$$
$$NH_2(g) \longrightarrow NH(g) + H(g) \qquad \Delta H_m^{\ominus}(298\,K) = +368\,kJ\,mol^{-1}$$
$$NH(g) \longrightarrow N(g) + H(g) \qquad \Delta H_m^{\ominus}(298\,K) = +356\,kJ\,mol^{-1}$$

None of these has a value equal to the average bond enthalpy, but the average bond enthalpy is equal to one-third of their total.

b Average bond enthalpies　Some typical values are summarised below for both single and multiple bonds.

Average single bond enthalpies/$kJ\,mol^{-1}$

		H	C	N	O	F	Si	P	S	Cl	Br	I
	H	436	413	391	464	566	323	322	344	431	366	299
Group 4B	C	413	346	305	358	485	301	264	272	339	284	218
	Si	323	301		368	582	226	213	226	391	310	234
Group 5B	N	391	305	163	201	272		209		193		
	P	322	264	209	351	450	213	209	230	319	264	184
Group 6B	O	464	358	201	146	190	368	351		205		201
	S	344	272			326	226	230	266	255	213	
Group 7B	F	566	485	272	190	158	582	490	326	255	239	
	Cl	431	339	193	205	255	391	319	255	242	218	209
	Br	366	284			239	310	264	213	218	193	180
	I	299	218		201		234	184		209	180	151
			1st short period				2nd short period					

Average multiple-bond enthalpies/$kJ\,mol^{-1}$

C=C	N=N	O=O	C=O	C≡C	N≡N
611	410	497	749	835	945
			C≡O	C≡N	P≡P
			1071	890	488

The figures must be used with care for they are average values, and they have been adjusted, in some cases, so that they can be used additively to the best advantage (p. 64). A bond enthalpy may also depend on the oxidation state of one of the atoms concerned and on the shape of the molecule in which the bond occurs. The Fe—Cl average in $FeCl_2(g)$, for example, is $401\,kJ\,mol^{-1}$, but that in $FeCl_3(g)$ is $345\,kJ\,mol^{-1}$. The average value for the P—Cl bond in $PCl_3(g)$ is $322\,kJ\,mol^{-1}$, whilst that in $PCl_5(g)$ is $259\,kJ\,mol^{-1}$. But in PCl_3 all three bonds are alike, whereas in PCl_5 two of them are different from the other three (p. 28).

6 Bond lengths

The value of the bond length or bond distance of a single covalent bond between two atoms, A and B, can be obtained, at least approximately, by adding the covalent radii of A and B (p. 130). In general, it is the strong bonds that are shortest.

Multiple-bond covalent radii can also be allotted to the elements to serve the same purpose for multiple bonding. Typical values are summarised, in nm, below.

C=	C≡	Si=	N=	N≡	O=	O≡	S=
0.067	0.060	0.107	0.060	0.055	0.060	0.050	0.094

A double-bond radius of an atom is smaller than its single-bond covalent radius, and a triple-bond radius is still smaller.

When there is a marked discrepancy between the measured bond distance in a molecule and that obtained by adding covalent radii it is usually an indication of delocalised bonding (p. 41). In CO_2, for example, the measured bond length is $0.115\,nm$, whereas the sum of the double-bond radii is $0.127\,nm$ and that of the triple-bond radii is $0.110\,nm$. On this evidence, the actual bond in CO_2 is something intermediate between a double and a triple bond. *The more multiple a bond is the shorter it is.*

7 Bond polarity

An ionic bond can have some covalent character (p. 16). Similarly, a covalent bond may have some ionic character; if so, it is said to be polar. The polarity arises because of unequal sharing of electrons between different atoms. The atom with the greatest attraction for electrons will 'pull' them in its direction, and the resulting displacement of charge will give the bond some ionic character, i.e. cause it to be polar. The effect can readily be demonstrated by bringing an electrically charged piece of plastic (simply rub it on the sleeve) close to a jet of water; the water, with polar H_2O molecules, is attracted quite strongly.

Fully ionic or covalent bonds are, in reality, extreme types of bonding.

In ionic bonds there is generally an incomplete transfer of electrons; in covalent bonds there is generally an unequal sharing of electrons. Most bonds are intermediate in type between ionic and covalent, and the gradation from the one to the other can be illustrated in terms of the shapes of the molecular orbitals, as in Fig. 24.

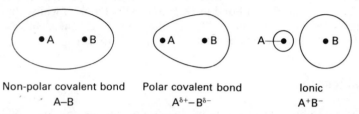

Non-polar covalent bond   Polar covalent bond   Ionic
    A—B $A^{\delta+}$—$B^{\delta-}$ A^+B^-

Fig. 24 *Representation of molecular orbital shapes for non-polar covalent bonds (as in Cl_2), polar bonds (as in HCl) and ionic bonds (as in NaCl).*

a Electronegativity   The degree of polarity in any covalent bond will depend on the relative electronegativities of the bonded atoms. *The electronegativity of an atom is a measure of the extent to which it attracts electrons in a covalent bond.* Various attempts have been made to allot numerical values to the electronegativities of atoms and, though they are all arbitrary methods, they give results which agree reasonably well. Pauling's values,* some of which are given below, are commonly used (see also pp. 127 and 345).

			H			
			2.1			

Li	Be	B	C	N	O	F
1.0	1.5	2.0	2.5	3.0	3.5	4.0

Na	Mg	Al	Si	P	S	Cl
0.9	1.2	1.5	1.8	2.1	2.5	3.0

A big difference in the electronegativities of two atoms bonded together is associated with a high percentage of ionic character in the bond between them. Estimates for typical bonds are

H—C	H—N	H—O	H—F	H—Cl	H—Br	H—I
4%	19%	39%	60%	19%	11%	4%

---

*Pauling took the geometric mean of the bond enthalpies of the A—A and X—X bonds as an estimate of the bond enthalpy of the A—X bond if it was *truly* covalent. The actual bond enthalpy of A—X is greater than the estimate, by an amount Δ kJ mol^{-1}. Electronegativity values, χ_A and χ_X were allotted on the basis of the relationship

$$96(\chi_A - \chi_X)^2 = \Delta$$

Mulliken based his electronegativity values on the arithmetic mean of the ionisation energy and the electron affinity, both values being expressed in eV.

b Effects of bond polarity A polar bond will have a dipole moment which is expressed in units of coulomb metre (C m) or in older Debye units (1 D = 3.34×10^{-30} C m). The dipole moment, in Debye units, is often approximately equal to the difference between the electronegativities of the two bonded atoms. The polarity also causes bond shortening, so that adding together the covalent radii (p. 131) does not give a good answer for a bond length when it is highly polar.

Polarity also builds up centres of negative and positive charge within a molecule, and these provide points at which a molecule can be attacked by other charged atoms or groups.

8 Delocalised bonds

In a σ bond, the concentration of charge is *between* the bonded atoms, and the bond is said to be localised (between the two atoms). It is reasonably well represented, at least positionally, by the traditional A—X or A⸫X symbolisms.

In a π bond between two atoms, the concentration of charge is *alongside* the atoms, though it is still localised between the two atoms. When more than two adjacent atoms have available p orbitals, however, the overlap can take place between *all* the atoms involved. The resulting distribution of charge over the whole molecule binds the atoms together. The bond is similar in type to a π bond but it extends over the whole molecule and is not localised between any two atoms. It is called a *delocalised* bond, and the electrons concerned are known as delocalised electrons.

The idea of delocalised bonds can account, in a reasonably simple way, for the properties of some molecules. In particular, unexpected bond lengths and enthalpies of formation can be accounted for, as in the following examples.

a Carbon dioxide The CO_2 molecule is collinear and was traditionally written as O=C=O. However, the measured bond length is 0.115 nm, whereas the expected value from covalent radii would be 0.127 nm and the C=O distance in propanone is 0.122 nm. Moreover, the measured enthalpy of formation from its atoms is -1605 kJ mol^{-1} but the value calculated from the bond enthalpies is -1498 kJ mol^{-1}. The actual carbon dioxide molecule is, therefore, more stable than the O=C=O molecule by 107 kJ mol^{-1}. This is known as the *delocalisation energy*.

The molecule is best represented as containing σ bonds between carbon and oxygen atoms, made up from sp hybrid orbitals of the carbon atom (as in ethyne, p. 34) overlapping with p orbitals of the oxygen atoms. All three atoms have remaining p orbitals, which can provide two π bond systems in perpendicular planes. Delocalisation occurs within the bond systems and it can be summarised as O⫤C⫥O.

b The nitrate, carbonate and sulphate ions All the bonds in the planar nitrate ion are of equal length (0.121 nm), between those calculated for the N—O bond (0.136 nm) and the N=O bond (0.115 nm); the bond angles are 120°. The nitrogen atom forms σ bonds with the three oxygen atoms, using sp^2 hybrid orbitals, and the remaining p orbitals on the nitrogen and oxygen atoms give delocalised π bonding; the delocalisation energy is 188 kJ mol^{-1}. The ion can be represented as in Fig. 25.

The carbonate ion and the sulphate ion also have delocalised bonding (Fig. 25), but the latter is tetrahedral in shape (p. 295).

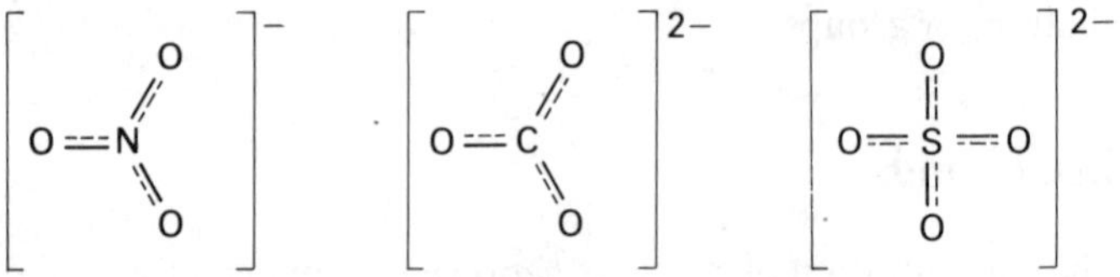

$$
\left[\ O\!=\!\!=\!N\!\!\underset{O}{\overset{O}{<}}\ \right]^{-}
\qquad
\left[\ O\!=\!\!=\!C\!\!\underset{O}{\overset{O}{<}}\ \right]^{2-}
\qquad
\left[\ O\!=\!\!=\!\underset{O}{\overset{O}{S}}\!\!=\!\!=\!O\ \right]^{2-}
$$

Fig. 25 *Delocalised bonding in the* NO_3^{-}, CO_3^{2-} *and* SO_4^{2-} *ions.*

c Graphite   The carbon atoms in the layers of the layer lattice structure (p. 202) are linked by σ bonds, using sp^2 hybrid orbitals. Each carbon atom has a p orbital remaining, and all these orbitals in any one layer can overlap to give extensive delocalised bonding, above and below the layer, throughout the structure.

This extensive delocalisation accounts for the electrical conductivity in the plane of the layers of graphite, and repulsion between the delocalised bonds in each layer is partially responsible for the large distance between the layers.

d Organic compounds of carbon   Delocalisation is particularly common in organic compounds, for it occurs whenever C=C and C—C bonds alternate in a molecule; such a system of bonding is said to be *conjugated*. That is why, for example, delocalisation is important in benzene and buta-1,3-diene. It also occurs whenever a C=C bond, or a benzene ring, is adjacent to an atom with available p orbitals, e.g. chlorine, oxygen or nitrogen.

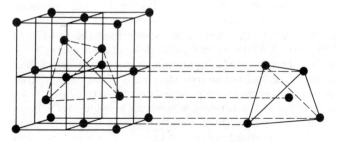

Fig. 26 *The crystal structure of diamond, showing (right) the tetrahedral arrangement of four carbon atoms around a central carbon atom.*

9 Atomic or covalent crystals

These are made up of atoms held together by covalent bonds. The structures are referred to as *giant lattices* or *giant molecules* because all the atoms are linked together.

The commonest completely covalent crystal structure occurs in diamond (Fig. 26). Each carbon atom is surrounded by four others arranged tetrahedrally. Silicon, germanium and grey tin have the same structure.

The *zinc blende* (ZnS) structure is closely related to that of diamond, the only difference being that adjacent atoms are different. Thus each zinc atom is surrounded by four sulphur atoms arranged tetrahedrally, and each sulphur atom by four zinc atoms similarly arranged (Fig. 27).

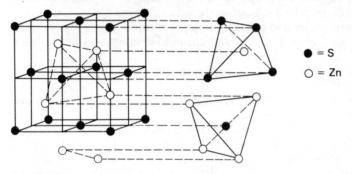

● = S

○ = Zn

Fig. 27 The crystal structure of zinc blende, showing (upper right) the tetrahedral arrangement of four sulphur atoms about a zinc atom, and (lower right) the tetrahedral arrangement of four zinc atoms about a sulphur atom. Compare the structure of diamond (Fig. 26).

Questions on Chapter 3

1 Give an account of the origin and development of the idea of valency.

2 (a) Distinguish in electronic terms between electrovalent (ionic), covalent and dative bonding. (b) Show how the results of the study of the structures of solids illustrate these types of chemical bonding and demonstrate the reality of the molecule and ion as chemical entities. (JMB)

3 What factors contribute most strongly to the formation of (a) a stable ionic compound and (b) a stable covalent compound between two elements?

4 If X, Y and Z represent elements of atomic numbers 9, 17 and 55 respectively, predict the type of bonding you would expect to occur between (a) X and Y, (b) X and Z, (c) Y and Z. Draw bond diagrams for the compounds formed, showing electronic configurations. Predict, giving your reasons, the relative (i) volatility, (ii) electrical conductance, (iii) solubility in various solvents, to be expected from these compounds. (SUJB)

5 Show the arrangement of bonds in any ten molecules not mentioned in this chapter that contain dative bonds.

6 What are covalency and electrovalency? Illustrate your answer with reference to chlorine and potassium chloride. Give the shapes of s and p orbitals and, in terms of orbitals, account for the shape of the carbon tetrachloride molecule. Explain why hydrogen chloride, which is covalent in the gaseous state, is ionic in water. (C)

7 'Each of the concepts electrovalency and covalency relates to an idealised state of chemical bonding which often does not exist in real compounds.' Discuss how far this statement is valid and give two examples, with suitable evidence, of cases where such 'non-ideality' in fact arises. (L)

8 In what ways are the properties of the following not typical of covalent substances: diamond, water, carbon dioxide, ammonia?

9 Discuss the order and principal features of topics you would adopt in explaining fundamental ideas about chemical combination and valency. (SUJB)

10 Discuss the statement 'Chemistry is the property of the outer shell of electrons.' (OC)

11 What are (a) atomic and (b) molecular orbitals? How do s, p, d, σ and π orbitals differ from each other?

12 Discuss the assumptions which are made when predicting the structures of molecules formed from non-transition elements by considering the repulsions between electron pairs in the molecule. Illustrate your answer (a) by reference to the structures of BF_3, NF_3 and PF_5, (b) by explaining why the bond angle in NH_3 is greater than that in H_2O. Comment on the structure of, and bonding in, XeF_4, PF_6^- and NH_4^+. Deduce the shapes of PF_3, H_3O^+ and ICl_2^-. (JMB)

13 Compare the shapes of the following, accounting for any differences; (a) molecules of carbon dioxide and sulphur dioxide, (b) molecules of hydrogen peroxide and ethyne, (c) ClO_3^- and NO_3^- ions.

14 What shapes would you predict for the following: (a) BrF_5, (b) IF_7, (c) ClF_3, (d) ICl_2^-, (e) ICl_4^-?

15 Predict the shapes of the following: (a) $TeCl_4$, (b) $COCl_2$, (c) $[Ag(CN)_2]^-$, (d) NO_2, (e) NH_2Cl, (f) PCl_6^-.

16 Draw diagrams to illustrate the shape and symmetry of s and p atomic orbitals, labelling them appropriately. Show how the idea of a hybridised orbital is necessary to describe the shape of molecules such as ethane, ethene and ethyne, and give a simple description of bonding in these molecules.

The shapes of many simple inorganic molecules can be explained in terms of the electrostatic repulsions of bonding and non-bonding pairs of electrons. Show how these ideas predict the shapes of the XeF_4 and SF_4 molecules. (O)

17 What do you understand by a 'chemical bond'? The 'dot-and-cross' diagram provides a useful elementary picture for many simple molecules and ions. Discuss the inadequacies of this picture, and how it has been supplemented by other models in order to meet the inadequacies.

Your answer should include general points, but be based on specific examples taken from your study of chemistry. You are free to choose your examples, but you may wish to discuss the nature of the bonding in hydrogen chloride, silver chloride, inorganic oxo-anions, and unsaturated hydrocarbons. (L)

18 (a) What are the shapes of the following species: $SO_3{}^{2-}$, $SO_4{}^{2-}$, H_2S, N_2O, H_2O_2? For each species your answer should consist of a clear sketch, appropriately labelled, and a brief description. Details of bonding or explanations are *not* required. (b) Describe the crystal structures of the following: graphite, sodium chloride. (c) Discuss the bonding in the complex $NH_3 . BF_3$. Your answers to (b) and (c) should include appropriate diagrams. (O)

19 Explain, by means of diagrams, what you understand by the following terms as they apply to the shapes of covalent molecules: (a) triangular planar, (b) triangular pyramidal, (c) tetrahedral, (d) octahedral. By referring to the appropriate theory, explain why (e) $BeCl_2$ is a linear molecule whereas H_2O is bent, (f) NH_3 is triangular pyramidal whereas BCl_3 is triangular planar, (g) the bond angle in NH_3 is less than that in CH_4, (h) the molecule of CO_2 has no dipole moment whereas the molecule of SO_2 possesses one. (L)

20 (a) State, with the aid of an example in each case, what is meant by each of the following types of lattice structures in solids: (i) covalent molecular, (ii) covalent macromolecular, (iii) ionic. For (iii) explain how the 'radius ratio' is related to the type of ionic lattice.

(b) Account for the shapes, stating the bond angles, of the molecules of the following compounds: (i) methane, (ii) ammonia, (iii) water, (iv) boron trichloride, (v) phosphorus pentafluoride. (AEB)

21 Study the bond enthalpy values given on p. 38, and comment on any significant points.

22 What differences are there between average bond enthalpy, average bond energy and bond dissociation enthalpy?

23 What differences are there between single, double and triple bonds between like elements?

24 What is meant by the electronegativity of an element, and how can its value be of use?

25 What is meant by delocalisation, and how does it affect the properties of a compound? Give illustrative examples.

26* Give examples of compounds in which the bond enthalpy of a particular bond is affected by (a) change in oxidation state of the atoms concerned, and (b) change in molecular shape.

27* Can you find any relationship between the electronegativity values of A and X and the dipole moment of the A—X bond?

28* Comment on the measured and calculated values for the bond distances in C_6H_6, H_2O, CO, N_2O and $CO_3{}^{2-}$.

4
Metallic, hydrogen and van der Waals' bonding

1 Metals and non-metals

There are many broad differences of property between metals and non-metals, as summarised below.

metals	non-metals
characteristic metallic crystals (p. 50) with high coordination number and metallic bonding *(closely bonded together)*	usually molecular crystals containing covalently-bound molecules held by van der Waals' forces (p. 55)
high electrical conductivity	non-conductors
high thermal conductivity	low thermal conductivity
exhibit thermionic emission	no thermionic emission
lustrous and can be polished	not lustrous
exhibit photoelectric emission	no photoelectric emission
high melting and boiling temperatures	low melting and boiling temperatures
high density	low density
malleable (can be hammered into shape); ductile (can be drawn out into wires)	not malleable (brittle); not ductile
form many alloys	minor components of alloys
form + ve ions by loss of electrons; electropositive	form − ve ions by gain of electrons; electronegative
form basic or amphoteric oxides when exerting their normal valency	form acidic or neutral oxides when exerting their normal valency
form basic or amphoteric hydroxides	form hydroxides which, by loss of water, become oxoacids
form solid, ionic halides	form liquid, covalent halides
do not generally form stable hydrides (p. 151)	form covalent hydrides which are usually stable gases

This summary is made up of broad generalisations and it is not difficult to find particular exceptions. Graphite, for instance, is a non-metal, but it is quite lustrous and a fair conductor of electricity; caesium, sodium and potassium have low enough densities to float on water; hydrogen, generally regarded as a non-metal, forms positive ions; boron, silicon and carbon have high melting and boiling temperatures; and so on.

Metals, nevertheless, do have very distinctive properties, and to account for them it is necessary to develop the idea of a special metallic bond.

2 The metallic bond

The simplest picture of the state of affairs in a metallic crystal is that positive ions of the metal are packed in a regular array within a 'sea' of electrons liberated from the atoms (Fig. 28). It is significant that only atoms with relatively low ionisation energies will form metallic crystals. It is the 'sea' of electrons that binds the positive atoms together, but the electrons are, also, relatively free and mobile so that they can move throughout the crystal under an applied potential difference.

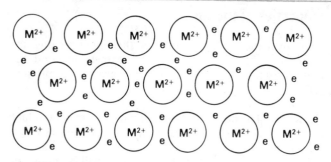

Fig. 28 *Representation of the 'sea' of delocalised electrons in the crystal of a metal with two available electrons per atom.*

The nature of the bonding in metals is peculiar, however, because there is such a shortage of electrons. Lithium, for example, has only three electrons per atom, and two of them, the 1s electrons, are very firmly held by the nucleus. In lithium, then, there is only one electron per atom available for any metallic bonding and yet each lithium has eight, or fourteen, near neighbours in its crystal (p. 50). It is impossible, then, for anything like individual bonds to exist in metals. The few electrons available for bonding purposes must be completely delocalised (p. 41). Molecular orbitals can be established for the metal as a whole, and this approach leads to the idea of energy bands within the crystal of a metal.

The arrangement of electrons in a single sodium atom is as follows:

1s	2s	2p	3s
2	2	6	1

each electron being in a particular energy level. If two sodium atoms are brought close to each other, the 3s orbital of the one can combine with the 3s orbital of the other to form two molecular orbitals as, for comparison, in the hydrogen molecule (p. 30). If a third sodium atom is introduced, three molecular orbitals will be possible; and for n atoms, n molecular orbitals can be formed. The final number of molecular orbitals will be equal to the number of atoms concerned, and this will be very large for any actual crystal of sodium. Each orbital will have its own particular energy level within a certain range, but the various energy levels will be so close together that the orbitals will make up an energy band containing n energy levels.

In a similar way it will be possible for 3p orbitals to combine to form another energy band and this band may overlap the s band. Even 2s and 2p orbitals may form energy bands but the electrons in the inner atomic orbitals generally stay in these orbitals and do not occupy molecular orbitals.

The existence of energy bands within crystals of metals is borne out by a study of X-ray spectra, but the details are beyond the scope of this book.

3 Physical properties of metals

The strong bonding in metals is responsible for their high melting and boiling temperatures (p. 121), and the idea of the metallic bond can account, reasonably satisfactorily, for many other characteristic physical properties of metals.

a Electrical conductivity The main physical hallmark of a metal is a high electrical conductivity which falls with a rise in temperature and which is greatly affected by impurities. (molecules are separated.)

In the 3s band of a sodium crystal, for example, there will be n possible molecular orbitals each capable of holding two electrons. The band, as a whole, can therefore hold $2n$ electrons, but as each sodium atom in the crystal provides only one 3s electron the total number of available electrons is only n so that the band is not full. As the energy differences between the orbitals in the 3s band are very small, the n 3s electrons can move about within the 3s band very easily. As the 3p band also overlaps the 3s band there are still greater possibilities for electron 'mobility'.

In a crystal under ordinary conditions, equal numbers of 3s electrons will move in all directions, but under an applied potential difference, the electrons will move preferentially in one direction and a current will flow. It is this movement of electrons, within what is sometimes known as a conduction band, which accounts for electrical conductivity.

For a metal with an even number of valency electrons, e.g. Mg, 2.8.2, it might be expected that all the levels in the 3s band would be full, because each atom contributes two 3s electrons. Magnesium is, however, still a good conductor and this is because some of the levels in the 3p band overlap those in the 3s band and thus provide opportunities for movement of electrons.

Even in aluminium, with $3n$ electrons available, the overlapping 3s and 3p bands are not full.

Electrical conductivity is associated, then, either with partially filled bands or with the overlapping of an unfilled band with a full one. The univalent metals, i.e. the alkali metals and the coinage metals, have the highest conductivities. They only have one outermost electron in the atomic orbitals, but the n electrons in the outermost energy bands have the greatest 'mobility'.

b Effect of temperature and impurities on electrical conductivity   It might be expected that an increase in temperature would cause more electrons to be 'promoted' into conducting bands with a consequent increase in conductivity or decrease in resistance. In general, however, the conductivity of a metal decreases with increasing temperature, i.e. the resistance increases. This is because increasing temperature produces increased thermal vibration within a metal crystal. This upsets the regularity within the crystal and interferes with the ease of movement of electrons through the crystal. It is rather like comparing movement through the ranks of a battalion of soldiers on parade with that through a London crowd. At very low temperatures, when vibration is negligible and the crystal assumes its most orderly array, metals may become superconducting. The regular array within a crystal may also be upset by the introduction of impurities. The resistance of copper, for example, is greatly increased by even traces of impurities.

c Thermal conductivity   The mobility of the electrons in a metal accounts for the high thermal conductivity, for energy absorbed as thermal energy in one part of the metal can easily be transferred right through the metal.

d Thermionic emission   At higher temperatures, metals will lose electrons in so-called thermal emission. The effect was first observed by Edison in 1883 and used by Fleming in 1904 in developing the valve.

e Lustre   The electrons in a metal can absorb energy from incident light and, in the process, they are promoted into excited levels within an energy band, or even into a higher energy band. On returning to their original level of energy, they re-emit light of the same wavelength as that which was absorbed. This accounts for the light reflection (lustre) of metals, and for the opacity, for no light is transmitted through even very thin films of metal.

f Photoelectric emission   If a metal is irradiated with light of the appropriate wavelength, electrons are sufficiently energised to be emitted. The rate of emission depends on the intensity of the light (so long as it is of the right wavelength); the energy of the emitted electrons depends on the wavelength of the irradiation. The alkali metals show the strongest photoelectric effect, and caesium, in particular, is used in photoelectric cells.

4 Metallic crystals

A metallic crystal contains metal atoms held together by metallic bonds. The atomic units in the crystal are all alike, so the radius ratio (p. 14) is 1. The structures have high coordination numbers (p. 14) and the great majority of metals (see below) crystallise with one of three structures:

a cubic close-packed, or face-centred cubic, structure
b hexagonal close-packed structure and
c body-centred cubic structure.

Many metals can crystallise with more than one structure.

Crystal structures of the metals

Li	Be									
c	b									
Na	Mg									
c	b									
K	Ca	Sc	Ti	V	Cr	Mn	Fe	Co	Ni	Cu
c	a,b	a,b	b,c	c	b,c	—	a,c	a,b	a,b	a
Rb	Sr	Y	Zr	Nb	Mo	Tc	Ru	Rh	Pd	Ag
c	a	b	b,c	c	b,c	b	a,b	a	a	a
Cs	Ba	La	Hf	Ta	W	Re	Os	Ir	Pt	Au
c	c	a,b	b,c	c	c	b	a,b	a	a	a

a Close-packed structures In close-packed structures the units are packed together in the tightest possible way to fill the maximum amount of available space. The tightest way of packing spheres in two dimensions is to have each sphere surrounded hexagonally by six others (Fig. 29).

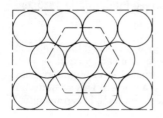 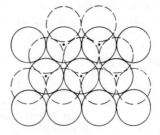

Fig. 29 Close-packing of spheres: in one layer (left) and in two layers (right).

A second layer of spheres fitted into the holes in the first layer gives the tightest packing for two layers (Fig. 29).

There are, now, two alternative ways of fitting in a third layer of spheres, each giving an equally close-packed structure. The spheres of the third

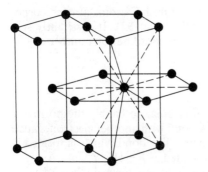

Fig. 30 Hexagonal close-packed structure showing a coordination number of 12.

layer may be placed so as to give a pattern that repeats itself every third layer, i.e. an ABABABA... structure. This is the *hexagonal close-packed* structure (Fig. 30). The coordination number is 12.

If, however, the third layer is placed so as to give a pattern that repeats itself every fourth layer, i.e. an ABCABCABC... structure, the *cubic close-packed* structure results. This structure is also known as the face-centred cubic structure for it has atoms at each corner and each face-centre of a cube; the coordination number is 12 (Fig. 31).

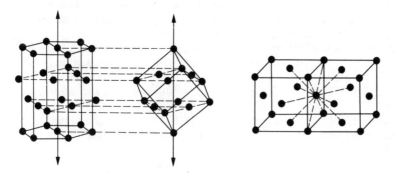

Fig. 31 Cubic close-packed structure, showing how the ABCA close-packing can be represented as a face-centred cubic structure (left), and the coordination number of 12 (right).

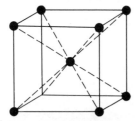

Fig. 32 Body-centred cubic structure.

b Body-centred cubic structure This structure has atoms at each corner of a cube and at the centre of the cube (Fig. 32). It is slightly less close-packed than the structures in **a**; it has a coordination number of 8.

5 Hydrogen bonding

A number of X—H bonds are sufficiently polar to enable the positively charged hydrogen atom to form a weak bond with another electronegative atom, Y, that has one or more lone pairs. The bond which is formed is known as a hydrogen bond and is represented as X—H···Y.

The strongest, and shortest, hydrogen bonds are formed when X and Y are both highly electronegative, and the commonest examples involve fluorine, oxygen or nitrogen atoms. Weaker hydrogen bonding can also occur when C, P, S, Cl, Br or I atoms are involved.

The strength of most hydrogen bonds $(15-30\,\text{kJ mol}^{-1})$ is some ten times smaller than that of normal covalent bonds, but the F—H···F bonding in the hydrogendifluoride ion, $[HF_2]^-$ is almost as strong (about $150\,\text{kJ mol}^{-1}$) as a weak covalent bond.

The bonding in $[HF_2]^-$ is also unusual because the hydrogen atom is equidistant $(0.113\,\text{nm})$ from the two fluorine atoms; the hydrogen bond is said to be *symmetrical*. In other substances, the hydrogen atom is not placed centrally, and the bond is said to be *unsymmetrical*. The X—H distance is smaller than the H—Y distance, but the values vary from substance to substance, even for the same atoms X and Y.

The nature of the hydrogen bonding in unsymmetrical cases is thought to be essentially electrostatic. The positively charged hydrogen atom in the X—H bond is very small, so it will be surrounded by a strong electrical field. Its small size will also allow other small atoms (Y) to approach closely.

In symmetrical hydrogen bonding, as in $[HF_2]^-$, it is suggested that two one-electron bonds, $[F\cdots H\cdots F]^-$ might occur, involving the delocalisation of two electrons over three centres.

a Hydrogen fluoride Solid hydrogen fluoride contains a zig-zag arrangement of F—H···F atoms in long chains; the bond angle is about 140° (Fig. 33). The zig-zag arrangement arises because the positively charged hydrogen atom in the F—H bond approaches the adjacent fluorine atom in the direction of one of its lone pairs.

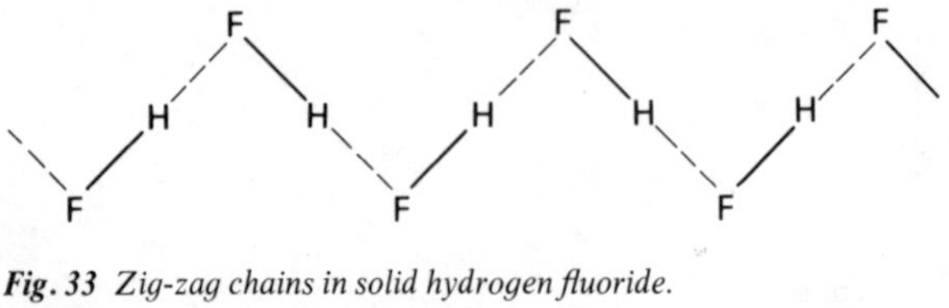

Fig. 33 Zig-zag chains in solid hydrogen fluoride.

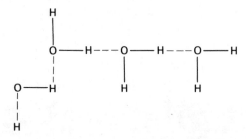

Fig. 34 *Hydrogen bonds causing association in water.*

b Hydrides of fluorine, oxygen, nitrogen and carbon Association, like that in hydrogen fluoride, also occurs in water (Fig. 34) and ammonia. It shows itself in the abnormally high melting and boiling temperatures of these hydrides compared with those of other hydrides in the same group of the periodic table (Fig. 35).

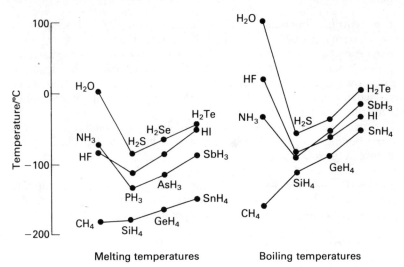

Fig. 35 *Melting and boiling temperatures of hydrides of Group 4B, 5B, 6B and 7B elements.*

Methane has 'normal' values because the C—H bond in methane is not sufficiently polar to participate in hydrogen bonding. The same bond in trichloromethane, $CHCl_3$, is much more polar so that it will form some weak hydrogen bonds.

c Ice The crystal structure of ice shows a tetrahedral arrangement of water molecules, similar to the arrangement of zinc and sulphur atoms in the wurtzite structure. Each oxygen atom is surrounded tetrahedrally by four others, hydrogen bonds linking pairs of oxygen atoms together, as shown in Fig. 36.

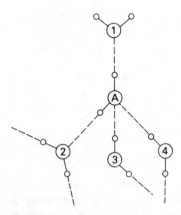

Fig. 36 *The crystal structure of ice. The central oxygen atom, A, is surrounded tetrahedrally by the oxygen atoms, 1, 2, 3 and 4. All other oxygen atoms are arranged similarly. The hydrogen atoms are shown as small circles.*

The hydrogen bonding is unsymmetrical, with the hydrogen atom 0.096 nm from one oxygen atom and 0.18 nm from the other.

The arrangement of the water molecules in ice is a very open structure and this explains the low density of ice. When ice melts, the structure breaks down and the molecules pack more closely together so that water has a higher density. But this breaking-down process is not complete until a temperature of 4 °C is reached.

Hydrogen sulphide shows no signs of hydrogen bonding; it has a normal boiling temperature (Fig. 35) and it has a close-packed crystal structure in the solid state.

d Copper(II) sulphate-5-water Gentle heating of copper(II) sulphate pentahydrate will convert it into the monohydrate: it is, in fact, efflorescent in hot, dry climates. A much higher temperature is required, however, to

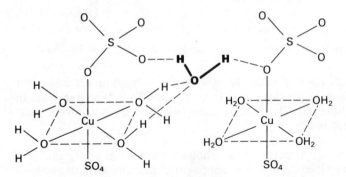

Fig. 37 *The arrangement in $CuSO_4.5H_2O$. Each Cu^{2+} ion is surrounded octahedrally by four H_2O molecules and two SO_4^{2-} ions. One H_2O molecule (shown in bold type) is linked between two such octahedral groups by hydrogen bonding.*

remove the final molecule of water of crystallisation from the mono-hydrate. This suggests that one of the molecules of water of crystallisation is different from the other four, and leads to the writing of the formula as $[Cu(H_2O)_4]SO_4.H_2O$. The crystal structure is shown in Fig. 37.

$[Cu(NH_3)_4]SO_4.H_2O$ exists and has a structure like that of $[Cu(H_2O)_4]SO_4.H_2O$, but $CuSO_4.5NH_3$ does not exist. This is probably due to the fact that NH_3 molecules will not form hydrogen bonds so readily as H_2O molecules.

e Sodium hydrogencarbonate, $NaHCO_3$ In sodium hydrogencarbonate crystals, $HCO_3{}^-$ groups are linked together into infinite chains by hydrogen bonds (Fig. 38). The chains are held laterally by Na^+ ions between them.

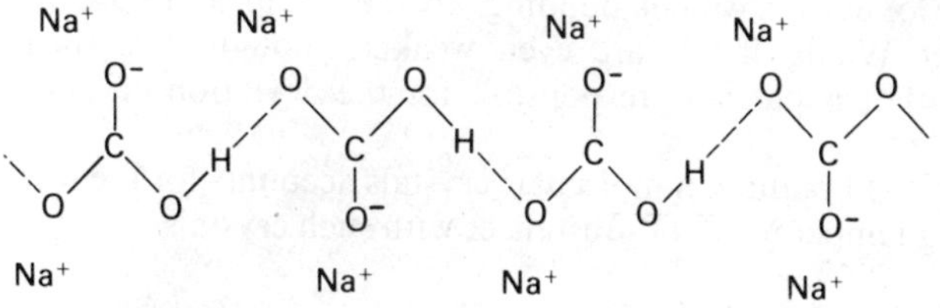

Fig. 38 The arrangement in $NaHCO_3$ crystals. Infinite chains of $HCO_3{}^-$ units, linked together by hydrogen bonds, are held together laterally by Na^+ ions.

f Boric acid, H_3BO_3 Crystals of boric acid are soft, and cleave easily. They are made up of infinite layers of $B(OH)_3$ units linked together by hydrogen bonds (Fig. 39).

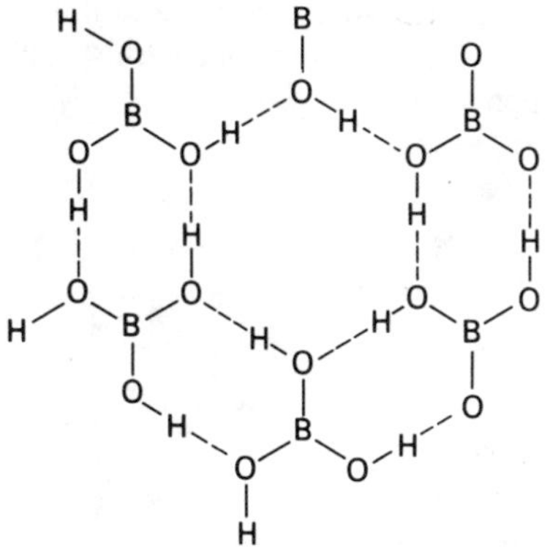

Fig. 39 Layers of $B(OH)_3$ units held together by hydrogen bonds in boric acid crystals.

6 Van der Waals' forces

Substances which are made up of individual, covalent molecules, e.g. oxygen and iodine, or of individual atoms, e.g. the noble gases, must have some cohesive forces holding the particles together in the solid state. These are the van der Waals' forces.

When the molecules held together in a molecular crystal are polar in nature, e.g. hydrogen chloride and ammonia, there will be a dipole–dipole attraction between molecules. Such attractive forces will also be increased by the fact that one dipole will induce another in a neighbouring molecule.

With non-polar substances such as iodine and the noble gases there are no initial dipoles; the molecules or atoms are electrically symmetrical. The forces in these cases are caused by slight and temporary displacements between nuclei and electrons in an atom, giving rise to temporary dipoles, which lead to attractive forces (sometimes referred to as London forces) in the same way as the more permanent dipoles in polar molecules.

Van der Waals' forces in solids are ten to twenty times weaker than those involved in ionic or covalent bonding. In the liquid and gaseous states the van der Waals' forces are even weaker, though it is their existence in gases that is partially responsible for the deviation of gases from the ideal gas laws.

The weakness of the bonding in molecular crystals accounts for the low melting and boiling temperatures of substances with such crystals.

dipole forces and Van der Waals' forces

7 Molecular crystals *are both weak.*

The structural units in a molecular crystal may be atoms, as in solidified noble gases, or molecules. With small molecules, which can be regarded as approximately spherical, the crystal structure is an approximately close-packed arrangement. Iodine and carbon dioxide (Fig. 40) provide typical examples and the other halogens, hydrogen, nitrogen, oxygen, ammonia, phosphine, hydrogen sulphide, carbon monoxide, hydrogen chloride, methane all exist as molecular crystals when solid.

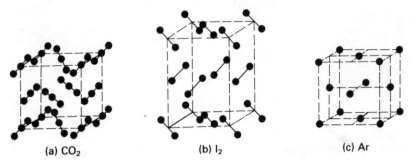

(a) CO_2 (b) I_2 (c) Ar

Fig. 40 *Molecular crystals of (a) CO_2, (b) I_2 and (c) Ar.*

With large molecules, the arrangement of the structural units becomes more complicated, although it is still as close-packed as the shape of the molecules will allow.

8 Layer lattice and chain structures

In a layer lattice structure, e.g. graphite (p. 202), H_3BO_3 (p. 55) and $AlCl_3$ (p. 193), atoms or ions are linked together in well-defined layers and adjacent layers are linked by weak van der Waals' forces or hydrogen bonds. This results in crystals with very clear-cut cleavage planes parallel to the planes of the layers. The crystals also tend to be anisotropic, i.e. they have physical properties which vary according to the direction in which they are measured.

In chain structures, the atoms are linked in long chains, and the chains are held to each other by van der Waals' bonding. Examples are provided by copper(I) chloride, silicon disulphide (p. 218), the catena forms of sulphur (p. 283) and solid hydrogen fluoride (p. 312).

Questions on Chapter 4

1 Solids may be classified, on the basis of the types of forces that hold them together, as (a) ionic, (b) covalent, (c) molecular or (d) metallic. Give one example each of (a), (b) and (c). In each case state what type of force operates to keep the substance in the solid state and give a critical comparison of the physical properties of the substances you select. List five properties which are characteristic of metals such as copper; show qualitatively why these properties cannot be accounted for by the types of force operative in (a), (b) and (c) and why a 'metallic' bond must be postulated. (JMB)

2 Discuss the bonding in (a) hydrogen chloride, (b) silica, (c) ice, (d) hexaamminecobalt(III) chloride and (e) tin(IV) chloride. In each case mention one property which is indicative of the type of bonding involved.

Describe the structures of sodium chloride and of caesium chloride in their solid state, and account for any differences which are observed. (JMB)

3 Give an account of the forces which hold together the particles (atoms, ions, molecules) in (a) a simple crystalline salt, (b) a metal, (c) water, (d) the coordination compound $[Cu(NH_3)_4]SO_4.H_2O$. Discuss how the physical properties of these materials are related to the binding forces operating in them. (L)

4 The basic crystal structures are body-centred and face-centred. Draw each structure and give some examples of substances with the structures.

5 What is anisotropy? Give some examples.

6 By reference to the bonding in solid copper(II) sulphate pentahydrate, or any other salt hydrate of your choice, show what you understand by the following terms: (a) covalent bond, (b) ionic bond, (c) coordinate bond, (d) hydrogen bond.

On gentle heating, copper(II) sulphate pentahydrate loses some of its water of crystallisation. In a typical experiment, 5.00 g of the pentahydrate gave 4.28 g of a residue. Determine the formula of the residue. On further very strong heating, a black solid was formed. What is this black solid? Calculate its mass. [Relative atomic masses: H = 1, Cu = 63.5, O = 16, S = 32.] (L)

7* Calculate the average volume occupied by one molecule in (a) ice at 0 °C, (b) water at 0 °C, (c) water at 4 °C and (d) steam at 100 °C.

8* Copper has a face-centred crystal. Use the density and relative atomic mass values to calculate (a) the apparent volume of a copper atom and (b) the radius of a copper atom.

5
Enthalpy changes

1 Heat changes

In most chemical reactions there is an associated energy change, shown by the evolution of heat in an exothermic process, or by an absorption of heat in an endothermic one. It is generally possible to measure what the *change* is, even though it may not be possible to determine the absolute energy levels before and after the change.

Heat changes measured at constant volume are known as *internal energy* changes (ΔU). Those measured at constant pressure are known as *enthalpy* changes (ΔH). If no volume change is involved, ΔU and ΔH are equal, but with a volume change, a difference between ΔH and ΔU arises because work is involved in expansion or contraction. As most chemical operations are carried out in open vessels, they take place at constant pressure. So it is the ΔH values that are concerned. The values depend on the conditions and on the amounts of the substances involved; they are commonly expressed as *standard molar enthalpy changes*. The standard conditions chosen are a pressure of $101.325\,kN\,m^{-2}$ and a temperature of $25\,°C$. If solutions are involved, concentrations of unit activity (approximately $1\,mol\,dm^{-3}$) are specified. Standard molar enthalpy changes are then symbolised by $\Delta H_m^{\ominus}(298\,K)$, but the simpler symbol, ΔH, may be used when there is no ambiguity. Subscripts may be used to indicate a particular type of enthalpy change; for example $\Delta H_{c,m}^{\ominus}(298\,K)$ refers to a standard molar enthalpy change of combustion. The units used for ΔH_m values are $kJ\,mol^{-1}$.

ΔH has negative values (heat evolved) for exothermic reactions, and positive values (heat absorbed) for endothermic reactions.

a Enthalpy of combustion   The standard molar enthalpy of combustion of a substance is the enthalpy change when 1 mol of it is completely burnt in oxygen under standard conditions, e.g.

	$\Delta H_{c,m}^{\ominus}(298\,K)$
$CH_4(g) + 2O_2(g) \longrightarrow CO_2(g) + 2H_2O(l)$	$-890.5\,kJ\,mol^{-1}$
$C(s) + O_2(g) \longrightarrow CO_2(g)$	$-393.5\,kJ\,mol^{-1}$

b Enthalpy of formation of compounds   The standard molar enthalpy of formation of a compound is the enthalpy change when 1 mol of the compound is formed from its elements under standard conditions, e.g.

$$H_2(g) + \tfrac{1}{2}O_2(g) \longrightarrow H_2O(l) \qquad \Delta H_{f,m}^{\ominus}(298\,K) = -285.9\,kJ\,mol^{-1}$$
$$2H_2(g) + O_2(g) \longrightarrow 2H_2O(l) \qquad \Delta H_{f,m}^{\ominus}(298\,K) = -571.8\,kJ\,mol^{-1}$$

As there is no energy change in forming an element under standard conditions the standard enthalpies of all pure elements in their standard states are zero.

The standard enthalpy of formation of an endothermic compound is positive and that of an exothermic compound negative, e.g.

exothermic	$NH_3(g)$	$HCl(g)$	$Fe_2O_3(s)$	$CO_2(g)$	$CH_4(g)$
$\Delta H_{f,m}{}^{\ominus}$(298 K)	-46	-92.3	-822	-393.5	-74.8

endothermic	$PH_3(g)$	$HI(g)$	$Au_2O_3(s)$	$CS_2(l)$	$SiH_4(g)$
$\Delta H_{f,m}{}^{\ominus}$(298 K)	5.4	26.5	80.8	87.9	34.3

It will be seen that compounds with similar formulae may be either exothermic or endothermic, and the underlying reasons are of great interest and significance.

c Enthalpy of formation of hydrated ions   This refers to the formation of 1 mol of hydrated ions from an element in its standard state, e.g.

$$Na(s) \longrightarrow Na^+(aq) + e^- \qquad \Delta H_{f,m}{}^{\ominus}(298\,K) = -240\,kJ\,mol^{-1}$$

The figures are usually quoted on an arbitrary scale on which the enthalpy of formation of $H^+(aq)$ is taken as zero, i.e.

$$\tfrac{1}{2}H_2(g) \longrightarrow H^+(aq) + e^- \qquad \Delta H_{f,m}{}^{\ominus}(298\,K) = 0$$

Absolute values can be calculated as explained on p. 17.

d Enthalpy of reaction   Tabulated values of the standard enthalpies of formation of compounds and/or hydrated ions can be used to calculate the standard enthalpy change for any reaction, for

$$\left\{\begin{array}{l}\text{standard}\\\text{enthalpy}\\\text{change of}\\\text{a reaction}\end{array}\right\} = \left\{\begin{array}{l}\text{sum of standard}\\\text{enthalpies of}\\\text{formation of}\\\text{products}\end{array}\right\} - \left\{\begin{array}{l}\text{sum of standard}\\\text{enthalpies of}\\\text{formation of}\\\text{reactants}\end{array}\right\}$$

Typical examples are summarised below.

$$\Delta H_m{}^{\ominus}(298\,K)/kJ\,mol^{-1}$$

$CO(g) + \tfrac{1}{2}O_2(g) \longrightarrow CO_2(g)$ $(-110.5)\ \ 0 \qquad\qquad (-393.5)$	$-393.5 - (-110.5) = -283.0$
$H_2O(g) + C(s) \longrightarrow CO(g) + H_2(g)$ $(-241.8)\ \ 0 \qquad\quad (-110.5)\ \ 0$	$-110.5 - (-241.8) = +131.3$
$Ag^+(aq) + Cl^-(aq) \longrightarrow AgCl(s)$ $(105.6) \qquad (-167.1) \qquad (-127.0)$	$-127 - (-167.1) - 105.6 = -65.5$

When more than 1 mol of a reactant or product is involved, this must be taken into account, e.g.

$$2SO_2(g) + O_2(g) \longrightarrow 2SO_3(g)$$
$$2(-297) \quad 0 \qquad\quad 2(-395)$$

The enthalpy change, in $kJ\,mol^{-1}$, is $2(-395) - 2(-297)$, i.e. $-196\,kJ\,mol^{-1}$.

e Enthalpy of atomisation This refers to the formation of 1 mol of gaseous atoms from an element in its standard state, e.g.

$$C(graphite) \longrightarrow C(g) \qquad \Delta H_{a,m}^{\ominus}(298\,K) = +715\,kJ\,mol^{-1}$$
$$\tfrac{1}{2}H_2(g) \longrightarrow H(g) \qquad \Delta H_{a,m}^{\ominus}(298\,K) = +218\,kJ\,mol^{-1}$$

When, as in the latter example, it is the bond in a covalent molecule that is being split, the enthalpy of atomisation is half the bond enthalpy. Thus the bond enthalpy for H—H is $436\,kJ\,mol^{-1}$, i.e.

$$H—H(g) \longrightarrow 2H(g) \qquad \Delta H_m^{\ominus}(298\,K) = +436\,kJ\,mol^{-1}$$

f Ionisation enthalpy or energy This refers (p. 9) to the conversion of 1 mol of free gaseous atoms into free gaseous cations, e.g.

$$Na(g) \longrightarrow Na^+(g) + e^- \qquad \Delta H_m(298\,K) = +502\,kJ\,mol^{-1}$$

ΔU values may be quoted instead of ΔH values (p. 9). The ΔU value for sodium is $+495.8\,kJ\,mol^{-1}$.

g Electron affinity This refers (p. 10) to the formation of 1 mol of free gaseous anions from 1 mol of free gaseous atoms, e.g.

$$Cl(g) + e^- \longrightarrow Cl^-(g) \qquad \Delta H_m(298\,K) = -354\,kJ\,mol^{-1}$$

The corresponding ΔU value (p. 10) is $-348\,kJ\,mol^{-1}$.

h Lattice enthalpy This is the name given to the enthalpy change for the reaction in which 1 mol of a crystalline solid is formed from its component ions in the gaseous phase, e.g.

$$Na^+(g) + Cl^-(g) \longrightarrow NaCl(s) \qquad \Delta H_m(298\,K) = -788\,kJ\,mol^{-1}$$

i Enthalpy of solution This most commonly refers to the formation of an infinitely dilute solution, and the molar enthalpy of solution at infinite dilution is defined as the enthalpy change when 1 mol of solute dissolves in such a large excess of solvent that addition of further solvent produces no further enthalpy change, e.g.

$$NaCl(s) + aq \longrightarrow Na^+(aq) + Cl^-(aq) \qquad \Delta H_{s,m}(298\,K) = +4\,kJ\,mol^{-1}$$
$$AgCl(s) + aq \longrightarrow Ag^+(aq) + Cl^-(aq) \qquad \Delta H_{s,m}(298\,K) = +66\,kJ\,mol^{-1}$$

j Enthalpy of hydration   This refers to the enthalpy change when 1 mol of gaseous ions is hydrated, e.g.

$$Na^+(g) + aq \longrightarrow Na^+(aq) \qquad \Delta H_{\text{hyd,m}}^{\ominus}(298\,\text{K}) = -406\,\text{kJ mol}^{-1}$$

The numerical values involved are discussed on p. 17.

2 Enthalpy diagrams

Hess's law states that the total enthalpy change for any process is independent of the way in which the process is carried out. The enthalpy change for the conversion of carbon into carbon dioxide, for example, is the same whether it is carried out in one stage

$$C(s) + O_2(g) \longrightarrow CO_2(g) \qquad \Delta H_m^{\ominus}(298\,\text{K}) = -393.5\,\text{kJ mol}^{-1}$$

or in two stages

$$C(s) + \tfrac{1}{2}O_2(g) \longrightarrow CO(g) \qquad \Delta H_m^{\ominus}(298\,\text{K}) = -110.5\,\text{kJ mol}^{-1}$$

$$CO(g) + \tfrac{1}{2}O_2(g) \longrightarrow CO_2(g) \qquad \Delta H_m^{\ominus}(298\,\text{K}) = -283.0\,\text{kJ mol}^{-1}$$

The relation between the various enthalpy changes is conveniently summarised in an enthalpy diagram (Fig. 41).

Such diagrams can also be used to obtain values for enthalpy changes

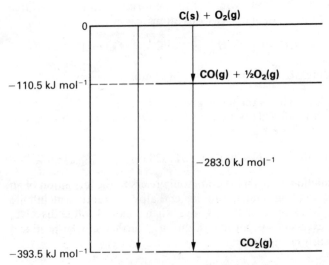

Fig. 41 *A simple enthalpy diagram showing the relationship between the molar enthalpies of formation of CO_2 from C and O_2 and from CO and O_2.*

that cannot be measured directly. The enthalpy of formation of methane, for example,

$$C(s) + 2H_2(g) \longrightarrow CH_4(g) \qquad \Delta H_m^{\ominus}(298\,K) = X\,kJ\,mol^{-1}$$

cannot be measured directly, for carbon and hydrogen do not react together. The enthalpies of combustion of carbon $(-393.5\,kJ\,mol^{-1})$, hydrogen $(-285.9\,kJ\,mol^{-1})$ and methane $(-890.5\,kJ\,mol^{-1})$ can, however, all be measured (p. 20) and the required enthalpy of formation is obtainable from them $(X = -74.8\,kJ\,mol^{-1})$ as summarised in Fig. 42.

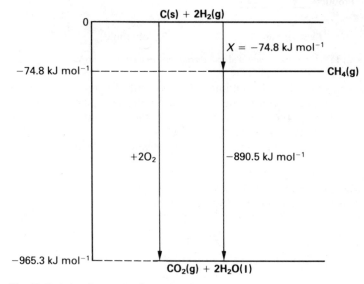

Fig. 42 *Enthalpy diagram to obtain the value for the standard molar enthalpy of formation of methane.*

3 Exothermic and endothermic reactions

Any chemical reaction involves a rearrangement of *atoms*. In the formation of carbon dioxide from carbon and oxygen, for example, the solid carbon is atomised to give gaseous carbon atoms and the oxygen molecules already in the gaseous state are split up into gaseous oxygen atoms. The carbon and oxygen atoms recombine to form gaseous CO_2 molecules.

In the formation of sodium chloride from sodium and chlorine, the solid sodium is atomised to give gaseous sodium atoms, which then lose electrons to form gaseous Na^+ ions. The bonds in chlorine molecules are split to give gaseous chlorine atoms, which gain electrons to form gaseous Cl^- ions. The Na^+ and Cl^- ions then pack together in an ionic crystal.

For an exothermic reaction the *total* binding of the atoms in the

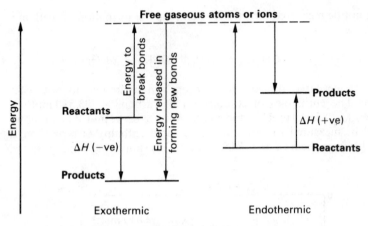

Fig. 43 *Energy changes in exothermic and endothermic reactions.*

products is stronger than the binding in the reactants. Energy has to be put in to convert the products into the reactants. For an endothermic reaction, the total binding of the atoms is stronger in the reactants than in the products. The situation is summarised in Fig. 43.

For reactions involving compounds with covalent bonding, it follows that

$$\begin{Bmatrix}\text{enthalpy}\\\text{change of}\\\text{reaction}\end{Bmatrix} = \begin{Bmatrix}\text{sum of bond}\\\text{enthalpies of}\\\text{reactants}\end{Bmatrix} - \begin{Bmatrix}\text{sum of bond}\\\text{enthalpies of}\\\text{products}\end{Bmatrix}$$

For example, the reaction

$$N_2(g) + 3H_2(g) \longrightarrow 2NH_3(g)$$
$$945 \qquad 3(436) \qquad 6(391)$$

involves the splitting of 1 mol of N≡N bond and 3 mol of H—H bonds, which requires 2253 kJ, and the formation of 6 mol of N—H bonds, which liberates 2346 kJ. The calculated enthalpy change for the reaction is, therefore, $(+2253 - 2346)$, i.e. -93 kJ. This is for the formation of 2 mol of ammonia, which is in good agreement with the measured $\Delta H_{f,m}$ value of -46 kJ mol^{-1}.

In an endothermic reaction, the energy required to break the bonds of the reactants is greater than the energy liberated in forming the bonds of the products, e.g.

$$N_2(g) + O_2(g) \longrightarrow 2NO(g)$$
$$945 \qquad 497 \qquad 2(627)$$

The calculated enthalpy change for the reaction is $+188$ kJ mol^{-1}, compared with a measured value of $+180.8$ kJ mol^{-1}.

Care must be taken in dealing with reactions involving solids and liquids, for energy is required to convert them into the gaseous state. Thus

$$H_2(g) + I_2(s) \longrightarrow 2HI(g)$$
$$436 \qquad 151 \qquad 2(299)$$

gives a value of $-11\,kJ\,mol^{-1}$ for the enthalpy change for the reaction. An energy term of $62.4\,kJ\,mol^{-1}$ must, however, be included for the conversion of $I_2(s)$ into $I_2(g)$, and this leads to a calculated enthalpy change for the reaction in the gaseous state of $+51.4\,kJ\,mol^{-1}$, as compared with a measured value of $+53.0\,kJ\,mol^{-1}$.

4 Feasibility of reaction

At one time it was thought that a negative value of ΔH was necessary for a reaction to be feasible, i.e. that only exothermic reactions would take place. Such an idea is no longer tenable, for endothermic reactions, with ΔH positive, though much less common than exothermic reactions, will take place under certain conditions.

As will be seen (p. 70), a reaction is only feasible if it has a negative *free energy change* (ΔG). For many reactions, particularly at low temperatures, it is the ΔH value that makes the largest contribution to the ΔG value for a reaction so that *some* indication as to whether or not a reaction is feasible can be deduced from the ΔH value. The more negative the ΔH value for a reaction, the more likely it is to be feasible.

Typical reactions that are highly feasible, and well known to be so, are

$$\Delta H_m^{\ominus}(298\,K)$$

$4Al(s) + 3O_2(g) \longrightarrow 2Al_2O_3(s)$ $0 \qquad\;\; 0 \qquad\qquad 2(-1675.7)$	$-3351.4\,kJ\,mol^{-1}$
$Ag^+(aq) + Cl^-(aq) \longrightarrow AgCl(s)$ $(105.6) \quad (-167.1) \quad\;\; (-127)$	$-65.5\,kJ\,mol^{-1}$
$Zn(s) + 2H^+(aq) \longrightarrow Zn^{2+}(aq) + H_2(g)$ $0 \qquad\;\; 0 \qquad\qquad (-152.3) \quad\;\; 0$	$-152.3\,kJ\,mol^{-1}$
$Zn(s) + Cu^{2+}(aq) \longrightarrow Zn^{2+}(aq) + Cu(s)$ $0 \qquad (64.4) \qquad\quad (-152.3) \quad\; 0$	$-216.7\,kJ\,mol^{-1}$

The reverse reactions, for which the ΔH values would be positive, are not feasible under normal conditions.

Deductions made from ΔH values can, however, only be approximations. They do not, moreover, give any indication about the rate of any reaction. Many reactions are thermodynamically, or energetically, feasible but they will only take place at reasonable rates under special conditions.

Hydrogen and oxygen, or hydrogen and chlorine, for example, can be kept together in the dark with no signs of any reaction, yet both reactions are feasible from the point of view of their ΔH values:

$$H_2(g) + \tfrac{1}{2}O_2(g) \longrightarrow H_2O(g) \qquad \Delta H_m^{\ominus}(298\,K) = -241.8\,kJ\,mol^{-1}$$
$$\tfrac{1}{2}H_2(g) + \tfrac{1}{2}Cl_2(g) \longrightarrow HCl(g) \qquad \Delta H_m^{\ominus}(298\,K) = -92.3\,kJ\,mol^{-1}$$

The gases will, indeed, react explosively under the right conditions. Nitrogen triiodide is a highly endothermic compound with a high positive enthalpy of formation. It will not decompose, however, until it is trodden on or hit with a hammer, when it does so explosively. The rate of any reaction is, in fact, controlled by its *energy of activation* and not by the enthalpy of the reaction, the meaning of the terms being shown diagrammatically in Fig. 44. It is only those molecules with energies greater than the activation energy that will participate in a reaction; the lower the activation energy, the more molecules will have enough energy to participate and, hence, the quicker the reaction. An increase in the concentration of the reacting molecules, and an increase in temperature, will also increase the reaction rate by providing more energised molecules.

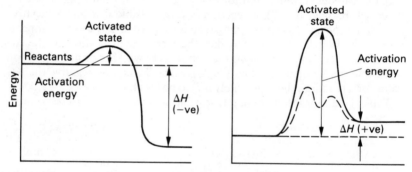

Fig. 44 *Reaction paths for (left) an exothermic reaction with a low activation energy and (right) an endothermic reaction with a high activation energy. The dotted line in the right-hand diagram shows an alternative path which might be possible in the presence of a catalyst.*

Catalysts may also increase the rate of a reaction by effectively lowering the activation energy. Many catalysts do this by providing alternative reaction paths with lower activation energies. Catalysts will not initiate reactions that are not thermodynamically feasible; they simply speed up reactions that are feasible.

5 Stability

A high negative enthalpy of formation for a compound suggests that it is very stable so far as decomposition into its elements is concerned. This is because the enthalpy change for its decomposition would be highly positive and, therefore, would be unlikely to take place.

Aluminium oxide, for example, is stable so far as decomposition into aluminium and oxygen is concerned

$$2Al_2O_3(s) \longrightarrow 4Al(s) + 3O_2(g) \quad \Delta H_m^{\ominus}(298\,K) = +3351.4\,kJ\,mol^{-1}$$

and the high input of electrical energy required to bring the change about is an indication of this. The oxide may, however, be unstable in relation to some other reaction, e.g.

$$Al_2O_3(s) + 3Mg(s) \longrightarrow 2Al(s) + 3MgO(s)$$
$$\Delta H_m^{\ominus}(298\,K) = -130\,kJ\,mol^{-1}$$

Whenever the idea of stability is used the terms of reference ought to be stated.

a Stability of compounds Typical examples of stable and unstable compounds so far as decomposition into the elements is concerned are summarised below, with the standard enthalpies of formation in $kJ\,mol^{-1}$ in parentheses.

stable compounds	unstable compounds
Al_2O_3 (-1675.7)	Cl_2O_7 ($+75.73$)
$NaCl$ (-411)	NCl_3 ($+230$)
HF (-271.1)	SnH_4 ($+162.8$)

The stable compounds require electrolytic methods to decompose them into the elements; the unstable ones decompose, possibly explosively, on very slight heating or even in the cold.

b Stability of allotropes By definition, the stable allotrope of an element under standard conditions is given an enthalpy of zero. Thus, graphite is more stable than diamond, red phosphorus more stable than white, and dioxygen more stable than trioxygen (ozone):

$$C(diamond) \longrightarrow C(graphite) \quad \Delta H_m^{\ominus}(298\,K) = -1.883\,kJ\,mol^{-1}$$
$$P_4(white) \longrightarrow P_4(red) \quad \Delta H_m^{\ominus}(298\,K) = -70.28\,kJ\,mol^{-1}$$
$$2O_3(g) \longrightarrow 3O_2(g) \quad \Delta H_m^{\ominus}(298\,K) = -284.6\,kJ\,mol^{-1}$$

Here again, however, the rate of change from the less to the more stable allotrope can, as with diamond, be immeasurably slow.

c Energetic (thermodynamic) and kinetic stability Because a feasible change may not take place because of unfavourable kinetic factors, it is sometimes convenient to refer to energetic or *thermodynamic stability* and to *kinetic stability* or *kinetic inertness*.

Diamond, for example, is energetically or thermodynamically unstable with respect to graphite, but it is kinetically stable for the same change. Similarly, petrol is energetically or thermodynamically unstable so far as its reaction with oxygen in air is concerned, but it is kinetically stable until the change is 'triggered off' by a spark or a flame.

The correct use of the words 'stable' or 'stability' must, then, refer to the change involved and must distinguish between energetic or thermodynamic stability and kinetic stability.

Questions on Chapter 5

1 (a) Explain the difference between enthalpy change (ΔH) and internal energy change (ΔU). (b) How are the two related? (c) Why are ΔH values generally used in chemistry?

2 Why are some reactions exothermic and some endothermic? Give illustrative examples.

3 Represent the following data on an enthalpy diagram:

$$A \longrightarrow B \quad \Delta H = +r\,kJ \qquad B \longrightarrow C \quad \Delta H = -s\,kJ$$
$$C \longrightarrow D \quad \Delta H = -t\,kJ \qquad D \longrightarrow A \quad \Delta H = +u\,kJ$$

What is the relationship between r, s, t and u?

4 What do you understand by the stability of a compound? Give examples.

5 Calculate the enthalpies of formation you would expect for the hydrogen halides using the bond enthalpies given on p. 38. Compare your results with the experimental values of -271.1, -92.3, -36.2 and $26.5\,kJ\,mol^{-1}$ for HF, HCl, HBr and HI respectively. Comment.

6 Distinguish between the enthalpy of formation of a hydrated ion, enthalpy of hydration and enthalpy of solution.

7 Why do some reactions 'go' whilst others do not?

8 Give examples of the correct and incorrect uses of the word 'stable' in chemistry.

9 If the enthalpy of atomisation of graphite is $715\,kJ\,mol^{-1}$ and the $O{=}O$ and $C{=}O$ bond enthalpies are 497 and $801\,kJ\,mol^{-1}$ respectively, calculate the enthalpy of combustion of graphite.

10* Use standard enthalpies of formation to compare the enthalpy of combustion of hydrogen with that of carbon monoxide.

11* Calculate the enthalpy change at 298 K for the reaction

$$AgNO_3(aq) + NaCl(aq) \longrightarrow AgCl(s) + NaNO_3(aq)$$

12* Compare the enthalpy changes for the reactions

$$2H^+(aq) + Zn(s) \longrightarrow H_2(g) + Zn^{2+}(aq)$$

$$2H^+(aq) + Cu(s) \longrightarrow H_2(g) + Cu^{2+}(aq)$$
$$Zn(s) + Cu^{2+}(aq) \longrightarrow Cu(s) + Zn^{2+}(aq)$$

and comment on the results.

13* Use standard enthalpies of formation to predict which of the following reactions are likely to be feasible:

(a) $2Fe^{3+}(aq) + Fe(s) \longrightarrow 3Fe^{2+}(aq)$

(b) $2HgO(s) \longrightarrow 2Hg(l) + O_2(g)$

(c) $2Fe^{2+}(aq) + Sn^{4+}(aq) \longrightarrow 2Fe^{3+}(aq) + Sn^{2+}(aq)$

14* Take a variety of fuels and comment on their enthalpies of combustion.

15* Use the necessary bond enthalpy values to calculate the enthalpy of formation from gaseous atoms of (a) ethane, (b) carbon dioxide, (c) silicon tetrachloride and (d) iodine trichloride.

6
Free energy changes

1 Enthalpy change and free energy change

In the preceding chapter the idea that a reaction was more likely to be feasible if it had a negative enthalpy change, i.e. ΔH was negative, was introduced as an approximation. Such an idea could not be absolutely true for, if it were, there could be no endothermic reactions.

To cover all situations it is necessary to replace enthalpy change (ΔH) by free energy change (ΔG), for it is universally true that *a process is only feasible if its free energy change is negative*. That ΔH gives any guidance is due to the fact that in many cases it provides the main contribution to the ΔG value. But ΔG is, in fact, made up of two terms,

$$\Delta G = \Delta H - T\Delta S$$

and it is only when the $T\Delta S$ term can be disregarded that ΔH can take the place of ΔG. The value of $T\Delta S$ may be equally significant, or more significant, than the ΔH value.

ΔS is the entropy change, and T the absolute temperature.

2 Entropy

Entropy is best considered (though this is an over-simplification) as a measure of disorderliness or randomness; what Gibbs once called 'mixed-upness'. Any increase in disorderliness gives an increase in entropy, i.e. ΔS is positive, and this causes a decrease in free energy, i.e. a negative ΔG.

ΔG can, therefore, be negative because of a negative value of ΔH or because of a positive value of ΔS, and it is the combination of ΔH and $T\Delta S$ values which decides the value and sign of ΔG. It is the interplay of ΔH and $T\Delta S$ values that accounts so elegantly for so many chemical facts.

ΔH is negative for exothermic reactions and positive for endothermic reactions (p. 64). Typical common ways in which entropy increases, i.e. ΔS is positive, include

a the change from solid to liquid
b the change from liquid to gas
c the solution of a solid in a liquid
d the conversion of a few molecules into a larger number
e the formation in a reaction of gases from liquids or solids, or of liquids from solids
f the mixing of two gases
g an increase in the gaseous volume in a reaction.

The reverse processes lead to a decrease in entropy, i.e. ΔS is negative.

a Standard molar entropies Tables of thermodynamic data (p. 401) give values for the standard molar entropies at 298 K, $S_m^{\ominus}(298\,K)$, of elements, compounds and hydrated ions. They are based on the conventions that the entropy of an element in its stable state is zero at 0 K, and that the entropy of $H^+(aq)$ is zero.

Values are quoted in $J\,K^{-1}\,mol^{-1}$ (*not* in $kJ\,K^{-1}\,mol^{-1}$). They can be used to calculate the standard entropy change in any reaction, for

$$\left\{\begin{array}{l}\text{standard entropy}\\\text{change in a}\\\text{reaction}\end{array}\right\} = \left\{\begin{array}{l}\text{sum of standard}\\\text{entropies of}\\\text{products}\end{array}\right\} - \left\{\begin{array}{l}\text{sum of standard}\\\text{entropies of}\\\text{reactants}\end{array}\right\}$$

Typical examples are

	$\Delta S_m^{\ominus}(298\,K)$
$H_2O(l) \longrightarrow H_2O(g)$ 70.0 $\qquad$ 188.7	$+118.7\,J\,K^{-1}\,mol^{-1}$
$2NH_3(g) \longrightarrow N_2(g) + 3H_2(g)$ 2(192.5) $\qquad$ 191.4 $\quad$ 3(130.6)	$+198.2\,J\,K^{-1}\,mol^{-1}$
$NH_3(g) + HCl(g) \longrightarrow NH_4Cl(s)$ 192.5 $\quad$ 186.7 $\qquad$ 94.6	$-284.6\,J\,K^{-1}\,mol^{-1}$
$Na(s) + H^+(aq) \longrightarrow Na^+(aq) + \frac{1}{2}H_2(g)$ 51.0 $\quad$ 0 $\qquad\qquad$ 59.0 $\qquad$ $\frac{1}{2}$(130.6)	$+73.3\,J\,K^{-1}\,mol^{-1}$
$NaCl(s) \longrightarrow Na^+(aq) + Cl^-(aq)$ 72.4 $\qquad$ 59 $\qquad$ 56.5	$+43.1\,J\,K^{-1}\,mol^{-1}$

3 Standard molar free energies of formation

Just as it is possible to draw up tables of standard molar enthalpies of formation (p. 59) on an arbitrary scale, so it is possible to allot standard molar free energies of formation. Thus the standard molar free energy of formation of a compound is *the free energy change when 1 mol of the compound is made from its elements in their standard states.* Standard molar free energies of formation can be treated in just the same way as standard molar enthalpies of formation (p. 60) to give the free energy change in a reaction, i.e.

$$\left\{\begin{array}{l}\text{standard free}\\\text{energy change in}\\\text{a reaction}\end{array}\right\} = \left\{\begin{array}{l}\text{sum of standard}\\\text{free energies of}\\\text{formation of}\\\text{products}\end{array}\right\} - \left\{\begin{array}{l}\text{sum of standard}\\\text{free energies}\\\text{of formation of}\\\text{reactants}\end{array}\right\}$$

For example,

a $CO(g) + \frac{1}{2}O_2(g) \longrightarrow CO_2(g)$ $\Delta H_m^{\ominus}(298\,K) = -283.0\,kJ\,mol^{-1}$
$(-110.5)\ 0 (-393.5)$

$CO(g) + \frac{1}{2}O_2(g) \longrightarrow CO_2(g)$ $\Delta G_m^{\ominus}(298\,K) = -257.1\,kJ\,mol^{-1}$
$(-137.3)\ 0 (-394.4)$

b $CO_2(g) + C(s) \longrightarrow 2CO(g)$ $\Delta H_m^{\ominus}(298\,K) = +172.5\,kJ\,mol^{-1}$
$(-393.5)\ 0 2(-110.5)$

$CO_2(g) + C(s) \longrightarrow 2CO(g)$ $\Delta G_m^{\ominus}(298\,K) = +119.8\,kJ\,mol^{-1}$
$(-394.4)\ 0 2(-137.3)$

Reaction **a** is feasible, for ΔG is negative, but reaction **b** is not feasible under standard conditions.

The difference between the ΔG_m and ΔH_m values for each reaction is due to the entropy change. For reaction **a**,

$$CO(g) + \frac{1}{2}O_2(g) \longrightarrow CO_2(g) \quad \Delta S_m^{\ominus}(298\,K) = -86.75\,J\,K^{-1}\,mol^{-1}$$
$$197.9 \quad \frac{1}{2}(204.9) \quad 213.6$$

the entropy change is $-86.75\,J\,K^{-1}\,mol^{-1}$, giving a $T\Delta S$ value of $-25.85\,kJ\,mol^{-1}$ at 298 K. The $T\Delta S$ term calculated from a ΔG_m value of $-257.1\,kJ\,mol^{-1}$ and a ΔH_m value of $-283\,kJ\,mol^{-1}$ is $-25.9\,kJ\,mol^{-1}$. The decrease in entropy for this reaction is due to the decrease in gas volume.

For reaction **b**,

$$CO_2(g) + C(s) \longrightarrow 2CO(g) \quad \Delta S_m^{\ominus}(298\,K) = +176.5\,J\,K^{-1}\,mol^{-1}$$
$$213.6 \quad 5.69 \quad 2(197.9)$$

the ΔS_m value is $176.5\,J\,K^{-1}\,mol^{-1}$, giving a $T\Delta S$ value of $52.6\,kJ\,mol^{-1}$ at 298 K. The value calculated from ΔG_m (119.8) and ΔH_m (172.5) is 52.7 $kJ\,mol^{-1}$. The increase in entropy, here, is due to an increase in gas volume.

In both the above examples, the ΔG_m and ΔH_m values do at least have the same sign, but this is not always so. For the solution of sodium chloride in water the relevant figures are

$$NaCl(s) \longrightarrow Na^+(aq) + Cl^-(aq)$$

	NaCl(s)	Na⁺(aq)	Cl⁻(aq)
$\Delta H_{f,m}^{\ominus}/kJ\,mol^{-1}$	-411	-240	-167.1
$\Delta G_{f,m}^{\ominus}/kJ\,mol^{-1}$	-384	-261.8	-131.3
$S_m^{\ominus}/J\,K^{-1}\,mol^{-1}$	72.4	59.0	56.5

giving a $\Delta H_m^{\ominus}$ value for the process of $+3.9\,kJ\,mol^{-1}$ and a $\Delta G_m^{\ominus}$ value of $-9.0\,kJ\,mol^{-1}$. The $\Delta S_m^{\ominus}$ value is $43.1\,J\,K^{-1}\,mol^{-1}$ giving $T\Delta S$, at 298 K, of $12.84\,kJ\,mol^{-1}$. The contribution of the $T\Delta S$ term to the value of

ΔG_m outweighs that of ΔH_m. This illustrates the dangers of drawing firm conclusions simply from ΔH_m values, i.e. of neglecting entropy changes.

4 The feasibility of reactions

There are six possible types of change, for both ΔH and ΔS can be zero and they can both be positive or negative.

a ΔS is zero It would be fortuitous if the ΔS value for any chemical reaction was zero, but it is zero for the situation in which a smooth ball runs down a smooth incline. Such a change takes place because the ΔH value is negative, giving a negative ΔG value. There is no entropy change, and $T\Delta S$ is zero.

$T\Delta S$ is also zero at $0\,K$, so that ΔH is equal to ΔG at that temperature. The old idea that only exothermic reactions were feasible is true at $0\,K$.

b ΔH is zero This situation is found on the mixing of two gases without reaction. They mix simply because there is an increase in entropy in the mixing process. As ΔS is positive, and ΔH zero, ΔG must be negative.

c Exothermic reactions ΔH is negative for exothermic reactions, so ΔG will be negative unless $T\Delta S$ has a higher negative value.

If there is an increase in entropy, i.e. ΔS is positive, ΔG will be negative at all temperatures, e.g.

	$\Delta H_m^{\ominus}(298\,K)$	$\Delta S_m^{\ominus}(298\,K)$	$\Delta G_m^{\ominus}(298\,K)$
$2C(s)+O_2(g)\longrightarrow 2CO(g)$	$-221\,kJ$	$+178.2\,J\,K^{-1}$	$-274.6\,kJ$
$S(s)+O_2(g)\longrightarrow SO_2(g)$	$-296.9\,kJ$	$+11.7\,J\,K^{-1}$	$-300.4\,kJ$

If there is a decrease in entropy, i.e. ΔS is negative, ΔG will become positive at temperatures for which $T\Delta S$ has a greater value than ΔH. Such reactions will take place at low temperatures but will reverse at high temperatures, e.g.

	$\Delta H_m^{\ominus}(298\,K)$	$\Delta S_m^{\ominus}(298\,K)$	$\Delta G_m^{\ominus}(298\,K)$
$N_2(g)+3H_2(g)\longrightarrow 2NH_3(g)$	$-92.4\,kJ$	$-198.2\,J\,K^{-1}$	$-33.3\,kJ$
$2SO_2(g)+O_2(g)\longrightarrow 2SO_3(g)$	$-197\,kJ$	$-189.7\,J\,K^{-1}$	$-139.8\,kJ$

d Endothermic reactions ΔH is positive for endothermic reactions, so a negative ΔG value can only arise if $T\Delta S$ is positive and greater in value than ΔH. The fact that this is more likely at high temperatures explains why most endothermic reactions only take place at high temperatures, e.g.

	$\Delta H_m^{\ominus}(298\,K)$	$\Delta S_m^{\ominus}(298\,K)$	$\Delta G_m^{\ominus}(298\,K)$
$N_2(g)+O_2(g)\longrightarrow 2NO(g)$	$+180.8\,kJ$	$+24.7\,J\,K^{-1}$	$+173.2\,kJ$
$C(s)+H_2O(g)\longrightarrow CO(g)+H_2(g)$	$+131.3\,kJ$	$+134.1\,J\,K^{-1}$	$+91.3\,kJ$

The ΔG values for these reactions are positive under standard conditions. It is only at higher temperatures that the $T\Delta S$ term becomes big enough to give negative ΔG values.

If ΔS is negative for an endothermic reaction, i.e. if there is an entropy decrease, ΔG can never be negative.

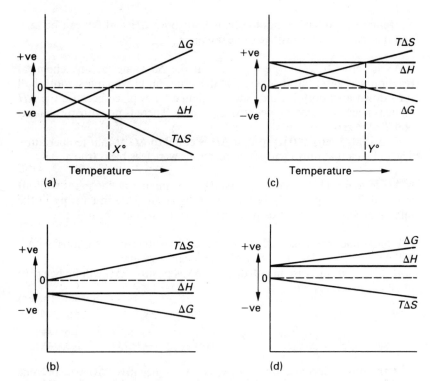

Fig. 45 *Four theoretically possible types of reaction, on the basis that ΔH and ΔS do not vary with temperature.*

(a) ΔH is $-ve$. ΔS is $-ve$. Reaction is feasible below $X\,°C$, e.g. $N_2 + 3H_2 \longrightarrow 2NH_3$.

(b) ΔH is $-ve$. ΔS is $+ve$. Reaction is feasible at any temperature, e.g. $2C + O_2 \longrightarrow 2CO$.

(c) ΔH is $+ve$. ΔS is $+ve$. Reaction is feasible above $Y\,°C$, e.g. $N_2 + O_2 \longrightarrow 2NO$.

(d) ΔH is $+ve$. ΔS is $-ve$. Reaction is never feasible.

e Summary   The four types of reaction given in **c** and **d** can be summarised as follows and as shown in Fig. 45. In this figure ΔH and ΔS are taken as being constant at different temperatures. This is an oversimplification though the change of ΔH with temperature may be small in relation to the change of $T\Delta S$.

It is rather like a pendulum which can be acted upon by two forces. Both may pull it to the right, or both to the left. Alternatively, the forces may counteract. When they balance each other exactly, the pendulum (or the reaction) is in an equilibrium position.

ΔH	ΔS	ΔG
negative	positive	feasible at all temperatures
negative	negative	feasible when $\Delta H > T\Delta S$, which is most likely at low temperatures
positive	positive	feasible when $T\Delta S > \Delta H$, which is most likely at high temperatures
positive	negative	never feasible

5 Free energy change and equilibrium constants

For practical purposes reactions may be classified as those which will not take place, those which go to completion, and those which are reversible, giving an equilibrium mixture which may contain an excess of the products or an excess of the reactants.

The value of the equilibrium constant for a reaction can, in fact, vary between 0 and infinity. If close to 0 the reaction is said not to take place; if close to infinity the reaction is said to go completely.

A feasible reaction, then, is one with a high equilibrium constant, and, as the feasibility depends on the ΔG_m value, it is not surprising that there is a relationship between the molar free energy change and the equilibrium constant (K), i.e.

$$\Delta G_m^{\ominus}(298\,\mathrm{K}) = -RT \ln K$$

where $\Delta G_m^{\ominus}$ is the standard molar free energy change at 298 K in $\mathrm{kJ\,mol^{-1}}$, R is $8.314 \times 10^{-3}\,\mathrm{kJ\,K^{-1}\,mol^{-1}}$ and T, at 25 °C, is 298 K. It follows that

$$\lg K = -0.1753\Delta G_m^{\ominus}(298\,\mathrm{K})$$

and the relationship between $\Delta G_m^{\ominus}$, K and $\lg K$ is

$\Delta G_m^{\ominus}(298\,\mathrm{K})$	-100	-57	-10	0	$+10$	$+57$	$+100$
K	3.4×10^{17}	10^{10}	57	1	0.018	10^{-10}	2.95×10^{-18}
$\lg K$	$+17.53$	10	1.753	0	-1.753	-10	-17.53

For practical purposes it is convenient to regard a 'reaction' with an equilibrium constant less than 10^{-10} ($\Delta G_m^{\ominus} > 57\,\mathrm{kJ\,mol^{-1}}$) as one which will not take place, whilst one with an equilibrium constant greater than 10^{10} ($\Delta G_m^{\ominus} < -57\,\mathrm{kJ\,mol^{-1}}$) will go completely. Reactions with intermediate values for the equilibrium constant can be regarded as reversible reactions.

6 Equilibrium constants at different temperatures

The use of standard molar free energy values enables values of equilibrium constants to be calculated at 25 °C as in the preceding section. Le Chatelier's theory implies that the value of an equilibrium constant increases with increase in temperature for an endothermic reaction but decreases for an exothermic reaction. The quantitative relation between temperature and equilibrium constant, for gas reaction, is

$$\frac{d(\ln K_p)}{dT} = \frac{\Delta H_m^{\ominus}}{RT^2}$$

In the simplest use of this expression it is assumed that $\Delta H_m^{\ominus}$ does not vary with temperature (which is a good approximation for most gas reactions over a limited range of temperature) so that the values of K_p at two temperatures, T_1 and T_2 are related by

$$\ln K_{p,T_2} - \ln K_{p,T_1} = \frac{-\Delta H_m^{\ominus}}{R}\left(\frac{1}{T_2} - \frac{1}{T_1}\right)$$

If, then, the K_p value is known at 25 °C, it can be calculated, with fair accuracy, for any other temperature, particularly if it is not far removed from 25 °C. In a more accurate treatment, the variation of $\Delta H_m^{\ominus}$ with temperature can be allowed for.

7 Free energy changes and standard electrode potentials

A standard electrode potential, $E^{\ominus}$ (p. 21), indicates how readily an ionisation process will take place and the standard molar free energy change for the process does likewise. It is not surprising, then, that the two are related. The expression is

$$\Delta G_m^{\ominus} = -zFE^{\ominus}$$

where z is the number of electrons taking part in the process, $\Delta G_m^{\ominus}$ is expressed in $J\,mol^{-1}$ and F is the Faraday constant ($96.487 \times 10^3\,C\,mol^{-1}$).

For the $Cu^{2+}(aq)/Cu$ couple, for example, the $E^{\ominus}$ value is 0.337 V and z is 2. The $\Delta G_m^{\ominus}$ value is, therefore, $-(2 \times 96\,487 \times 0.337)\,J\,mol^{-1}$, i.e. $-65\,kJ\,mol^{-1}$. For the $Zn^{2+}(aq)/Zn$ couple it is $-(2 \times 96\,487 \times (-0.763))$ $J\,mol^{-1}$, i.e. $+147.2\,kJ\,mol^{-1}$. Similar values for other couples are given on p. 400.

An overall reaction will only take place if the combined $E^{\ominus}$ value, made up of the two electrode potentials involved, is positive (p. 88). This, as can be seen in the following example, means a negative $\Delta G_m^{\ominus}$ value:

	$E^\ominus$	$\Delta G_m^\ominus$
$Cu^{2+}(aq) + 2e^- \longrightarrow Cu(s)$	$+0.337\,V$	$-65\,kJ$
$Zn^{2+}(aq) + 2e^- \longrightarrow Zn(s)$	$-0.763\,V$	$+147.2\,kJ$
$Cu^{2+}(aq) + Zn(s) \longrightarrow Zn^{2+}(aq) + Cu(s)$	$+1.1\,V$	$-212.2\,kJ$

The $\Delta G_m^\ominus$ value of $-212.2\,kJ\,mol^{-1}$ enables the equilibrium constant to be calculated:

$$\lg K = -0.175 \times \Delta G_m^\ominus = -0.175(-212.2) = 37.13,$$
$$\therefore K = 1.35 \times 10^{37}$$

The very large value of K means that the reaction can be considered as going to completion.

The same equilibrium constant can be calculated direct from the $E^\ominus$ value, without introducing $\Delta G_m^\ominus$, for

$$\lg K = -0.175 \times \Delta G_m^\ominus = \frac{E^\ominus z}{0.0592}$$

Thus $\quad \lg K = \dfrac{1.1 \times 2}{0.0592} = 37.16$

$$\therefore K = 1.45 \times 10^{37}$$

For the reaction

$$Ag^+(aq) + Fe^{2+}(aq) \longrightarrow Fe^{3+}(aq) + Ag(s)$$

the $E^\ominus$ value is $(0.799 - 0.771)$, i.e. $0.028\,V$. This gives $\lg K$ as $0.028/0.059$, i.e. 0.4745, and K as 2.982. The low value for K means that the reaction is reversible.

The relationship between $E^\ominus$, $\Delta G_m^\ominus$ and K values for processes involving one electron is as follows:

K	10^{50}	10^{10}	10	1	10^{-1}	10^{-10}	10^{-50}
$E^\ominus/V$	$+2.96$	$+0.592$	$+0.0592$	0	-0.0592	-0.592	-2.96
$\Delta G_m^\ominus/kJ\,mol^{-1}$	-285	-57	-5.7	0	$+5.7$	$+57$	$+285$

In general, an effective $E^\ominus$ value greater than $+0.592\,V$ (corresponding to a $\Delta G_m^\ominus$ value of $-57\,kJ\,mol^{-1}$) means that the reaction may be regarded as going completely ($K > 10^{10}$); an $E^\ominus$ value more negative than $-0.592\,V$ (corresponding to a $\Delta G_m^\ominus$ of $+57\,kJ\,mol^{-1}$) means that the reaction does not, for practical purposes, take place ($K < 10^{-10}$). For example

	$E^\ominus/V$	$\Delta G_m^\ominus/kJ\,mol^{-1}$	K
$Zn(s) + Pb^{2+}(aq) \longrightarrow Zn^{2+}(aq) + Pb(s)$	$+0.637$	-122.9	3.2×10^{21}
$Sn(s) + Pb^{2+}(aq) \longrightarrow Sn^{2+}(aq) + Pb(s)$	$+0.01$	-1.93	2.18
$Cu(s) + Pb^{2+}(aq) \longrightarrow Cu^{2+}(aq) + Pb(s)$	-0.463	$+89.3$	2.3×10^{-16}

Questions on Chapter 6

1 Choose some examples of endothermic reactions and comment on the conditions under which they will take place.

2 How would you explain the meaning of the term activation energy to an intelligent person who knew no chemistry?

3 If chemical change always proceeds in the direction of greater stability, why is it that endothermic reactions occur at all? Illustrate your answer with some specific examples. (OS)

4 Use the free energy values given on p. 402 to calculate the equilibrium constants for the following reactions:

(a) $NO(g) \longrightarrow \frac{1}{2}N_2(g) + \frac{1}{2}O_2(g)$

(b) $CO(g) + H_2O(g) \longrightarrow H_2(g) + CO_2(g)$

(c) $H_2(g) + \frac{1}{2}O_2(g) \longrightarrow H_2O(l)$

5 The ΔG value for the reaction

$$Fe^{2+}(aq) + Ag^+(aq) \longrightarrow Fe^{3+}(aq) + Ag(s)$$

is $2.7 \, kJ \, mol^{-1}$ at 298 K. What is the equilibrium constant and what does this value represent?

6 The enthalpy and entropy changes, at 25 °C, for the reaction

$$H^+(aq) + OH^-(aq) \longrightarrow H_2O(l)$$

are $-56 \, kJ \, mol^{-1}$ and $80.5 \, J \, K^{-1} \, mol^{-1}$ respectively. Calculate (a) the $\Delta G^\ominus$ value, and (b) the equilibrium constant at 25 °C. Comment on the answer you get in (b).

7 'Not all exothermic reactions are spontaneous; some endothermic changes are spontaneous.' Discuss the extent to which a knowledge of the enthalpy change for a reaction is a guide to its ability to proceed. Illustrate your answer by considering *four* reactions of varied type, *two* of which are exothermic, and *two* endothermic. You may wish to consider some or all of the following reactions or others of your own choice.

$$N_2O_4(g) \rightleftharpoons 2NO_2(g) \qquad\qquad NH_4Cl(s) \rightleftharpoons NH_3(g) + HCl(g)$$
$$N_2(g) + 3H_2(g) \rightleftharpoons 2NH_3(g) \qquad CaCO_3(s) \rightleftharpoons CaO(s) + CO_2(g)$$

(L)

8* Calculate the free energy change at 298 K for the reaction

$$H^+(aq) + OH^-(aq) \longrightarrow H_2O(l)$$

Hence calculate the value of the equilibrium constant and comment on its significance.

9* Use standard molar free energies of formation to decide which of the following reactions may be regarded as reversible at room temperature:

(a) $N_2O_4(g) \rightleftharpoons 2NO_2(g)$

(b) $2H_2O(g) \rightleftharpoons 2H_2(g) + O_2(g)$

(c) $2Ag(s)+H_2S(g)\rightleftharpoons H_2(g)+Ag_2S(s)$
(d) $BiCl_3(s)+H_2O(l)\rightleftharpoons BiOCl(s)+2HCl(aq)$
(e) $Pb(s)+2H^+(aq)\rightleftharpoons Pb^{2+}(aq)+H_2(g)$
(f) $Ag^+(aq)+Fe^{2+}(aq)\rightleftharpoons Fe^{3+}(aq)+Ag(s)$
10* The K_p value for the following reaction

$$N_2(g)+O_2(g)\longrightarrow 2NO(g) \qquad \Delta H^\ominus(298\,K) = 180\,kJ\,mol^{-1}$$

at 298 K is 4×10^{-31}. Calculate the value at 1500 K. How would you expect this value to agree with the measured value?
11* The K_p values for the reaction

$$H_2(g)+I_2(g)\longrightarrow 2HI(g)$$

are 794 and 160 at 298 and 500 K respectively. Calculate the enthalpy change for the reaction at 298 K. Why does your answer differ from the measured value of $-9.6\,kJ\,mol^{-1}$?
12* Use standard molar free energies of formation to predict which of the following processes is feasible:

$$2Ag(s)+Hg_2Cl_2(s)\longrightarrow 2AgCl(s)+2Hg(l)$$
$$2AgCl(s)+2Hg(l)\longrightarrow Hg_2Cl_2(s)+2Ag(s)$$

Compare, and explain, the answer you get using standard molar enthalpies of formation.
13* Compare the free energy and the enthalpy changes for (a) the freezing of water at 0 °C and (b) the boiling of water at 100 °C. Comment on the values and calculate the entropy changes for the two processes.
14* The enthalpy change for the reaction:

$$CaCl_2(aq)+Na_2CO_3(aq)\longrightarrow CaCO_3(s)+2NaCl(aq)$$

is positive $(+13.1\,kJ\,mol^{-1})$. Why, then, is a precipitate of calcium carbonate formed?

7
Oxidation and reduction

1 Introduction

Oxidation involves the addition of oxygen atoms, the removal of hydrogen atoms, or the loss of electrons; the chemical that brings about the change is called an *oxidising agent* (O.A. for short) or an oxidant. Examples are the oxidations of carbon, hydrogen chloride, iron, and Fe^{2+}(aq) ions, which are summarised in the following equations.

$$C(s) + O_2(g) \longrightarrow CO_2(g)$$
$$4HCl(g) + O_2(g) \longrightarrow 2Cl_2(g) + 2H_2O(g)$$
$$Fe(s) \longrightarrow Fe^{2+}(aq) + 2e^-$$
$$Fe^{2+}(aq) \longrightarrow Fe^{3+}(aq) + e^-$$

Reduction occurs when hydrogen atoms are added, oxygen atoms are removed, or when there is a gain of electrons; it is brought about by a *reducing agent* (R.A. for short), or a reductant. Chlorine, copper(II) oxide, Fe^{3+}(aq) and Fe^{2+}(aq) ions, for example, are reduced in the changes represented by the following equations.

$$Cl_2(g) + H_2(g) \longrightarrow 2HCl(g)$$
$$CuO(s) + C(s) \longrightarrow Cu(s) + CO(g)$$
$$Fe^{3+}(aq) + e^- \longrightarrow Fe^{2+}(aq)$$
$$Fe^{2+}(aq) + 2e^- \longrightarrow Fe(s)$$

Oxidising agents, therefore, give oxygen atoms, take hydrogen atoms, or take electrons. Reducing agents give hydrogen atoms, take oxygen atoms, or give electrons. In terms of electrons, **O.A. Take** and **R.A. Give** (**OAT** and **RAG**) are good mnemonics to distinguish between oxidising agents and reducing agents. Similarly, **OIL** (**O**xidation **I**s **L**oss) and **RIG** (**R**eduction **I**s **G**ain) may be useful for oxidation and reduction.

Oxidation and reduction are complementary processes, for neither can take place without the other. The oxidising agent is reduced by the reducing agent; the reducing agent is oxidised by the oxidising agent. Many oxidising agents will, therefore, react with many reducing agents; the processes are called *redox reactions*.

They are best understood by consideration of the oxidation state or number of an element, and of how it might change in the course of a reaction.

2 Oxidation number or oxidation state

All elements in their elementary, uncombined state are given an oxidation number of zero. Whenever the element can be considered as having lost some of its electrons, it is said to be oxidised. When it gains some electrons it is said to be reduced.

a Reactions involving simple ions   When sodium, for example, forms the Na^+ ion it loses one electron. The sodium is being oxidised, with a change in oxidation state from 0 for Na to $+1$ for Na^+. Similarly, the oxidation numbers of the Cu^{2+} and Al^{3+} ions are $+2$ and $+3$, respectively.

When chlorine atoms are reduced to Cl^- ions they gain electrons and the oxidation number changes from 0 to -1. This is a reduction. Similarly, the oxidation number of O^{2-} is -2. Thus

$$\underset{0}{Na} \xrightarrow{\text{oxidation}} \underset{+1}{Na^+} + e^- \qquad \underset{0}{Cl} + e^- \xrightarrow{\text{reduction}} \underset{-1}{Cl^-}$$

$$\underset{0}{Cu} \xrightarrow{\text{oxidation}} \underset{+2}{Cu^{2+}} + 2e^- \qquad \underset{0}{O} + 2e^- \xrightarrow{\text{reduction}} \underset{-2}{O^{2-}}$$

For simple ions, the oxidation number is equal to the charge on the ion. Oxidation involves an increase in the oxidation number; reduction, a decrease.

$$\underset{\substack{+2 \\ R.A.}}{Fe^{2+}(aq)} \underset{\text{reduction}}{\overset{\text{oxidation}}{\rightleftharpoons}} \underset{\substack{+3 \\ O.A.}}{Fe^{3+}(aq)} + e^-$$

b Reactions involving covalent molecules   The loss or gain of electrons when a covalent molecule is oxidised or reduced is less obvious. The oxidation number of an element in a covalent compound is, therefore, decided by taking it to be equal to the charge that the element would carry if all the bonds in the compound were ionic instead of covalent. In doing this, a shared pair of electrons between two atoms is assigned to the atom with the higher electronegativity (p. 40).

Thus, in the oxidation of carbon to carbon dioxide by oxygen, the dioxide is to be regarded as $C^{4+}(O^{2-})_2$. The carbon has been oxidised, because its oxidation number has increased from 0 to $+4$; the oxygen has been reduced, because its oxidation number has decreased from 0 to -2. In the reduction of nitrogen to ammonia by hydrogen, the ammonia is regarded as $N^{3-}(H^+)_3$, so that the oxidation number of the nitrogen changes from 0 to -3, and that of the hydrogen from 0 to $+1$. In summary:

$$\underset{0}{C} + \underset{0}{O_2} \xrightarrow[\text{oxidation}]{\text{reduction}} \underset{\substack{+4 \\ -2}}{C O_2} \qquad \underset{0}{N_2} + 3\underset{0}{H_2} \xrightarrow[\text{reduction}]{\text{oxidation}} 2\underset{\substack{+1 \\ -3}}{NH_3}$$

3 Further examples of oxidation numbers

Even though they are somewhat artificial, oxidation numbers are useful in naming compounds (p. 278), in deciding whether oxidation and reduction have taken place, and in balancing equations for redox reactions (p. 84).

There are no real problems in allotting oxidation numbers to the atoms in a molecule unless the molecule contains bonds between like atoms (see section **g** below). The following points should be noted.

a The algebraic sum of the oxidation numbers of all the atoms in an uncharged compound is zero. In an ion, the algebraic sum is equal to the charge on the ion.

b As fluorine is the most electronegative element it always has an oxidation number of -1. Hence

$$\underset{+1\ -1}{H\ F} \qquad \underset{+1\ -1}{Cl\ F} \qquad \underset{+5\ -1}{I\ F_5} \qquad \underset{+3\ -1}{[Al\ F_6]^{3-}} \qquad \underset{+2\ -1}{O\ F_2}$$

c In ionic metallic hydrides, hydrogen has an oxidation number of -1; in all other compounds it is $+1$. Hence

$$\underset{+1\ -1}{Li\ H} \qquad \underset{+2\ -1}{Ca\ H_2} \qquad \underset{+1\ -1}{H\ Cl} \qquad \underset{+1\ -2}{H_2O} \qquad \underset{-3\ +1}{N\ H_3} \qquad \underset{-3\ +1}{N\ H_4{}^+}$$

d Oxygen, second only to fluorine in electronegativity, has an oxidation number of -2, except in OF_2 (when it is $+2$) or unless the compound contains $-O-O-$ bonds (see **g** below). Hence

$$\underset{+2\ -2}{MgO} \qquad \underset{+4\ -2}{C\ O_2} \qquad \underset{+7\ \ -2}{Mn_2O_7} \qquad \underset{+6\ -2}{Cr\ O_3} \qquad \underset{-2\ +1}{O\ H^-} \qquad \underset{+6\ -2}{S\ O_4{}^{2-}}$$

e In compounds containing more than two elements, the oxidation number of any one of them may be obtained by first assigning reasonable oxidation numbers to the other elements. In sulphuric acid, for example, the most reasonable oxidation numbers for hydrogen and oxygen are $+1$ and -2 respectively, which gives sulphur an oxidation number of $+6$. Hence

$$\underset{+1\ \ +6\ -2}{H_2S\ O_4} \qquad \underset{+1\ \ +4\ -2}{H_2S\ O_3} \qquad \underset{+1\ \ +7\ -2}{H\ Cl\ O_4} \qquad \underset{+1\ +1\ -2}{H\ Cl\ O}$$

$$\underset{-4\ +1}{C\ H_4} \qquad \underset{-2\ +1\ \ -1}{C\ H_3\ Cl} \qquad \underset{0\ \ +1\ -1}{C\ H_2\ Cl_2} \qquad \underset{+4\ -1}{C\ Cl_4}$$

f Some elements may have widely different oxidation numbers in different compounds, as is shown by the following compounds of manganese, chromium, nitrogen and chlorine.

−3	−2	−1	0	+1	+2	+3	+4	+5	+6	+7
			Mn		$MnCl_2$	$MnCl_3$ Mn_2O_3	MnO_2		MnO_4^{2-}	MnO_4^- Mn_2O_7
			Cr		$CrCl_2$	$CrCl_3$ Cr_2O_3			CrO_3 CrO_4^{2-} $Cr_2O_7^{2-}$	
NH_3 NH_4^+	N_2H_4	NH_2OH	N_2	N_2O	NO	N_2O_3 HNO_2	NO_2	N_2O_5 HNO_3		
		HCl	Cl_2	HClO		$HClO_2$	ClO_2	$HClO_3$		$HClO_4$

g In molecules containing —X—X— bonds, the oxidation numbers of the atoms can be obtained by considering only the other bonds. For example

$$H—O—O—H \qquad F—O—O—F \qquad Cl—O—O—O—Cl$$
$$\text{+1 \ −1 \ −1 \ +1} \qquad \text{−1 \ +1 \ +1 \ −1} \qquad \text{+1 \ −1 \ \ 0 \ \ −1 \ +1}$$

In the thiosulphate ion, $S_2O_3^{2-}$, one sulphur atom is linked to three oxygen atoms, and the other is linked to sulphur. By direct analogy with SO_4^{2-}, sulphur might be regarded as having an oxidation number of +6. In fact, one sulphur atom has oxidation number of +4 and the other of zero. The 'average' value is +2.

$$S \ O_4^{2-} \qquad SS \ O_3^{2-}$$
$$\text{+6 −2} \qquad \text{0 +4 −2}$$

Similarly, in $S_4O_6^{2-}$, two of the sulphur atoms have oxidation numbers of +5 and two of zero:

$$O_3^-S \ SSS \ O_3^-$$
$$\text{−2 \ \ +5 0 0 \ +5 −2}$$

The 'average' is +2.5.

4 Balancing equations for redox reactions

This can be done by using half-equations and electron transfer, or by using oxidation numbers.

a Half-equations The functioning of oxidising and reducing agents can be summarised in half-equations which show the number of electrons involved. These equations do not represent reactions which will take place on their own; they will only take place when suitable oxidising and reducing agents are present to make up a redox reaction.

oxidising agent	half-equation
O_2	$\frac{1}{2}O_2 + 2e^- \longrightarrow O^{2-}$
Cl_2	$\frac{1}{2}Cl_2 + e^- \longrightarrow Cl^-$
conc. HNO_3	$NO_3^- + 2H^+ + e^- \longrightarrow NO_2 + H_2O$
conc. H_2SO_4	$2H_2SO_4 + 2e^- \longrightarrow SO_2 + SO_4^{2-} + 2H_2O$
$KMnO_4$ in acid soln.	$MnO_4^- + 8H^+ + 5e^- \longrightarrow Mn^{2+} + 4H_2O$
$KMnO_4$ in alk. soln.	$MnO_4^- + 4H^+ + 3e^- \longrightarrow MnO_2 + 2H_2O$
$K_2Cr_2O_7$ in acid soln.	$Cr_2O_7^{2-} + 14H^+ + 6e^- \longrightarrow 2Cr^{3+} + 7H_2O$
Fe^{3+}	$Fe^{3+} + e^- \longrightarrow Fe^{2+}$
H_2O_2	$H_2O_2 + 2H^+ + 2e^- \longrightarrow 2H_2O$

reducing agent	half-equation
H_2	$\frac{1}{2}H_2 \longrightarrow H^+ + e^-$
I^-	$I^- \longrightarrow \frac{1}{2}I_2 + e^-$
Na	$Na \longrightarrow Na^+ + e^-$
Fe^{2+}	$Fe^{2+} \longrightarrow Fe^{3+} + e^-$
Sn^{2+}	$Sn^{2+} \longrightarrow Sn^{4+} + 2e^-$

b Electron transfer To get a balanced equation for a redox reaction, it is necessary to balance the number of electrons provided by the reducing agent against the number required by the oxidising agent. The electrons provided by $Fe^{2+}(aq)$, for example,

$$5Fe^{2+}(aq) \longrightarrow 5Fe^{3+}(aq) + 5e^-$$

can be taken up by $MnO_4^-(aq)$ ions in acid solution,

$$MnO_4^-(aq) + 8H^+(aq) + 5e^- \longrightarrow Mn^{2+}(aq) + 4H_2O(l)$$

to give an overall equation of

$$5Fe^{2+} + MnO_4^- + 8H^+ \longrightarrow 5Fe^{3+} + Mn^{2+} + 4H_2O$$

c Use of oxidation numbers For the reaction between manganate(VII) and iron(II) ions in acid solution, the two half-equations give

$$MnO_4^-(aq) + Fe^{2+}(aq) \longrightarrow Mn^{2+}(aq) + Fe^{3+}(aq)$$
$${+7}\phantom{(aq) + Fe^{2}}{+2}{+2}{+3}$$

and show that the oxidation number of the oxidising agent has decreased by 5 whilst that of the reducing agent has increased by 1. The equation must, therefore, be balanced as follows:

$$MnO_4{}^-(aq) + 5Fe^{2+}(aq) \longrightarrow Mn^{2+}(aq) + 5Fe^{3+}(aq)$$

The oxygen atoms must now be balanced by introducing water,

$$MnO_4{}^-(aq) + 5Fe^{2+}(aq) \longrightarrow Mn^{2+}(aq) + 5Fe^{3+}(aq) + 4H_2O(l)$$

and a final balancing of the hydrogen by introducing H^+ ions gives the balanced equation:

$$MnO_4{}^- + 5Fe^{2+} + 8H^+ \longrightarrow Mn^{2+} + 5Fe^{3+} + 4H_2O$$

5 The relative strengths of oxidising and reducing agents

When any oxidising agent reacts with any reducing agent, weaker oxidising and reducing agents are produced. For example

$$Zn(s) + Cu^{2+}(aq) \longrightarrow Zn^{2+}(aq) + Cu(s)$$
R.A.   O.A.   weaker O.A. weaker R.A.

It is only because the zinc metal and the $Cu^{2+}(aq)$ ions are the stronger pair of oxidising and reducing agents that the reaction takes place, and, in this example, they are so much the stronger that the reaction takes place almost fully (p. 77).

It is, then, the relative strengths of oxidising and reducing agents that determine which redox reactions will take place, and how fully they will do so. The strengths can be measured on an arbitrary scale where the $H^+(aq)/\frac{1}{2}H_2(g)$ couple, under standard conditions, is given a standard electrode potential ($E^\ominus$) of 0 V (p. 21). The $E^\ominus$ values for other redox couples can be measured against the standard hydrogen couple in cells. The values obtained, some of which are listed below, give the relative strengths of oxidising and reducing agents under standard conditions (see also Appendix III, p. 400). These values are for aqua-ions, with water molecules as the ligand; with other ligands the values may be very different (p. 337), and many of them are affected by changes in pH value (pp. 88 and 348).

oxidising agents (take electrons)	reducing agents (give electrons)	$E^{\ominus}/V$
$Li^+(aq) + e^- \longleftrightarrow$	$Li(s)$	-3.04
$Na^+(aq) + e^- \longleftrightarrow$	$Na(s)$	-2.71
$Zn^{2+}(aq) + 2e^- \longleftrightarrow$	$Zn(s)$	-0.76
$H^+(aq) + e^- \longleftrightarrow$	$\frac{1}{2}H_2(g)$	0.00
$Cu^{2+}(aq) + 2e^- \longleftrightarrow$	$Cu(s)$	$+0.34$
$\frac{1}{2}I_2(aq) + e^- \longleftrightarrow$	$I^-(aq)$	$+0.54$
$Fe^{3+}(aq) + e^- \longleftrightarrow$	$Fe^{2+}(aq)$	$+0.77$
$\frac{1}{2}Br_2(l) + e^- \longleftrightarrow$	$Br^-(aq)$	$+1.07$
$Cr_2O_7{}^{2-}(aq) + 14H^+(aq) + 6e^- \longleftrightarrow$	$2Cr^{3+} + 7H_2O(l)$	$+1.33$
$\frac{1}{2}Cl_2(aq) + e^- \longleftrightarrow$	$Cl^-(aq)$	$+1.36$
$MnO_4{}^-(aq) + 8H^+(aq) + 5e^- \longleftrightarrow$	$Mn^{2+}(aq) + 4H_2O(l)$	$+1.52$
$\frac{1}{2}F_2(aq) + e^- \longleftrightarrow$	$F^-(aq)$	$+2.87$

(left margin, vertical) increasing strength as O.A.

(centre, vertical) increasing strength as R.A.

$E^{\ominus}$ values and the associated oxidation state changes can be summarised in diagrams, as in Fig. 46. An alternative way of doing the same thing is described on p. 92.

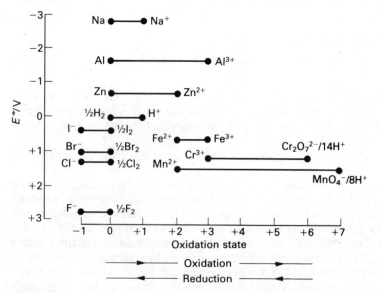

Fig. 46 *Diagrammatic method of summarising electrode potentials and oxidation states (see also Fig. 47, p. 92).*

6 Factors affecting $E^\ominus$ values

Standard electrode potential values, $E^\ominus$, refer to ionic concentrations of $1 \, mol \, dm^{-3}$ (strictly, unit activity) at 298 K; when a gas, e.g. hydrogen, is involved it is at a pressure of 101.325 kPa. The values normally quoted also refer to aqueous solutions, i.e. they are the values for the hydrated ions. At different ionic concentrations, at different pH values, and with different ligands, the electrode potential values will be different from the $E^\ominus$ values.

a Effect of ionic concentration Any redox couple can be generalised as

$$\text{oxidised form} + ze^- \rightleftharpoons \text{reduced form}$$

and the electrode potential for the couple at different ionic concentrations, E, is related to the standard electrode potential, $E^\ominus$, by the Nernst equation,

$$E = E^\ominus + \frac{RT}{zF} \ln \left(\frac{[\text{oxidised form}]}{[\text{reduced form}]} \right)$$

where R is the gas constant ($8.313 \, J \, K^{-1} \, mol^{-1}$), T is the temperature in K, F is the Faraday constant ($96\,485 \, C \, mol^{-1}$), z is the number of electrons concerned and [oxidised form] and [reduced form] are the respective concentrations of the ions involved, in $mol \, dm^{-3}$. Conversion to logarithms to base 10 and including the numerical values of the constants gives, at 298 K,

$$E = E^\ominus + \frac{0.059}{z} \lg \left(\frac{[\text{oxidised form}]}{[\text{reduced form}]} \right)$$

For the Zn^{2+}/Zn couple, the concentration of solid zinc is taken as unity so that

$$E = -0.76 + \frac{0.059}{2} \lg [Zn^{2+}]$$

It follows that the redox potential changes by 0.0295 V for every ten-fold change in Zn^{2+} ion concentration. The redox potential is, therefore, $(-0.76 - 0.0295) \, V$ in M/10 solution, or $(-0.76 + 0.0295) \, V$ in 10M solution. As the ionic concentration increases the value of the redox potential gets less negative or more positive.

For the Fe^{3+}/Fe^{2+} couple, the relationship is

$$E = +0.76 + 0.059 \lg \frac{[Fe^{3+}]}{[Fe^{2+}]}$$

As the $[Fe^{3+}]/[Fe^{2+}]$ ratio increases, the redox potential gets larger. When the $[Fe^{3+}]$ is 100 times greater than the $[Fe^{2+}]$ the redox potential is $(0.76+0.118)$ V, i.e. 0.878 V.

The Nernst equation also enables electrode potentials at different temperatures to be calculated from $E^{\ominus}$ values.

b Effect of acidity Hydrogen ions are concerned in some redox processes, e.g.

$$MnO_4^-(aq)+8H^+(aq)+5e^- \longrightarrow Mn^{2+}(aq)+4H_2O(l)$$

$$E^{\ominus} = +1.52 \text{ V}$$

and they must be regarded as part of the oxidising agent so that

$$E = E^{\ominus} + \frac{0.059}{5}\lg\frac{[MnO_4^-][H^+]^8}{[Mn^{2+}]}$$

For equal concentrations of MnO_4^- and Mn^{2+} ions, the value of E changes with the pH value as follows:

pH	0	1	2	5	6
E/V	1.52	1.43	1.33	1.05	0.96

so that the oxidising power of the MnO_4^- ion is greatly affected by acidity.

Thus, at a pH of 1, potassium manganate(VII) will oxidise Cl^-, Br^- and I^- ions; at a pH of 2, Br^- and I^- will be oxidised; at a pH of 6, only I^- will be oxidised. The halides can be selectively oxidised in this way.

c Effect of change of ligand See p. 337.

7 The feasibility of redox reactions

Lists of $E^{\ominus}$ values, such as the one in section **5**, are useful in predicting feasible redox reactions. Under standard conditions, any oxidising agent on the left-hand side of the list will, theoretically, react with any reducing agent higher up the list on the right-hand side.

Thus, lithium, the strongest reducing agent shown, will react with any of the oxidising agents below it. Fluorine is the strongest oxidising agent given and it will oxidise any reducing agent above it.

A redox reaction will, therefore, only be feasible, under standard conditions, if its combined $E^{\ominus}$ value, made up from the two electrode potentials concerned, has a *positive* value (p. 76). For example, $Cu^{2+}(aq)$ with react with zinc,

$$
\begin{array}{ll}
Cu^{2+}(aq)+2e^- \longrightarrow Cu(s) & E^{\ominus} = +0.34 \text{ V} \\
Zn^{2+}(aq)+2e^- \longrightarrow Zn(s) & E^{\ominus} = -0.76 \text{ V} \\
\hline
Cu^{2+}(aq)+Zn(s) \longrightarrow Zn^{2+}(aq)+Cu(s) & E^{\ominus} = +0.34-(-0.76) \\
& \quad\quad = +1.1 \text{ V}
\end{array}
$$

but $Fe^{2+}(aq)$ will not react with $Sn^{2+}(aq)$,

$$
\begin{array}{lll}
Fe^{2+}(aq) + 2e^- & \longrightarrow Fe(s) & E^{\ominus} = -0.44\,V \\
Sn^{4+}(aq) + 2e^- & \longrightarrow Sn^{2+}(aq) & E^{\ominus} = +0.15\,V \\
\hline
Fe^{2+}(aq) + Sn^{2+}(aq) \longrightarrow Fe(s) + Sn^{4+}(aq) & & E^{\ominus} = -0.44 - 0.15 \\
& & \phantom{E^{\ominus} } = -0.59\,V
\end{array}
$$

It would, in fact, be the reverse reaction, with an $E^{\ominus}$ value of $+0.59$ V, that would be feasible. Other examples are given on p. 77. Standard electrode potentials must, however, be used with care. They refer only to standard conditions, although the values can be converted to other temperatures, pressures or concentrations. Moreover, electrode potentials only indicate which reactions are thermodynamically feasible; they give no indication of the rate at which any reaction will take place. The combined $E^{\ominus}$ value for the reaction

$$H_2(g) + 2Ag^+(aq) \longrightarrow 2H^+(aq) + 2Ag(s) \qquad E^{\ominus} = +0.8\,V$$

is positive, but bubbling hydrogen gas into silver nitrate solution will not produce any silver. Nor will hydrogen gas reduce $Fe^{3+}(aq)$ to $Fe^{2+}(aq)$ (p. 148).

8 Disproportionation

In some reactions an element in a compound or an ion appears to undergo simultaneous oxidation and reduction, i.e. its oxidation number is both increased and decreased in the same reaction. Such a change is known as disproportionation. Typical examples are described below.

a Copper(I) compounds in aqueous solution   Copper(I) ions can act either as reducing agents, as in i, or as oxidising agents, as in ii:

$$
\begin{array}{llll}
i & Cu^{2+}(aq) + e^- \longrightarrow Cu^+(aq) & & E^{\ominus} = +0.153\,V \\
ii & Cu^+(aq) + e^- \longrightarrow Cu(s) & & E^{\ominus} = +0.521\,V \\
\hline
iii & 2Cu^+(aq) \longrightarrow Cu^{2+}(aq) + Cu(s) & & E^{\ominus} = -0.153 + 0.521 \\
& {+1} {+2} 0 & & \phantom{E^{\ominus} } = +0.368\,V
\end{array}
$$

In the overall change, as in iii, which will take place because the $E^{\ominus}$ value is positive, the Cu^+ ions acting as oxidising agents in ii have oxidised those acting as reducing agents in i. The oxidation states are as shown.

Copper(I) compounds are, therefore, only stable in the presence of water when they are insoluble and the Cu^+ ion concentration is necessarily very low. Under these circumstances, the $E^{\ominus}$ figures quoted above are not valid.

b Iron(II) ions in aqueous solution Fe^{2+} ions might be expected to disproportionate in the same way as Cu^+ ions:

$$3Fe^{2+}(aq) \longrightarrow 2Fe^{3+}(aq) + Fe(s)$$
$$\;\;+2 \qquad\qquad\quad +3 \qquad\qquad 0$$

Such disproportionation will not, however, take place for the change would have a negative $E^\ominus$ value:

$$
\begin{array}{ll}
Fe^{2+}(aq) + 2e^- \longrightarrow Fe(s) & E^\ominus = -0.440\,V \\
\underline{2Fe^{3+}(aq) + 2e^- \longrightarrow 2Fe^{2+}(aq)} & E^\ominus = +0.771\,V \\
3Fe^{2+}(aq) \qquad\; \longrightarrow 2Fe^{3+}(aq) + Fe(s) & E^\ominus = -0.440 - 0.771 \\
& \quad\;\; = -1.211\,V
\end{array}
$$

Indeed, iron will reduce Fe^{3+} ions to Fe^{2+} ions, the $E^\ominus$ value for this change being $+1.211\,V$.

9 Tests for oxidising and reducing agents

To test for an oxidising agent it is necessary to choose a strong reducing agent which undergoes some easily observable change when oxidised.

For insoluble solids, the evolution of chlorine on warming with concentrated hydrochloric acid is used.

For solutions, the liberation of iodine (which will turn starch blue) from an acidified solution of potassium iodide is best. Most oxidising agents will also oxidise iron(II) salts to iron(III) salts; ammonium iron(II) sulphate is commonly used.

To test for a reducing agent, a strong oxidising agent which undergoes easily detectable changes is required.

For an insoluble solid, the evolution of nitrogen dioxide on warming with concentrated nitric acid can be used.

For aqueous solutions, the decolorisation of acidified potassium manganate(VII) or the conversion of an acidified solution of potassium dichromate(VI) from orange to green are satisfactory tests. Reduction of iron(III) chloride to iron(II) chloride can also be used.

10 Redox titrations

Many oxidising agents will react with many reducing agents both quantitatively and rapidly so that they can be titrated together. The reactions and the measurements can be used to estimate an oxidising or a reducing agent, as in acid–base titrations.

A standard solution of some oxidising or reducing agent is required, and the substances used as primary standards must be available in a pure form, readily soluble in water, and stable.

a The use of oxidising agents as standards   Potassium manganate(VII) and potassium dichromate(VI) are the most commonly used oxidising agents. They are generally used in acid solution, i.e.

$$MnO_4^-(aq)+8H^+(aq)+5e^- \longrightarrow Mn^{2+}(aq)+4H_2O(l)$$
$$E^\ominus = +1.52\,V$$

$$Cr_2O_7^{2-}(aq)+14H^+(aq)+6e^- \longrightarrow 2Cr^{3+}(aq)+7H_2O(l)$$
$$E^\ominus = +1.33\,V$$

Potassium manganate(VII) is the more convenient to use, though it is not easy to obtain it in a very pure form, and it dissolves only slowly in water. It is convenient, however, because the very marked colour change from the deep purple MnO_4^- ion to the almost colourless Mn^{2+} ion means that no indicator is required. With potassium dichromate(VI), a redox indicator has to be used; this consists of some other oxidising or reducing agent which does have a marked colour change between its oxidised and reduced forms.

Other oxidising agents that are used in redox titrations include potassium bromate(V), potassium iodate(V) and cerium(IV) sulphate:

$$BrO_3^-(aq)+6H^+(aq)+5e^- \longrightarrow \tfrac{1}{2}Br_2(aq)+3H_2O(l) \quad E^\ominus = +1.52\,V$$
$$IO_3^-(aq)+6H^+(aq)+5e^- \longrightarrow \tfrac{1}{2}I_2(aq)+3H_2O(l) \quad E^\ominus = +1.19\,V$$
$$Ce^{4+}(aq)+e^- \longrightarrow Ce^{3+}(aq) \qquad\qquad\qquad\quad E^\ominus = +1.45\,V$$

b The use of reducing agents as standards   Reducing agents are less satisfactory than oxidising agents for use as standards because they do not store well, particularly in solution, due to atmospheric oxidation. If a solution of a reducing agent is to be used as a standard, it is generally necessary to standardise it before use against some oxidising agent solution.

In estimating an oxidising agent it may be preferable to reduce it completely by adding, say, excess zinc amalgam and an acid, and then titrate the resulting solution against a standard oxidising agent solution.

Reducing agents that may be used directly as standard solutions include iron(II) sulphate, tin(II) sulphate, sodium ethanedioate, and titanium(III) chloride or sulphate:

$$Fe^{3+}(aq)+e^- \longrightarrow Fe^{2+}(aq) \qquad\qquad\qquad\qquad E^\ominus = +0.77\,V$$
$$Sn^{4+}(aq)+2e^- \longrightarrow Sn^{2+}(aq) \qquad\qquad\qquad\quad E^\ominus = +0.15\,V$$
$$2CO_2(g)+2e^- \longrightarrow C_2O_4^{2-}(aq) \qquad\qquad\qquad E^\ominus = -0.1\,V$$
$$TiO_2^{2+}(aq)+2H^+(aq)+e^- \longrightarrow Ti^{3+}(aq)+H_2O(l) \quad E^\ominus = +0.1\,V$$

c Iodimetry   Many oxidising agents react quantitatively and rapidly with excess potassium iodide to form iodine, and this can readily be estimated by titration with sodium thiosulphate solution. The thiosulphate reduces the iodine:

$$I_2(s)+2S_2O_3^{2-}(aq) \longrightarrow 2I^-(aq)+S_4O_6^{2-}(aq)$$

The colour change at the end-point, from brown iodine to colourless iodide ions, is generally made much clearer by adding starch solution; this gives a colour change from blue–black to colourless.

Typical substances that can be estimated in this way include hydrogen peroxide, iodates(V) and Cu^{2+} ions (and hence, copper in alloys):

$$H_2O_2(aq) + 2H^+(aq) + 2I^-(aq) \longrightarrow I_2(aq) + 2H_2O(l)$$
$$IO_3^-(aq) + 6H^+(aq) + 5I^-(aq) \longrightarrow 3I_2(aq) + 3H_2O(l)$$
$$2Cu^{2+}(aq) + 4I^-(aq) \longrightarrow 3I_2(aq) + Cu_2I_2(s)$$

11 Oxidation state diagrams

Oxidation state diagrams provide an alternative method of summarising a lot of electrode potential data to that shown in Fig. 46, p. 86. The oxidation state of an element (n) in a compound or ion, is plotted against $nE^{\ominus}$.

The *gradient* of the line joining two points in the diagram is, then, the electrode potential of the couple made up of the two species represented by the points (Fig. 47(a)). A high positive gradient for any couple means that it is strongly oxidising, e.g. F_2/F^- ($+2.87$ V). The more negative the gradient for a couple, the more reducing it will be, e.g. Na^+/Na (-2.71 V). As it is the gradients that matter, they can be collected together, as in Fig. 47(b), for comparison.

The gradients can also be used to predict feasible reactions, but give no indication of rates. A couple with a high positive gradient will oxidise one

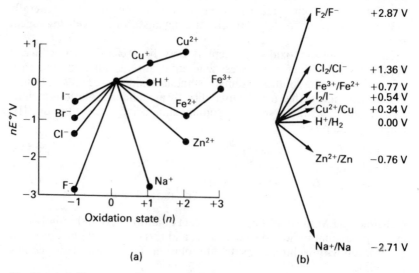

Fig. 47 (**a**) *Oxidation state diagram summarising some electrode potentials.* (**b**) *Summary of the gradients from oxidation state diagrams, giving electrode potentials in numerical order.*

with a lower positive (or a negative) gradient. Alternatively, a couple with a high negative gradient will reduce one with a lower negative (or a positive) gradient. F_2, for example, will oxidise Cl^-, Br^- and I^-, or Na and Zn. Na will reduce Zn^{2+} or Fe^{2+} (to Fe). I^- (but not Br^-, Cl^- or F^-) will reduce Fe^{3+} to Fe^{2+}.

a Disproportionation This occurs when the point representing the middle one of any three oxidation states lies *above* the line joining the higher and lower states. Thus, Cu^+ ions disproportionate into Cu^{2+} and Cu, but Fe^{2+} ions do not disproportionate into Fe^{3+} and Fe.

b Oxidation state diagram for the halogens Figure 48 shows how it is possible to collect together a lot of data in a way that makes it readily comparable. The diagram shows, for example,

 i that the halogen ion couples are oxidising, with the $\frac{1}{2}I_2/I^-$ couple being the weakest
 ii that Cl_2 would be expected to disproportionate, into Cl^- and ClO^-, more readily than Br_2 or I_2
 iii that the ClO^- ion would be expected to disproportionate, into Cl^- and ClO_3^-, more readily than BrO^- or IO^-
 iv that F_2 would be expected to oxidise water to oxygen.

All these predictions turn out to be accurate (see Chapter 23).

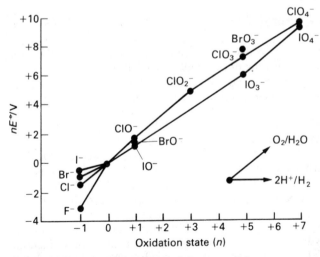

Fig. 48 *Oxidation state diagram for the halogens at pH 0.*

c Oxidation state diagram for manganese Figure 49 gives the $E^\ominus$ values for various manganese couples in both acidic and alkaline solutions, emphasising the dependence on pH. It shows, amongst other things, that the +2 oxidation state is the most stable state, that the MnO_4^-/Mn^{2+}

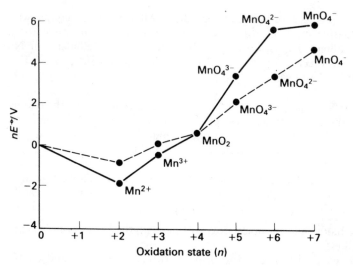

Fig. 49 *Oxidation state diagram for manganese. The bold line refers to pH 0; the dotted line to pH 14.*

couple in acidic solution is more strongly oxidising than the MnO_4^-/MnO_2 couple in alkaline solution, and that MnO_4^{2-} will disproportionate, into MnO_2 and MnO_4^-, in acidic, but not in alkaline solution (p. 348).

Questions on Chapter 7

1 Taking the tests for oxidising and reducing agents summarised on p. 90, write equations showing one example of each test in operation.

2 Give, with one example of each, three different ways, other than by the addition of oxygen, in which you could recognise that oxidation had taken place in a reaction.

Sulphur dioxide, manganese dioxide and hydrogen peroxide can, in different reactions, be oxidising or reducing agents. Explain why this is so and give one example of each type of reaction for each substance. (OC)

3 Define the terms 'oxidation' and 'reduction' in as many ways as you can. Identify the oxidising and reducing agents in each of the following processes, and discuss the application of your definition to them: (a) the reaction of hydrogen peroxide with acidified potassium iodide; (b) the reaction of hydrogen peroxide with acidified potassium permanganate; (c) the passage of hydrogen over heated sodium; (d) the addition of metallic calcium to water; (e) the reaction of hydrogen sulphide with moist sulphur dioxide. (L)

4 Define *oxidation* in terms of electron transfer. Rewrite each of the following molecular equations first as a single ionic equation then as a pair of ion–electron half-reaction equations, and show in each case which species is oxidised and which reduced.

(a) $Fe_2(SO_4)_3 + 2KI = 2FeSO_4 + I_2 + K_2SO_4$
(b) $2KMnO_4 + 16HCl = 2KCl + 2MnCl_2 + 5Cl_2 + 8H_2O$
(c) $3Cu + 8HNO_3 = 3Cu(NO_3)_2 + 2NO + 4H_2O$
(d) $4Zn + 10HNO_3 = 4Zn(NO_3)_2 + NH_4NO_3 + 3H_2O$
(e) $3CdS + 8HNO_3 = 3Cd(NO_3)_2 + 3S + 2NO + 4H_2O$ (JMB)

5 Which of the following conversions involves oxidation?
(a) $[Cr(H_2O)_6]^{3+} \longrightarrow CrO_4^{2-}$; (b) $Cr_2O_7^{2-} \longrightarrow CrO_4^{2-}$;
(c) $CrO_3 \longrightarrow CrO_4^{2-}$; (d) $[Cr(H_2O)_6]^{3+} \longrightarrow Cr_2O_7^{2-}$;
(e) $H_2O_2 \longrightarrow 2OH^-$; (f) $IO_3^- \longrightarrow I^-$; (g) $2H^- \longrightarrow H_2$;
(h) $[Fe(CN)_6]^{3-} \longrightarrow [Fe(CN)_6]^{4-}$; (i) $MnO_2 \longrightarrow MnO_4^-$. Give reasons
for your answers.

6 What is the change in the oxidation state of the metal in the following
conversions?
(a) $[Co(H_2O)_6]^{2+} \longrightarrow [CoCl_4]^{2-}$
(b) $[Fe(H_2O)_6]^{2+} + 6CN^- \longrightarrow [Fe(CN)_6]^{4-} + 6H_2O$
(c) $Ni + 4CO \longrightarrow Ni(CO)_4$
(d) $2MnO_4^{2-} + Cl_2 \longrightarrow 2MnO_4^- + 2Cl^-$
(e) $Zn + 2H^+ \longrightarrow Zn^{2+} + H_2$

7 What is meant by disproportionation? Discuss the following reactions
from this point of view:

$2H_2O_2 \longrightarrow 2H_2O + O_2$ $\qquad$ $HIO_3 + 5HI \longrightarrow 3I_2 + 3H_2O$
$3HIO \longrightarrow HIO_3 + 2HI$ $\qquad$ $Cl_2 + H_2O \longrightarrow HCl + HClO$

8 What is meant by the term 'oxidation number' ('oxidation state')?
Discuss the rules for assigning oxidation number, and the applications of
the concept in organic and inorganic chemistry. Place a suitable selection
of the compounds of *either* nitrogen *or* sulphur on a scale of oxidation
number. Comment on the oxidation number of *either* nitrogen in HN_3
(hydrazoic acid) *or* sulphur in $Na_2S_2O_3$, in relation to the structures of the
compounds. (L)

9 Explain, with examples, how changes in (a) concentration, (b) temperature, (c) acidity and (d) ligand affect the electrode potential of a redox
couple.

10 There are three common ways of expressing the tendencies of metals to
lose electrons: (a) ionisation energy, (b) electrode potential, and (c)
electropositivity. Explain the meaning of each of these terms and discuss
the ways in which they are related. Describe how *either* (a) *or* (b) could be
measured.

Which of these properties would you use in order to compile a reactivity
series of metals? Comment on the positions of calcium and sodium in such
a series, in the light of your knowledge of the general chemistry of these
elements. (L)

11 (a) Describe, with the aid of a labelled diagram, an experiment to
measure the standard electrode potential of silver and write an equation
representing the cell reaction. (b) Construct a cycle of the Born–Haber
type for the formation of silver ions in aqueous solution from solid silver.

Name the enthalpy change in each step and indicate its sign. (c) By
reference to the following data, discuss possible methods of preparation of
fluorine and chlorine.

$$\frac{1}{2}F_2(g) + e^- \longrightarrow F^-(aq) \qquad\qquad\qquad\qquad\qquad E^\ominus = +2.87\,V$$

$$\frac{1}{2}Cl_2(g) + e^- \longrightarrow Cl^-(aq) \qquad\qquad\qquad\qquad\qquad E^\ominus = +1.36\,V$$

$$MnO_4^-(aq) + 8H^+(aq) + 5e^- \longrightarrow Mn^{2+}(aq) + 4H_2O \quad E^\ominus = +1.51\,V$$

$$MnO_2(s) + 4H^+(aq) + 2e^- \longrightarrow Mn^{2+}(aq) + 2H_2O \quad E^\ominus = +1.23\,V$$

(JMB)

8
Some industrial processes

1 Introduction

One of the main purposes of chemistry is to make the best use of the limited raw materials that occur naturally. Sometimes, a raw material can be used on its own after a simple purification process as, for example, in making table salt from rock salt, in the treatment of water for drinking, in the production of noble gases from air, and in the use of North Sea gas for burning. Otherwise, raw materials are converted into specific chemicals which are then used to make the required end-products.

A large part of chemical industry, the so-called petrochemical sector, is concerned with crude oil as the basic raw material and, in the main, with organic compounds. But inorganic chemicals also play a large part in industrial processes, as shown by Chart 1 (p. 98), which summarises some of the processes. Both large-scale processes, such as making steel, ammonia and sulphuric acid, and small-scale processes, such as making very pure metals and pyrographite, are involved. The terms 'heavy' industry and 'heavy' chemicals are used for the large-scale processes; for smaller quantities the terms 'light' industry and 'fine' chemicals are used.

Necessity is the mother of invention and, over the years, it has been possible to manufacture chemicals to serve many varied needs. The history of the chemical industry has, therefore, been one of constant challenge and change, with the continual introduction of new processes and the replacement of old processes by more successful ones; and so it is today.

The development and operation of a successful process depends on a full understanding and application of basic chemical principles. But important changes may have to be made in transferring a small-scale process, which can be operated successfully in a laboratory or on a pilot scale, into a full-scale operation. A chemical engineer has to be both a good chemist and a good engineer. Nor is it just a matter of chemistry and engineering, because the success of a modern industrial process depends on many political and socio-economic factors. Some of these are outlined below; more details are given in the discussions of some individual processes later in the chapter.

a Economics It may sometimes be necessary for a basic process, e.g. steel making or coal mining, to be subsidised by the State, but the general aim is to operate profitably; the normal rules of supply and demand apply. It is necessary, too, to compete internationally because every advanced nation has its own extensive chemical industry.

One of the major factors in operating successfully is an adequate supply of a cheap raw material, and the purer it is, the better. Air is ideal because it is free and has no delivery charges; water and natural gas are also good.

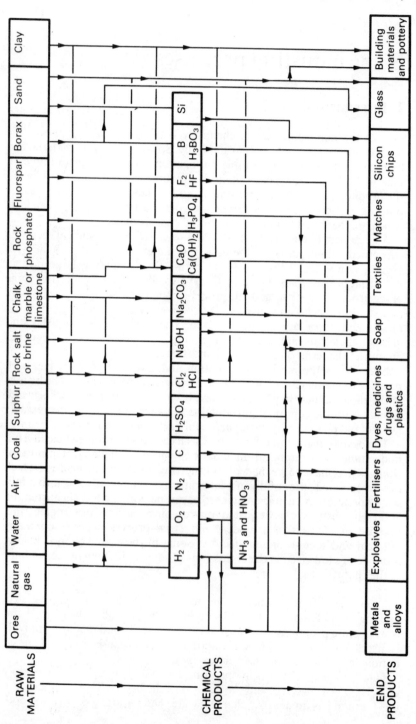

Chart 1 Industrial inorganic chemistry.

At the other end of the scale, any process that has to rely on imported materials is at a disadvantage. This applies to many operations in the United Kingdom, which has poor supplies of some important raw materials, including sulphur and metallic ores.

Once a supply of raw material is obtained, it is necessary to convert a high percentage of it into the required end-product, i.e. the process must have a high yield. The conversion operations must also be easy and rapid, so using the best catalyst is of particular importance. Other factors involved are fuel and transport costs (particularly in 'heavy' industry), the efficient recycling of any unreacted chemicals and the efficient use of heat-exchangers so that maximum use is made of the fuel.

The economics of any process can be broken down into capital costs and running costs. A very complex plant, operated by computers but with a small work-force, will have a high capital, but a low running cost; a large work-force will put up the running cost. For any process, it is advantageous to build as large a plant as possible, in order to achieve 'economies of scale'. But the most important thing is to make a good product at the right price. Making something that no-one wants to buy, or making something that is so expensive that no-one can afford it, is of little use. It is *possible*, for example, to make gold from sea water, and this has been tried on an industrial scale. But this superficially attractive process has, so far, been a quick way to bankruptcy.

b Location Chemical plants are best sited close to a supply of their raw materials and their required fuel, close to their customers, and in areas with favourable sea, road, rail or air transport and good supplies of labour and housing. That is why there is a heavy concentration of chemical industry around the coal and salt supplies and good transport facilities of South Lancashire and North Cheshire; why industries requiring a lot of electricity are located close to big power stations or hydro-electric schemes, as at Niagara Falls; and why many modern plants, particularly in 'heavy' industry, are sited close to ports. It is easier to find good sites for 'light' industry because their transport costs are much smaller.

It is also important to introduce new industries into regions where old ones are dying out; the development of the silicon chip industry in Scotland provides a good example.

c The environment In days gone by, little attention was paid to the environment: 'where there's muck there's money' had some truth in it. An increasing public awareness and government legislation have, however, changed that for the better. Much more care has to be taken, for example, in the design of chemical plants, in the disposal of waste products, in avoiding the escape of deleterious products into the atmosphere, and in protecting any workers who might be at risk within a plant.

The problems of 'acid rain' and of the disposal of radioactive waste show, however, that the pollution issues have not yet been entirely solved.

2 Manufacture of ammonia

Ammonia is manufactured by the Haber process involving the exothermic reaction between hydrogen and nitrogen:

$$N_2(g) + 3H_2(g) \rightleftharpoons 2NH_3(g) \qquad \Delta H_m^{\ominus}(298\,K) = -92.4\,kJ\,mol^{-1}$$

The initial $1:3$ mixture of nitrogen and hydrogen is obtained from methane, naphtha, or other hydrocarbon mixtures. They are treated with steam (steam reforming, p. 145) and air; the amount of air used being chosen to give the correct proportion of nitrogen. The reactions taking place may be outlined as follows:

$$C_xH_y + xH_2O \rightleftharpoons xCO + (x + \tfrac{y}{2})H_2$$
$$C_xH_y + \tfrac{x}{2}O_2 + N_2 \rightleftharpoons xCO + \tfrac{y}{2}H_2 + N_2$$
$$CO + H_2O \rightleftharpoons CO_2 + 2H_2$$

The carbon dioxide in the final gas mixture is removed by solution in a concentrated solution of potassium carbonate or a solution of an amine. Traces of carbon monoxide, which is a catalyst poison, are converted into methane by passing the mixture over a nickel catalyst at 350 °C:

$$CO(g) + 3H_2(g) \rightleftharpoons CH_4(g) + H_2O(g)$$

a The background For many years farmers were able to grow adequate amounts of food by using natural manures and crop rotation. But towards the end of the nineteenth century it became clear that artificial fertilisers would be needed, to allow the more intensive use of agricultural land, at least in heavily populated countries. Fortunately, the role of fertilisers, and the necessity of providing soluble compounds of nitrogen, phosphorus and potassium, had already been established in the latter half of that century.

The vast deposits of Chile saltpetre, containing $NaNO_3$, which had long been used in making gunpowder, became widely used as the main nitrogenous fertiliser, and the $NaNO_3$ also provided a good source of nitric acid. Ammonium compounds, obtained as by-products from coal gas manufacture, were also used as fertilisers.

Early processes for making artificial fertilisers involved the manufacture of ammonia from cyanamide (p. 186) and of nitric acid from air (p. 242). But both these processes were very expensive, so the introduction of the Haber process around 1910 was particularly timely and important. It was of great significance, too, that the ammonia that could be manufactured by it provided an alternative supply of nitric acid. During the First World War a lot of the acid was made in Germany in this way, in order to make explosives. At the end of the war, every country switched to the Haber process for making both ammonia and nitric acid, and the method is still in use today.

The raw materials for the process are relatively cheap and widely available, but high temperatures and pressures are needed. Much engineering skill is required to build the plant (parts of it need to be built with special alloys to resist chemical attack) and the very large compressors. It is, however, the use of special catalysts that makes the operation successful by providing economic yields of ammonia at reasonable rates.

In Britain, the main ammonia plants are sited at Billingham on the River Tees in Cleveland. They were, originally, close to the Durham coalfields and to good supplies of electricity; now, by chance, they are conveniently close to North Sea oil and gas.

b Optimum conditions for reaction The conversion of nitrogen and hydrogen into ammonia is exothermic and results in a decrease in gas volume. Le Chatelier's principle shows, therefore, that the yield of ammonia will be greatest at low temperature and high pressure (Fig. 50).

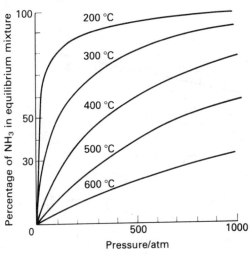

Fig. 50 *The percentage of ammonia in the equilibrium mixture obtained from a 3:1 mixture of hydrogen and nitrogen at different temperatures and pressures.*

At low temperature, however, equilibrium is reached slowly; at high pressure, the cost of equipment is high. In practice a compromise has to be struck. Pressures between 150 and 350 atm are used at temperatures of 400–500 °C and the rate of reaction is increased by using a catalyst of iron with added promoters.

To reach a temperature of 400–500 °C the initial mixture of gases must be heated, but once started, the reaction produces enough heat to maintain the temperature. Indeed, the catalyst bed is cooled down in the later stages of the reaction, for lower temperatures give the best results as the amount of ammonia in the mixture rises. The gases leaving the catalyst bed are, therefore, at a temperature between 100 and 200 °C. The ammonia is condensed into a liquid by refrigeration at $-20\,°C$. The little ammonia remaining as a gas, together with unreacted hydrogen and nitrogen, is recycled.

c Effect of temperature

The relevant thermodynamic data for the Haber process are summarised below.

temperature/K	298	400	500	600	800
ΔH/kJ mol^{-1}	-92.4	-96.9	-101.3	-105.8	-114.6
ΔG/kJ mol^{-1}	-33.3	-12.3	13.9	31.9	79.1
$T\Delta S$/kJ mol^{-1}	-59.1	-84.6	-115.2	-137.7	-193.7
K_p/atm^{-2}	6.73×10^5	40.7	3.55×10^{-2}	1.66×10^{-3}	6.92×10^{-6}
$\lg K_p$	5.83	1.61	-1.45	-2.78	-5.16

The values of ΔH, ΔG and $T\Delta S$ are plotted on Fig. 51. It will be seen that ΔG has a zero value at approximately 450 K and this corresponds to a $\lg K_p$ value of 0 and a K_p value of 1.

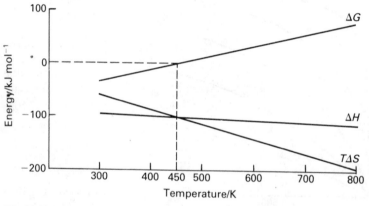

Fig. 51 *Showing how ΔG, ΔH and $T\Delta S$ vary with temperature for the reaction*

$$N_2(g) + 3H_2(g) \longrightarrow 2NH_3(g)$$

ΔG is zero at approximately 450 K.

The $\lg K_p$ values at different temperatures can also be calculated from the value at 298 K, as explained on p. 76, and some calculated and measured values are compared below.

temperature/K	298	400	500	600	700	800
$\lg K_p$ (calculated)	5.83	1.69	-0.73	-2.34	-3.45	-4.35
$\lg K_p$ (measured)	5.83	1.61	-1.45	-2.78	-4.11	-5.16

It will be seen that there is some discrepancy between the calculated and measured values, which increases as the temperature rises. This occurs because the calculation assumes (p. 76) that ΔH does not change with temperature; but this is not quite true, as the figures quoted at the start of this section show.

d Effect of pressure By assuming that the gases behave as ideal gases, and by considering an initial $3:1$ mixture of hydrogen and nitrogen, it is also possible to calculate approximate values for the yield of ammonia at different pressures.

K_p is taken as being independent of temperature, though it does change slightly, particularly at high pressures, due to the non-ideal behaviour of the gases. If the partial pressures in the equilibrium mixtures are p_{H_2}, p_{N_2} and p_{NH_3}, then

$$p_{N_2} + p_{H_2} + p_{NH_3} = P(\text{total pressure}) \quad \text{and} \quad K_p = \frac{(p_{NH_3})^2}{(p_{N_2})(p_{H_2})^3}$$

As p_{H_2} is three times as big as p_{N_2} it follows that

$$p_{N_2} = \tfrac{1}{4}(P - p_{NH_3}) \qquad p_{H_2} = \tfrac{3}{4}(P - p_{NH_3}) \qquad K_p = \frac{9.48(p_{NH_3})^2}{(P - p_{NH_3})^4}$$

At 700 K, K_p is 7.76×10^{-5}, and for a 50 per cent yield of ammonia in the equilibrium mixture p_{NH_3} must be equal to $\tfrac{1}{2}P$; P will have a value of 700 atm. For a yield of 20 per cent at the same temperature, p_{NH_3} must be equal to $\tfrac{1}{5}P$ and a pressure of 110 atm is required. These figures and others which can be obtained similarly are in good agreement with those plotted in Fig. 50.

3 Manufacture of sulphuric acid

Sulphuric acid is mainly manufactured by the contact process, in which sulphur dioxide is oxidised to sulphur(VI) oxide, which is then reacted with water:

$$2SO_2(g) + O_2(g) \longrightarrow 2SO_3(g)$$
$$SO_3(g) + H_2O(l) \longrightarrow H_2SO_4(l)$$

a The source of sulphur dioxide Most of the sulphur dioxide used in the contact process in this country is made by burning sulphur in dry air. Liquid sulphur is sprayed into a combustion furnace where it burns with a hot flame to produce a mixture of sulphur dioxide (10%), oxygen (10%) and nitrogen (80%), which is pumped out from the end of the furnace. In other parts of the world, roasting sulphide ores (p. 116), e.g.

$$2ZnS(s) + 3O_2(g) \longrightarrow 2ZnO(s) + 2SO_2(g)$$

provides an alternative source of sulphur dioxide.

b The contact process   The conversion of sulphur dioxide and oxygen into sulphur(VI) oxide is an exothermic reaction with a decrease in gas volume:

$$2SO_2(g) + O_2(g) \rightleftharpoons 2SO_3(g) \qquad \Delta H_m^{\ominus}(298\,K) = -197.6\,kJ\,mol^{-1}$$

Increase in pressure would give an increased yield of sulphur(VI) oxide, but the effect is small and the yield is good even at low pressures just above atmospheric which are, therefore, used in practice.

The greatest yield of sulphur(VI) oxide would be obtained at low temperature, but the attainment of the equilibrium would be very slow. A compromise temperature, initially about 420 °C, is used, together with a catalyst of vanadium(V) oxide with potassium sulphate as a promoter, on a support of silicon(IV) oxide.

Yields above 95.5 per cent are obtained, but only by carrying out the process in stages. The mixture of gases from the sulphur burner is passed into the first converter at 420 °C; a 63 per cent conversion is achieved but the temperature rises as the reaction is exothermic. Above 600 °C, the catalyst deteriorates and the sulphur(VI) oxide begins to dissociate. The partially converted gas mixture is therefore cooled before passing into a second converter, and then, likewise, into a third.

The sulphur(VI) oxide in the mixture is now removed by absorption in 98 per cent sulphuric acid. Water is added to maintain that concentration or, alternatively, the sulphur(VI) oxide is allowed to go on dissolving in the acid, forming *oleum* (p. 294). 98 per cent sulphuric acid is used to absorb the sulphur(VI) oxide in this way, as it does it more effectively than pure water, which forms a mist.

The unreacted gas mixture, containing only about 0.7 per cent of sulphur dioxide, is passed into a fourth converter, and the extra sulphur(VI) oxide formed is again absorbed. In this way all but about 0.05 per cent of the original sulphur dioxide is converted into sulphur(VI) oxide.

c The background   Sulphuric acid has been known for over 1000 years. It was originally made from the SO_3 and SO_2 obtained by heating iron(II) sulphate, $FeSO_4$ (green vitriol):

$$2FeSO_4(s) \longrightarrow Fe_2O_3(s) + SO_2(g) + SO_3(g)$$

hence the old name for the acid, *oil of vitriol*.

Larger-scale manufacture began about 200 years ago, the main demands for it being for bleaching textiles, and for making nitric acid and chlorine. The process used became known as the *lead chamber process* because SO_2, steam, air and NO_2 were heated in lead-lined containers. The details of the process were complicated, the capital cost of the plant was high and, though the yields were good, the acid made was only dilute and not very pure. The introduction of the contact process about 100 years ago has, by now, made the lead chamber process almost obsolete; it is not operated at all in the United Kingdom.

The capital cost of the contact process plant is relatively low, and it produces pure, concentrated acid. It was, however, only the development of good catalysts (at first platinum, but now V_2O_5), and the control of temperature in the various converters, that enabled reasonable yields to be obtained.

The overall conversion of sulphur into sulphur(VI) oxide is highly exothermic,

$$2S(s) + 3O_2(g) \longrightarrow 2SO_3(g) \qquad \Delta H_m^{\ominus}(298\,K) = -790\,kJ\,mol^{-1}$$

and by careful heat exchange in the converters it is possible to make a lot of steam as a by-product. It is the selling of this steam that makes the contact process very economic and allows the sulphuric acid to be sold very cheaply. The main disadvantage of the process, in the United Kingdom is the necessity to use imported sulphur. Various alternative supplies of SO_2 have been sought, but not with complete success.

d Thermodynamic considerations Some data for the contact process reaction is summarised below and plotted on Fig. 52.

temperature/K	298	800	900	1000	1100
$\Delta H/kJ\,mol^{-1}$	−197.6	−193	−191	−189	−187
$\Delta G/kJ\,mol^{-1}$	−139.8	−46	−28	−10	+8
$T\Delta S/kJ\,mol^{-1}$	−55.8	−147	−163	−178	−195
K_p/atm^{-1}	33×10^{24}	1000	42.7	3.39	0.4
$lg\,K_p$	24.52	3.0	1.63	0.53	−0.4

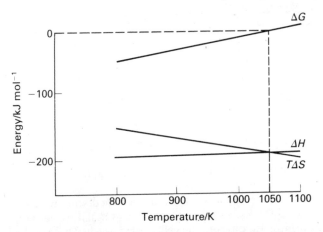

Fig. 52 *Showing how* ΔG, ΔH *and* $T\Delta S$ *vary with temperature for the reaction*

$$2SO_2(g) + O_2(g) \longrightarrow 2SO_3(g)$$

ΔG *is zero at approximately 1050 K.*

The approximate temperature at which ΔG is zero, i.e. at which ΔH is equal to $T\Delta S$, is 1050 K (p. 73).

The value of K_p falls rapidly with rising temperature. The entropy change is negative, due to the decrease in the number of gas molecules in the course of the reaction.

4 Manufacture of chlorine

Electrolysis of a dilute, aqueous solution of sodium chloride using carbon or platinum electrodes produces hydrogen at the cathode and oxygen at the anode. By using brine (a saturated solution of sodium chloride in water) and special conditions, however, chlorine can be made, together with the useful by-products, sodium hydroxide and hydrogen.

a Use of a flowing mercury cathode (Fig. 53)   This is the main process used in the United Kingdom; it is sometimes known as the Castner–Kellner process.

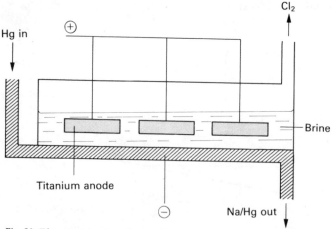

Fig. 53 *The principle of a mercury cell.*

The electrolysis is carried out in a slightly tilted steel trough lined with rubber. The anodes are adjustable titanium plates, which are maintained about 2 mm above a stream of mercury that flows across the bottom of the trough and acts as the cathode. Purified brine flows through the trough in the same direction as the mercury.

Chlorine is produced at the anode, for $Cl^-(aq)$ ions are discharged there in preference to $OH^-(aq)$ ions, as they are present in much higher concentration. The chlorine is led off from the cell, cooled, washed with water, dried, and liquefied.

At the cathode, $Na^+(aq)$ ions are discharged instead of $H^+(aq)$ ions, mainly because of the high over-voltage of hydrogen at a mercury surface, and sodium amalgam, a liquid alloy of sodium and mercury, is formed. This amalgam runs off from the cell continuously, and passes into a decomposer. This is a slightly-tilted steel cell containing activated graphite blocks. The amalgam flows down the cell over the graphite blocks and against a counter-flow of water. The amalgam and water react on the graphite surface to produce sodium hydroxide, hydrogen and mercury:

$$2Na/Hg(l) + 2H_2O(l) \longrightarrow 2NaOH(aq) + H_2(g) + 2Hg(l)$$

The sodium hydroxide can be used as a solution or evaporated to give the solid; the mercury is recycled.

Mercury cells are operated at about 4.4 V, and currents up to 400 kA can be used.

b Use of a diaphragm cell (Fig. 54)   In this type of cell, a titanium anode is separated from a steel cathode by a porous asbestos diaphragm.

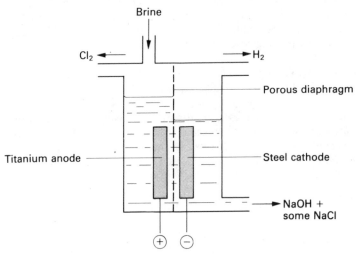

Fig. 54 The principle of a diaphragm cell.

Pure saturated brine, containing hydrated Cl^-, Na^+, H^+ and OH^- ions, is fed into the anode compartment; the $Cl^-(aq)$ ions are preferentially discharged, and chlorine gas bubbles off. The solution, depleted in $Cl^-(aq)$ ions, passes through the diaphragm into the cathode compartment, where $H^+(aq)$ ions are preferentially discharged, and hydrogen bubbles off.

The solution remaining in the cathode compartment contains sodium hydroxide with some chloride impurity. On concentration by evaporation, sodium chloride crystallises out, and a concentrated solution of fairly pure sodium hydroxide remains. A typical cell operates at 3.8 V with currents up to 150 kA.

c Use of a membrane cell This new type of cell is very similar to the diaphragm cell; it uses a titanium anode and a nickel cathode separated by an ion exchange membrane. The reactions taking place in the cell are identical to those in the diaphragm cell, but the ion exchange membrane prevents the passage of Cl^- ions (but allows the passage of Na^+ ions) into the cathode compartment. The sodium hydroxide solution that forms in the cathode compartment is, therefore, not contaminated with sodium chloride, as it is in the diaphragm cell. The solution still has to be concentrated but the sodium hydroxide that is obtained is purer than that from a diaphragm cell. A typical cell operates at 3.1 V and currents up to 75 kA.

d The background Chlorine was originally mainly used for bleaching textiles and, in the United Kingdom, the textile industry developed very greatly in the last century. At first, cloth was bleached by heating with chlorine water, but later, bleaching powder (p 322) or solutions of chlorates(I) were used. Prior to the use of chlorine, bleaching had been a very lengthy and uncertain business, involving repeated boiling, washing and exposure to sunlight for up to six months.

Carl Scheele discovered chlorine in 1774, and it was manufactured shortly after that by heating NaCl, H_2SO_4 and MnO_2,

$$2NaCl(s) + H_2SO_4(aq) \longrightarrow Na_2SO_4(s) + 2HCl(g)$$
$$4HCl(aq) + MnO_2(s) \longrightarrow MnCl_2(aq) + 2H_2O(l) + Cl_2(g)$$

but large quantities of manganese(II) salts were left as by-products. Many ingenious attempts were made to reconvert these salts into MnO_2, but by 1868 the *Deacon process* had taken over. In this process, the HCl from the reaction of NaCl with H_2SO_4 was oxidised by air, using a catalyst of copper(II) chloride at 450 °C, and not by MnO_2;

$$4HCl(g) + O_2(g) \longrightarrow 2H_2O(l) + 2Cl_2(g)$$

The electrolytic processes started around 1900 as supplies of electricity became available. They had the advantage of producing valuable products other than chlorine, and they came to make up what is now known as the *chlor-alkali* industry. The electricity they use is not cheap, and it is not easy to ensure a balanced demand for all the products, but the raw materials are cheap and readily available. In this country, the plants are sited around Runcorn on Merseyside, where they are close to the salt supplies of North Cheshire and the Lancashire coalfields.

Mercury cells produce very pure sodium hydroxide, but they are expensive to construct, and the mercury used is both expensive and toxic. Diaphragm cells are much cheaper to build, but the diaphragms need frequent replacement, and it is costly to evaporate the solution from the cathode compartment to purify and concentrate it. Membrane cells require pure brine but they yield a high purity product, and the membrane only needs to be changed every two to three years. As they have lower capital and operating costs than either the mercury cell or the diaphragm cell, they are likely to be used increasingly in the future.

Nowadays, much of the chlorine made by the electrolytic processes is used in the chlorination of organic compounds, and this produces a lot of by-product hydrogen chloride. There has, as a result, been some revival of the older Deacon process for oxidising this back to chlorine, for re-use.

5 Manufacture of sodium carbonate

a The ammonia–soda or Solvay process (Fig. 55)   In this process, sodium chloride and calcium carbonate are the raw materials,

$$2NaCl(aq) + CaCO_3(s) \longrightarrow Na_2CO_3(aq) + CaCl_2(aq)$$

but they will not react directly because of the insolubility of calcium carbonate. A roundabout method has to be adopted.

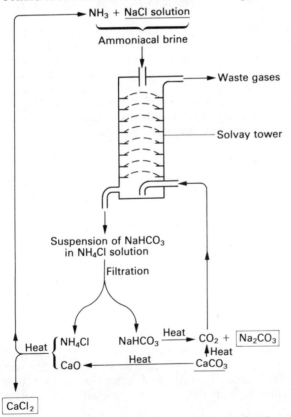

Fig. 55 *The Solvay or ammonia–soda process (simplified). The starting materials are underlined; the final products enclosed in rectangles.*

In practice, purified brine (a strong solution of sodium chloride) is saturated with ammonia and the mixture is trickled down a tower fitted with baffle-plates whilst carbon dioxide, obtained by heating limestone,

i $$CaCO_3(s) \longrightarrow CaO(s) + CO_2(g)$$

is passed up the tower. The carbon dioxide and ammonia react to form HCO_3^- and NH_4^+ ions:

ii $$CO_2(g) + H_2O(l) \rightleftharpoons HCO_3^-(aq) + H^+(aq)$$
iii $$NH_3(g) + H^+(aq) \longrightarrow NH_4^+(aq)$$

Na^+ and Cl^- ions are also present, but the least soluble product is sodium hydrogencarbonate and this separates out as a suspension in a solution of ammonium chloride:

iv $$Na^+(aq) + Cl^-(aq) + NH_4^+(aq) + HCO_3^-(aq) \longrightarrow$$
$$NaHCO_3(s) + NH_4^+(aq) + Cl^-(aq)$$

The reactions produce heat so that the bottom of the tower has to be water-cooled to lower the solubility of sodium hydrogencarbonate.

Sodium hydrogencarbonate would normally be regarded as soluble (0.122 mol per 100 g of water at 25 °C) but the high concentration of Na^+ ions in this mixture lowers the solubility by a common ion effect. Potassium hydrogencarbonate is more soluble (0.362 mol per 100 g of water at 25 °C) and it will not separate out as the sodium salt will. It is for this reason that the Solvay process cannot be used directly for making potassium carbonate.

The sodium hydrogencarbonate is removed from the suspension by rotary filters and heated in large rotary calciners to convert it into anhydrous sodium carbonate (soda ash):

v $$2NaHCO_3(s) \longrightarrow Na_2CO_3(s) + H_2O(g) + CO_2(g)$$

Recrystallisation gives sodium carbonate-10-water, $Na_2CO_3.10H_2O$ (washing soda).

The carbon dioxide required in ii is provided by i and v, whilst the ammonia required for iii is obtained by heating the calcium oxide from i with the ammonium chloride from iv. Calcium chloride is something of a waste product for which there is, as yet, little use.

b Sodium carbonate from sodium hydroxide   Electrolysis of brine produces both sodium hydroxide and chlorine, and the demand for the two products is not always in balance. If a surplus of sodium hydroxide arises, it can be converted into sodium carbonate by reaction with carbon dioxide:

$$2NaOH(aq) + CO_2(g) \longrightarrow Na_2CO_3(aq) + H_2O(l)$$

c The background   Sodium carbonate and potassium carbonate have been used for a very long time in making soap and glass. Until the end of the eighteenth century, supplies were obtained from the ashes of plants: wood provided K_2CO_3 (hence the name potash); and seaweed, and a Spanish plant called barilla, provided Na_2CO_3 (soda ash). The supplies were always inadequate and expensive, and this led the French Academy of Science to offer a prize of 12 000 francs, in 1775, for the best new method of manufacture; the prize was won by Nicolas Leblanc.

His process involved heating sodium chloride with sulphuric acid to make sodium sulphate, and the subsequent roasting of this with calcium carbonate and coal:

$$2NaCl(s) + H_2SO_4(aq) \longrightarrow Na_2SO_4(aq) + 2HCl(g)$$
$$Na_2SO_4(s) + CaCO_3(s) + 4C(s) \longrightarrow Na_2CO_3(s) + CaS(s) + 4CO(g)$$

The sodium carbonate was extracted by dissolving in water.

This French process was, eventually, much used in the United Kingdom. It had the advantage that the raw materials (except the sulphuric acid) were readily available, and the by-product hydrogen chloride provided cheap acid and was used in making chlorine. However, the sodium carbonate produced was not very pure, and the process caused such extensive pollution that it eventually led to the introduction of the first Alkali Act (1863), which attempted to control emissions from factories and which set up an inspectorate to supervise them.

The pollution arose both from the emission of hydrogen chloride fumes into the atmosphere, and from the insoluble calcium sulphide. Originally, the latter was stored in heaps, or dumped into the sea, but it reacted slowly with water to produce hydrogen sulphide. Eventually a means of converting it into sulphur, which could be used in making sulphuric acid, was developed, but by then much damage had been done.

The Solvay process was invented in 1863 and introduced into the United Kingdom, at Northwich in North Cheshire, by Ludwig Mond in 1872. The Leblanc process began to decline, but it lasted in some parts of the world until around 1925, mainly because of the valuable hydrogen chloride and sulphur by-products. The Solvay process is still operated at Northwich, which is close to natural supplies of brine and not far from limestone deposits in Derbyshire. Transport facilities are good, and the nearby manufacturers of glass, soap and textiles are the main customers.

6 Ellingham diagrams

Metallic oxides may be reduced by heating with carbon,

$$i \qquad 2MO(s) + 2C(s) \longrightarrow 2M(s) + 2CO(g)$$

so long as a temperature at which the ΔG for the reaction is negative can be achieved. This overall reaction is really a combination of two reactions,

ii $2C(s) + O_2(g) \longrightarrow 2CO(g)$
iii $2M(s) + O_2(g) \longrightarrow 2MO(s)$

and the ΔG for reaction *i* will be equal to that for reaction *ii* minus that for reaction *iii*.

In reaction *ii* there is an entropy increase (p. 70) because one mole of gas is converted into two; ΔS is, therefore, positive. At 298 K

	$\Delta H_m^{\ominus}/$ kJ mol^{-1}	$\Delta G_m^{\ominus}/$ kJ mol^{-1}	$\Delta S_m^{\ominus}/$ J K^{-1} mol^{-1}
$2C(s) + O_2(g) \longrightarrow 2CO(g)$	-221	-274.6	$+178.2$

As the value of $T\Delta S$ increases with rise in temperature, the ΔG value gets more and more negative at higher temperatures (Fig. 56).

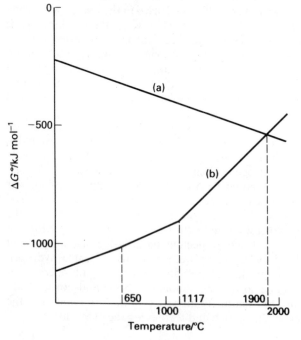

Fig. 56 *Changes of $\Delta G^{\ominus}$ with temperature for (**a**) $2C(s) + O_2(g) \longrightarrow 2CO(g)$ and (**b**) $2Mg(s) + O_2(g) \longrightarrow 2MgO(s)$.*

Conversely, the entropy change in reaction *iii* is negative because a solid is obtained from a gas. For magnesium oxide, for instance, at 298 K

	$\Delta H_m^{\ominus}$	$\Delta G_m^{\ominus}$	$\Delta S_m^{\ominus}$
$2Mg(s) + O_2(g) \longrightarrow 2MgO(s)$	-1203.4	-1138.8	-226.7

As a result, the ΔG value becomes less negative as temperature rises (Fig. 56). The change of ΔG with temperature is not regular because entropy changes at the melting temperature and, particularly, the boiling temperature of magnesium (650 °C and 1117 °C) cause abrupt changes in the ΔG value.

From Fig. 56, which is known as an Ellingham diagram, it will be seen that the ΔG value for reaction i will only become negative at a temperature of approximately 1900 °C. Magnesium oxide can, then, be reduced by carbon above that temperature. Below it, the ΔG value would be positive and the oxide would not be reduced by carbon; indeed it would be the magnesium that would reduce the carbon monoxide.

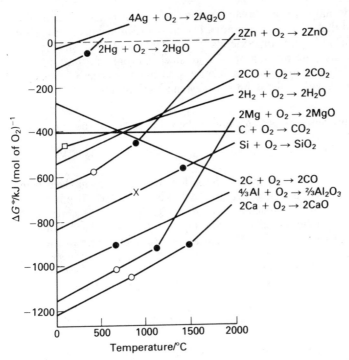

Fig. 57 *Ellingham diagrams. 'O' signifies melting temperature of element; '●' boiling temperature of element; '□' boiling temperature of oxide; 'x' transition temperature.*

Data for a number of oxides is plotted on the Ellingham diagrams in Fig. 57, and it will be seen that discontinuities occur at the melting temperatures and, particularly, the boiling temperatures of the elements or oxides concerned. There may also be discontinuities at transition temperatures.

Ellingham diagrams provide an extremely useful method of summarising a lot of data. The ones given refer to the reduction of oxides; similar ones can also be drawn relating to the reduction of sulphides or chlorides.

7 Methods of extracting metals

a Use of carbon to reduce an oxide   The ΔG–T plot for the reaction

$$2C(s) + O_2(g) \longrightarrow 2CO(g) \qquad \Delta G_m^{\ominus}(298\,K) = -274.6\,kJ\,mol^{-1}$$

slopes downwards and cuts across the lines for many metallic oxides, which slope upwards, the point of cutting giving the temperature at which the reduction of the oxide will begin to take place. Thus, zinc oxide is reduced by carbon above approximately 950 °C, and magnesium oxide above about 1900 °C; aluminium and calcium oxides require temperatures above 2000 °C.

It is lucky that carbon is available (as coke or anthracite) reasonably cheaply, and that it is easy to handle. It is also advantageous that carbon will form carbides with some metals; the formation of such carbides, for instance, is vital in making iron and steel. Carbide formation may, however, be disadvantageous for it may spoil the properties of the metal as, for example, with molybdenum, tungsten and titanium.

b Use of hydrogen to reduce an oxide   The value of ΔG for the reaction

$$2H_2(g) + O_2(g) \longrightarrow 2H_2O(g) \qquad \Delta G_m^{\ominus}(298\,K) = -457.2\,kJ\,mol^{-1}$$

gets less negative as temperature rises, for two moles of gas are being formed from three, giving a negative ΔS value. The ΔG–T plot slopes upwards (with a marked discontinuity at 100 °C) in much the same way as the lines for metallic oxides (Fig. 57). Because of this upward slope, the ΔG–T line for hydrogen does not cut across the metal lines in the same way as the carbon line does, and hydrogen will reduce many fewer oxides than carbon. It is, also, a much less convenient industrial reagent.

Hydrogen is used in making pure molybdenum and tungsten.

c Reduction of oxides by aluminium   The ΔG value for the reaction

$$\tfrac{4}{3}Al(s) + O_2(g) \longrightarrow \tfrac{2}{3}Al_2O_3(s) \qquad \Delta G_m^{\ominus}(298\,K) = -1055\,kJ\,mol^{-1}$$

is highly negative, so aluminium is a powerful reducing agent for many other metallic oxides. It is used, for example, in the thermit process for making iron, and in making chromium, manganese and cobalt:

	$\Delta G_m^{\ominus}(298\,K)$
$Fe_2O_3(s) + 2Al(s) \longrightarrow 2Fe(s) + Al_2O_3(s)$	$-861.4\,kJ\,mol^{-1}$
$Cr_2O_3(s) + 2Al(s) \longrightarrow Al_2O_3(s) + 2Cr(s)$	$-535.6\,kJ\,mol^{-1}$
$3Mn_3O_4(s) + 8Al(s) \longrightarrow 4Al_2O_3(s) + 9Mn(s)$	$-2489\,kJ\,mol^{-1}$
$3Co_3O_4(s) + 8Al(s) \longrightarrow 4Al_2O_3(s) + 9Co(s)$	$-4653\,kJ\,mol^{-1}$

d Reduction of oxides on heating alone   Some oxides, e.g. HgO and Ag$_2$O decompose into the metal and oxygen simply on heating. The ΔG values for the decompositions

$$2HgO(s) \longrightarrow 2Hg(l) + O_2(g) \qquad \Delta G_m^{\ominus}(298\,K) = +117\,kJ\,mol^{-1}$$
$$2Ag_2O(s) \longrightarrow 4Ag(s) + O_2(g) \qquad \Delta G_m^{\ominus}(298\,K) = +21.6\,kJ\,mol^{-1}$$

have low positive values even at 25 °C and they become negative at relatively low temperatures, as shown in the Ellingham diagrams (Fig. 57). Silver oxide begins to decompose at about 160 °C and mercury(II) oxide at 360 °C.

e Reduction of sulphides   Many metals occur naturally as sulphides and it would be convenient if the sulphide could be reduced directly to the metal. Sulphides of very unreactive metals, e.g. HgS and Ag$_2$S, can, like their oxides, be split up into the metal simply by heating, but reduction of other sulphides is not easily feasible, as neither carbon nor hydrogen are effective reducing agents in this situation, mainly because the ΔG values for the reactions

$$C(s) + 2S(s) \longrightarrow CS_2(l) \qquad \Delta G_m^{\ominus}(298\,K) = +63.6\,kJ\,mol^{-1}$$
$$H_2(g) + 2S(s) \longrightarrow H_2S(g) \qquad \Delta G_m^{\ominus}(298\,K) = -33.6\,kJ\,mol^{-1}$$

are so much less negative than the corresponding reactions involving oxygen and because of the non-existence of CS to correspond with CO. That is why the naturally occurring sulphides have to be roasted in air to convert them into the oxide, prior to reduction with carbon.

f Reduction of chlorides   As with sulphides, it would be convenient if metallic chlorides could be reduced easily by carbon or hydrogen but the ΔG values for the reactions

$$C(s) + 2Cl_2(g) \longrightarrow CCl_4(l) \qquad \Delta G_m^{\ominus}(298\,K) = -69.3\,kJ\,mol^{-1}$$
$$H_2(g) + Cl_2(g) \longrightarrow 2HCl(g) \qquad \Delta G_m^{\ominus}(298\,K) = -190.6\,kJ\,mol^{-1}$$

are not negative enough.

Metallic chlorides can be reduced by reactive metals. Before the development of the electrical process, aluminium was made from sodium and aluminium chloride, and titanium is made, in the Kroll process, by reducing titanium(IV) chloride with magnesium or sodium:

	$\Delta G_m^{\ominus}(298\,K)$
AlCl$_3$(s) + 3Na(s) $\longrightarrow$ 3NaCl(s) + Al(s)	$-647\,kJ\,mol^{-1}$
TiCl$_4$(s) + 2Mg(s) $\longrightarrow$ 2MgCl$_2$(s) + Ti(s)	$-469.1\,kJ\,mol^{-1}$
TiCl$_4$(s) + 4Na(s) $\longrightarrow$ 4NaCl(s) + Ti(s)	$-820.5\,kJ\,mol^{-1}$

g Electrolysis of oxides Electrical energy can be used to force 'difficult' reactions to take place, but it is not easy to get oxides into a state in which they will conduct, for they have such high melting points. Aluminium oxide, however, can be dissolved in molten cryolite, $Na_3[AlF_6]$, to give a conducting solution. A high input of energy is needed to decompose the oxide,

$$2Al_2O_3(s) \longrightarrow 4Al(s) + 3O_2(g) \qquad \Delta G_m^{\ominus}(298\,K) = 3164.8\,kJ\,mol^{-1}$$

but this is to some extent offset by using a graphite anode, so that the overall reaction is really

$$2Al_2O_3(s) + 3C(s) \longrightarrow 4Al(s) + 3CO_2(g)$$
$$\Delta G_m^{\ominus}(298\,K) = 1981.6\,kJ\,mol^{-1}$$

with carbon playing a very indirect role as a reducing agent.

h Electrolysis of chlorides The chlorides of reactive metals, like the oxides, require a high input of energy to decompose them, e.g.

$$2NaCl(s) \longrightarrow 2Na(s) + Cl_2(g) \qquad \Delta G_m^{\ominus}(298\,K) = 768\,kJ\,mol^{-1}$$
$$MgCl_2(s) \longrightarrow Mg(s) + Cl_2(g) \qquad \Delta G_m^{\ominus}(298\,K) = 592.3\,kJ\,mol^{-1}$$

but they melt at reasonably low temperatures to give conducting liquids, and the temperature of the melt can be lowered by adding other metallic halides. It is important to get as low a working temperature as possible to lower the reactivity of the chlorine and to limit loss of the metal by volatilisation. This is done by using mixtures of halides as the electrolytes.

Many of the Group 1A and Group 2A metals are manufactured by electrolysis of their molten chlorides.

i Summary

general method of extraction	metal and typical ore
electrolysis of molten chloride with added halides to lower temperature	Na (rock salt, NaCl) Mg (carnallite) Ca (limestone, $CaCO_3$) Sr (strontianite, $SrCO_3$) Ba (witherite, $BaCO_3$)
electrolysis of solution of oxide (in molten Na_3AlF_6)	Al (bauxite, Al_2O_3)
high temperature reduction of oxide by carbon	Zn (zinc blende, ZnS) Pb (galena, PbS) Sn (cassiterite, SnO_2) Fe (haematite, Fe_2O_3) Ni (pentlandite) Cr and Mn (see below)

high temperature reduction of oxide by hydrogen	Mo (molybdenite, MoS_2) W (scheelite, $CaWO_4$)
high temperature reduction of oxide by aluminium	Cr (chromite, $FeCr_2O_4$) Mn (pyrolusite, MnO_2) Co (smaltite, $CoAs_2$)
indirect reduction of sulphide by roasting with oxide	Cu (copper pyrites, $CuFeS_2$) Pb (galena, PbS)
chemical reduction of chloride	Ti (rutile, TiO_2)
reduction of oxide by heating	Hg (cinnabar, HgS)

Questions on Chapter 8

1 Compare the reducing powers of (a) carbon and carbon monoxide, and (b) aluminium and magnesium, so far as other oxides are concerned.

2 Why is it that both hydrogen and carbon are very ineffective reducing agents so far as metallic sulphides are concerned?

3 Explain why (a) zinc is not obtained by direct reduction of its sulphide, (b) titanium is not made by reducing its oxide with carbon, (c) calcium is not made by electrolysis of its molten oxide, and (d) mercury is not made by reducing its sulphide with carbon.

4 Describe fully the manufacture of (a) iron, (b) aluminium or (c) titanium.

5 Why is it true to say that hydrogen is a better reducing agent for metallic oxides than carbon at low temperatures but a poorer one at high temperatures?

6 Iron, aluminium and titanium all occur as their oxides. Why are the processes for manufacturing the three metals so different?

7* Plot the standard free energies of formation per atom of oxygen for the oxides of some common metals against the position of the metal in the electrochemical series. Comment on the result.

8* Carbon is a better reducing agent than hydrogen for metallic oxides, but hydrogen is better than carbon for metallic chlorides. Explain why this should be so.

9* Use standard enthalpies of formation and standard entropies to calculate the approximate temperature at which the free energy change for the reaction

$$C(s) + CO_2(g) \longrightarrow 2CO(g)$$

becomes zero. What is the significance of this figure?

10* Calculate the free energy changes for the reduction of (a) silver sulphide and (b) iron(II) sulphide by hydrogen.

11* Why is it that carbon monoxide is a better reducing agent at low temperatures than carbon, but that the situation is reversed at high temperatures? Calculate the approximate temperatures at which they have equal reducing powers.

12* Why do you think that copper(II) oxide will split up into copper(I) oxide and oxygen at a considerably lower temperature than iron(III) oxide will split into iron(II) oxide and oxygen?

13* Use the Ellingham diagrams to compare the reducing powers of carbon and silicon.

14* Some metals will reduce water or steam, i.e.

$$M(s) + H_2O(g) \longrightarrow MO(s) + H_2(g)$$

Calculate the standard free energy changes for this type of reaction for Ca, Mg, Fe, Sn, Pb, Cu and Zn, and hence list the metals in order of reducing activity.

9
General aspects of periodicity

Mendeleev stated that the elements exhibited a periodicity in their properties when arranged in the order of their relative atomic masses, and such periodicity applies both to physical and to chemical properties. An appreciation of such periodic trends, and the reasons for them, plays an essential part in the study of inorganic chemistry.

1 Periodicity of physical properties

a Atomic volume The 'atomic volume' of an element is the volume occupied by 1 mol of atoms of the element. It is equal to the mass of 1 mol of atoms divided by the density. If the mass is expressed in g, and the density in $g\,cm^{-3}$, the units of atomic volume are $cm^3\,mol^{-1}$.

Shortly after Mendeleev had published his first periodic table, Lothar Meyer pointed out the periodicity that was apparent on plotting the atomic volumes of elements against their atomic numbers (Fig. 58). The alkali metals, for example, all have high values, compared with those of the other elements. (because they have the lowest density).

If all atoms packed together in the same way in their crystals, the atomic volume would give a good comparative measure of the size of the atoms, but as the packing is very varied, the atomic volume of an element is not, nowadays, of much significance. Much more precise information about atomic sizes is available (p. 130).

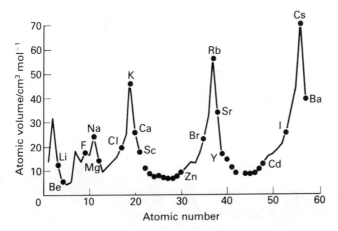

Fig. 58 *The atomic volumes of the elements.*

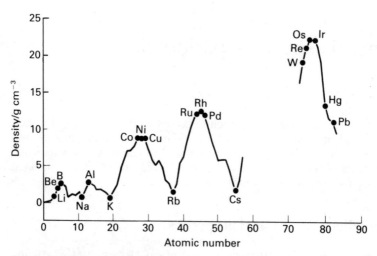

Fig. 59 *The densities of the elements. (The values for elements that are gaseous at room temperature are given in the liquid state.)*

b Density   The density of an element is like the atomic volume in that it depends, partially, on the size of the atoms. But it also depends on the mass of the atoms and on their packing in crystals. The variation of density with atomic number is, therefore, not of great significance. There is, nevertheless, evident periodicity (Fig. 59). The alkali metals, for example, have low densities, and lithium, sodium and potassium are able to float on water. The elements in the middle of a long period have the highest densities; rhenium and osmium, for example, have densities twice as large as that of lead. It was the wide variation in the density of metals

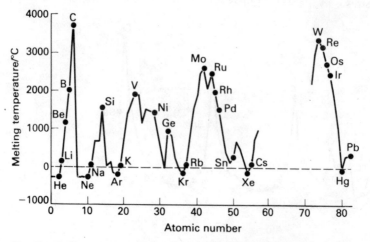

Fig. 60 *Variation of melting temperature of the elements with atomic number.*

(from $0.53\,\text{g cm}^{-3}$ for lithium to $22.48\,\text{g cm}^{-3}$ for osmium) that led to the use of the old terms 'light' and 'heavy' metal.

c Melting temperature The variation of melting temperature with atomic number is shown in Fig. 60. As both melting temperature and density depend, to some extent, on the binding forces within a crystal, it is not surprising that the variations of melting temperature and density with atomic number have some similarities. Most of the transition metals, for example, have high melting temperatures and high densities (p. 344).

The metallic elements have higher melting temperatures than most of the non-metallic elements, but there is a wide range from $-38.9\,°\text{C}$ for mercury (the only metal which is a liquid at room temperature) to $3377\,°\text{C}$ for tungsten. The alkali metals have low melting temperatures.

d Enthalpy of fusion This is the enthalpy change when 1 mol of a solid is converted into a liquid at its melting point. It applies both to compounds and to elements, e.g.

$$H_2O(s) \longrightarrow H_2O(l) \qquad \Delta H_{\text{fus, m}} = +6\,\text{kJ mol}^{-1}$$
$$Na(s) \longrightarrow Na(l) \qquad \Delta H_{\text{fus, m}} = +2.6\,\text{kJ mol}^{-1}$$
$$W(s) \longrightarrow W(l) \qquad \Delta H_{\text{fus, m}} = +35\,\text{kJ mol}^{-1}$$

It is a good measure of the difference between the binding forces in a solid at its melting point, and those in the liquid at the same temperature.

For elements, there is a reasonably close relationship between the melting temperatures and the enthalpies of fusion, a plot of the two being approximately linear.

e Boiling temperature The boiling temperatures of the elements change with atomic number (Fig. 61) in much the same way as the melting temperatures.

f Enthalpy of vaporisation This is the enthalpy change when 1 mol of a liquid is vaporised at its boiling point, e.g.

$$H_2O(l) \longrightarrow H_2O(g) \qquad \Delta H_{\text{vap, m}} = +41.1\,\text{kJ mol}^{-1}$$
$$Na(l) \longrightarrow Na(g) \qquad \Delta H_{\text{vap, m}} = +99\,\text{kJ mol}^{-1}$$
$$W(l) \longrightarrow W(g) \qquad \Delta H_{\text{vap, m}} = +824\,\text{kJ mol}^{-1}$$

It is a measure of the difference between the binding forces in a liquid and those in a vapour. Enthalpies of vaporisation are considerably higher than enthalpies of fusion. There is a small lowering in the binding forces in passing from a solid to a liquid, but those in the liquid are still high. In contrast, the forces between particles in a vapour are very low.

For elements, a plot of boiling temperature against enthalpy of vaporisation is approximately linear.

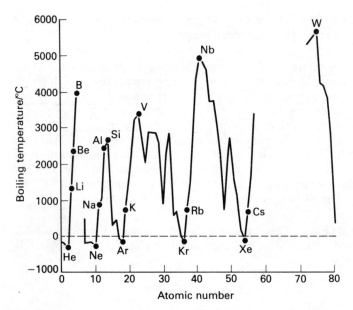

Fig. 61 *Variation of boiling temperature of the elements with atomic number.*

g Enthalpy of atomisation This is the enthalpy change when 1 mol of gaseous atoms of an element are formed from the element in its standard state, e.g.

$$C(\text{graphite}) \longrightarrow C(g) \qquad \Delta H_{a,m}^{\ominus}(298\,\text{K}) = +715\,\text{kJ mol}^{-1}$$
$$\tfrac{1}{2}H_2(g) \longrightarrow H(g) \qquad \Delta H_{a,m}^{\ominus}(298\,\text{K}) = +218\,\text{kJ mol}^{-1}$$

The values for the noble gases are zero because they exist as single gaseous atoms in their standard states.

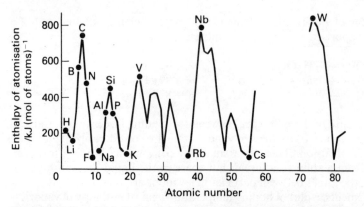

Fig. 62 *Variation of enthalpy of atomisation of the elements with atomic number.*

The variation of enthalpy of atomisation with atomic number (Fig. 62) follows closely that of boiling temperature, and a plot of boiling temperature against enthalpy of atomisation is almost linear.

Elements with high enthalpies of atomisation, e.g. carbon, tend to be unreactive, because it is difficult to convert them into free gaseous atoms and this is an essential stage in any reaction. Fluorine, which has a low enthalpy of atomisation, is very reactive. The strength of the new bonds formed in a reaction is also very important. Reactions take place most easily when it is easy to break the bonds in the reactants and to form stronger bonds in the products.

2 Periodicity of ionisation energy

The variation of first ionisation energy (p. 9) with atomic number is shown in Fig. 63. The following points should be noted.

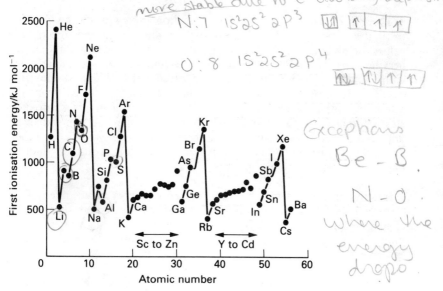

Fig. 63 Variation of first ionisation energy of the elements with atomic number.

a The noble gases have the highest ionisation energies and are least likely to form X^+ ions. This is because of the very high stability of their electron arrangements.

b The alkali metals (Group 1A), with one electron more than a noble gas, have the lowest ionisation energies; they form X^+ ions most readily.

c There is a decrease in the value of the ionisation energy in passing down Group 1A, so that caesium, the biggest atom, forms an ion more easily than lithium, the smallest. The same trend is noticeable in other groups, all the more so if the energy required to form X^{n+} ions is considered, e.g.

H^+	1311						
Li^+	520.0	Be^{2+}	2657	(B^{3+})	6887	(C^{4+})	14290
Na^+	495.8	Mg^{2+}	2189	Al^{3+}	5140	(Si^{4+})	9946
K^+	418.7	Ca^{2+}	1735	Ga^{3+}	5520	Ge^{4+}	10000
Rb^+	402.9	Sr^{2+}	1614	In^{3+}	5083	Sn^{4+}	8989
Cs^+	375.5	Ba^{2+}	1468	Tl^{3+}	5440	Pb^{4+}	9327

The values are quoted in $kJ\,mol^{-1}$.

This result is to be expected, for the outermost electrons are further away from the positively charged nucleus in the bigger atoms.

In Group 1A, caesium ionises most readily; in Group 2A, barium ionises most readily. In Group 3B, boron is not big enough to form B^{3+} ions, and in Group 4B neither C^{4+} nor Si^{4+} exist. because the charge is large enough for the ion to form.

d There is a general rise in the first ionisation energy in passing from left to right across a period but boron, oxygen, aluminium and silicon have values below the trend. The value for boron is lower than that for beryllium because a 2p, and not a 2s, electron is involved; similarly, for aluminium, it is a 3p electron that is involved rather than a 3s one. (It is easier to remove the p electrons because they are 'shielded' from the attraction of the nucleus by the s electrons.) For oxygen, it is a paired 2p electron that is being removed, which is easier than the removal of an unpaired 2p electron, as in nitrogen, because electrons repel each other. Similarly, it is easier to remove a paired 3p electron in sulphur than an unpaired 3p electron in phosphorus.

e It is important to remember that ionisation energy values refer to the formation of free, gaseous ions, and not to hydrated ions. The energy change in the latter process is measured by the standard electrode potential (p. 21) or by the enthalpy of formation of the hydrated ion (p. 20), e.g.

$$Na(s) \longrightarrow Na^+(g) + e^- \qquad \Delta H_m = +502\,kJ\,mol^{-1}$$
$$Na(s) + aq \longrightarrow Na^+(aq) + e^- \qquad \Delta H_m = +204\,kJ\,mol^{-1}$$

Generally, it is the hydrated ion that is formed as the final product of a reaction, so the ionisation energy value is of limited significance.

3 Types of element

Once the arrangement of electrons in atoms had been worked out, the significance of the periodic table became much clearer. In particular, all the atoms in any one group of the table were found to have the same number of electrons in their outermost orbital. Moreover, five general types of element were revealed, as follows.

a The noble gases   The arrangement of electrons in the noble gases is as follows:

	1	2	3	4	5	6
	s	s p	s p d	s p d f	s p d	s p
2 He	2					
10 Ne	2	2 6				
18 Ar	2	2 6	2 6			
36 Kr	2	2 6	2 6 10	2 6		
54 Xe	2	2 6	2 6 10	2 6 10	2 6	
86 Rn	2	2 6	2 6 10	2 6 10 14	2 6 10	2 6

Every orbital which is occupied at all is fully occupied and it is this unique arrangement of electrons which accounts for the chemical inactivity of the noble gases. They were known as the inert gases for many years, but it is now known that they do form a few compounds, particularly with fluorine and oxygen (p. 143). The noble gases have particularly high ionisation energies.

b s-block elements   These elements, found in Groups 1A and 2A, have atoms containing either one or two electrons in the outermost s orbital. All the other orbitals which contain electrons are fully occupied. The structures of the atoms concerned may be summarised as follows:

Li 2.1   Be 2.2
Na 2.8.1   Mg 2.8.2
K   2.8.8.1   Ca 2.8.8.2
Rb 2.8.18.8.1   Sr 2.8.18.8.2
Cs 2.8.18.18.8.1   Ba 2.8.18.18.8.2
Fr 2.8.18.32.18.8.1   Ra 2.8.18.32.18.8.2

The chemical similarity within the groups is due to the fact that all the elements in the same group have the same number of electrons in the outermost orbital, this number being equal to the group number.

It is the ready loss of the outer electrons, particularly in the larger atoms, that facilitates the formation of positive ions; the elements are said to be strongly electropositive. They have low ionisation energies and high negative standard electrode potentials.

The s-block elements form ionic compounds, which are colourless unless they contain a coloured anion. The metals exhibit fixed oxidation states equal to the group number; +1 for Group 1A and +2 for Group 2A.

A fuller discussion of the characteristics of s-block elements is given in Chapter 12 (pp. 154–63). Summaries for Group 1A and Group 2A elements will be found on p. 166 and p. 179, respectively.

Both hydrogen and helium can be regarded as s-block elements, but the chemistry of hydrogen is unique (p. 146), and it is best to link helium with the other noble gases.

c p-block elements These elements occur in Groups 3B–7B. They have atoms in which the outermost p orbitals are filling up, containing from one to five electrons, with the total number of electrons in the outermost shell rising from three for Group 3B elements to seven for Group 7B elements.

Of the 25 elements concerned, 15 would generally be regarded as non-metallic and 10 as metallic, though the diagonal dividing line must not be taken absolutely rigidly because the distinction between a metal and a non-metal is not absolutely clear cut.

group number	3B	4B	5B	6B	7B	
	B	C	N	O	F	
	Al	Si	P	S	Cl	
	Ga	Ge	As	Se	Br	non-metals
metals	In	Sn	Sb	Te	I	
	Tl	Pb	Bi	Po	At	

The p-block elements generally have oxidation states equal to the group number, the group number minus two, or eight minus the group number. The s- and p-blocks together are sometimes referred to as *main group* elements.

d d-block elements In these elements, inner 3d, 4d or 5d orbitals fill up, from one to ten electrons, so that there are three series of ten elements – Sc to Zn, Y to Cd and La to Hg (omitting the lanthanoids). The first nine elements in each of these series are called the *transition elements* (p. 341). They form coloured ions, which are paramagnetic (p. 352); exhibit variable valencies, which may differ from each other by one or more units; form many complexes; and often possess marked catalytic activity.

e f-block elements There are two series of f-block elements in which the 4f and 5f orbitals fill up, from one to fourteen electrons. The first series contains the lanthanoids; it is because there are the same number of electrons in the two outermost shells of all these fourteen elements that they are so very much alike.

The second series contains the actinoids and, again, there is remarkable chemical similarity.

4 The main group elements

The s- and p-block elements make up what are known as the main group elements. The periodic trends centre mainly on the change from metals to non-metals in passing from left to right along the horizontal periods, and on the increase in metallic character in passing down the vertical groups. These changes are reflected in many of the properties of the elements, and in the nature of their compounds.

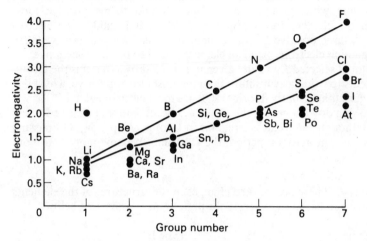

Fig. 64 *Electronegativities of s- and p-block elements.*

a Electronegativity values The electronegativity values (p. 40) are plotted in Fig. 64 and summarised below.

		H 2.1				
Li 1.0	Be 1.5	B 2.0	C 2.5	N 3.0	O 3.5	F 4.0
Na 0.9	Mg 1.2	Al 1.5	Si 1.8	P 2.1	S 2.5	Cl 3.0
K 0.8	Ca 1.0	Ga 1.6	Ge 1.8	As 2.0	Se 2.4	Br 2.8
Rb 0.8	Sr 1.0	In 1.7	Sn 1.8	Sb 1.8	Te 2.1	I 2.5
Cs 0.7	Ba 0.9	Tl 1.8	Pb 1.8	Bi 1.8	Po 2.0	

There is a general fall in value in passing down a group, and a rise in passing from left to right along a period. The halogens (particularly fluorine) have high, and the alkali metals (particularly caesium) low, values.

The high electronegativities of fluorine, oxygen and nitrogen mean that the F—H, N—H and O—H bonds have high polarities, which causes hydrogen bonding (p. 52). As a result, many compounds containing these bonds have 'unusual' properties.

b Standard electrode potentials, $E^{\ominus}$ An element with a high negative $E^{\ominus}$ value is one which readily loses electrons and forms positive ions; although the element has a negative $E^{\ominus}$ value it is said to be electropositive. A large positive $E^{\ominus}$ value indicates a readiness to form negative ions by gaining electrons; such an element is said to be electronegative.

The plot of selected $E^{\ominus}$ values given in Fig. 65 shows that it is the Group 1A elements (particularly lithium) that are most electropositive, whilst it is the halogens (particularly fluorine) that are most electronegative.

On passing down a group, the $E^{\ominus}$ values become more negative (or less positive), though the trend is irregular in Group 1A (p. 154). On passing from left to right across a period, the $E^{\ominus}$ values become more and more positive.

c Structures   There are marked changes in the structures of the elements in passing from left to right across the periods, as summarised below.

Type I		Type II		Type III		
				H		
Li	Be	B	C	N	O	F
Na	Mg	Al	Si	P	S	Cl
K	Ca	Ga	Ge	As	Se	Br
Rb	Sr	In	Sn	Sb	Te	I
Cs	Ba	Tl	Pb	Bi	Po	

Type I   Metallic crystals
Type II   Giant covalent lattices, e.g. diamond
          Layer structures, e.g. graphite, black phosphorus
          Chains, e.g. red phosphorus, plastic sulphur
Type III Molecular crystals

The structures of the elements on the borderlines between the types vary according to the allotrope being considered, or have other irregularities; hence the dotted lines.

The elements forming metallic crystals do so with a variety of structures (p. 50), but those of aluminium, gallium, indium, thallium and lead have some irregularities compared with those of the 'true' metals of Groups 1A and 2A. The β- and γ-allotropes of tin (p. 223) have metallic structures, but the α-allotrope has a covalent lattice like that of diamond, silicon and germanium. On the other borderline, phosphorus, sulphur and selenium can have chain structures or molecular structures, again depending on the allotropic form.

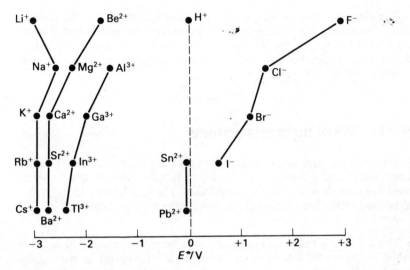

Fig. 65 *Some standard electrode potentials.*

In the elements with molecular crystals, the molecules are held together by van der Waals' forces. A variety of molecular species is involved, depending on the atomicity of the molecule as follows:

diatomic molecules	H_2 N_2 O_2 F_2 Cl_2 Br_2 I_2
triatomic molecules	O_3
tetra-atomic molecules	P_4
octa-atomic molecules	S_8 Se_8

d Physical properties The differences in structure between the elements show up in their physical properties. These are summarised below for the elements lithium to chlorine. In this table, *i* gives the melting temperature in °C, *ii* gives the boiling temperature in °C, *iii* gives the enthalpy of fusion in kJ mol^{-1} and *iv* gives the enthalpy of vaporisation in kJ mol^{-1}.

	Li	Be	B	C	N	O	F
i	180	1283	2027	3700	−210	−219	−220
ii	1331	2477	3927	c.4800	−196	−183	−188
iii	3.0	11.7	22.2		0.36	0.22	0.26
iv	13.4	309	314		2.8	3.4	3.2

	Na	Mg	Al	Si	P	S	Cl
i	98	650	660	1423	44	119	−101
ii	890	1120	2447	2680	280	444	−34
iii	2.6	8.9	11	46	0.63	1.4	3.2
iv	100	136	284	170	12	13	10

The elements with molecular crystals have low values for the physical constants quoted because the forces holding the molecules together in the crystals are weak (p. 56). The other elements have much higher values, because the forces holding their crystals together (metallic or covalent bonds) are much stronger.

5 The sizes of main group atoms

It is not easy to measure or define the exact size of an atom for it has no sharp edges, and atoms are usually found in molecules or crystals where their sizes are affected by the binding forces. The term 'atomic radius' tends to be used rather loosely to indicate three different quantities.

a Van der Waals' radius   Half the distance between the nuclei of adjacent atoms in the crystal of a noble gas measures what is known as the van der Waals' radius of the atom concerned. It gives the best value for the actual size of an atom for it measures how close two atoms will approach when not attracted by any very strong bond.

For a crystal (such as iodine) containing diatomic covalent molecules, the closest distance between the nuclei of atoms in adjacent molecules will be equal to twice the van der Waals' radius (in this case, for iodine).

Some typical values for van der Waals' radii are given below, in nm.*

H 0.12				He 0.12
	N 0.15	O 0.140	F 0.135	Ne 0.160
	P 0.19	S 0.185	Cl 0.180	Ar 0.192
	As 0.20	Se 0.200	Br 0.195	Kr 0.197
	Sb 0.22	Te 0.220	I 0.215	Xe 0.217

The values increase on passing down a group. They decrease on passing from left to right across the periods, until they come to the noble gas, when they increase. This is because of the very weak forces within noble gas crystals (p. 142).

b Covalent radius   Half the bond distance between two like atoms joined by a single covalent bond is known as the covalent radius of the atom concerned, and it is this value that is most commonly used as the 'atomic

---

*Values may also be quoted in pm (1 nm = 10^3 pm) and this has the advantage of involving fewer decimal points in the quoted values. The older (non SI) unit, the Ångstrom, also still occurs (1 nm = 10 Å).

radius'. Typical values are given below, in nm.

			H 0.030			

Li 0.123	Be 0.089	B 0.080	C 0.077	N 0.070	O 0.066	F 0.064
Na 0.157	Mg 0.136	Al 0.125	Si 0.117	P 0.110	S 0.104	Cl 0.099
K 0.203	Ca 0.174	Ga 0.125	Ge 0.122	As 0.121	Se 0.117	Br 0.114
Rb 0.216	Sr 0.191	In 0.150	Sn 0.140	Sb 0.141	Te 0.137	I 0.133
Cs 0.235	Ba 0.198	Tl 0.155	Pb 0.154	Bi 0.152		

The covalent radii decrease in passing from left to right across the periods. Each succeeding atom contains one more electron, but the increased positive charge on the nucleus attracts the electrons more strongly, and causes the atom to contract in size. In passing down the groups, there is an increase in size as successive outer shells become occupied.

The covalent radii are useful because they can be added together to give the approximate value of the length of a single covalent bond between the two atoms concerned (p. 39).

c **Metallic radius**   Half the internuclear distance between the atoms in a metallic crystal is called the metallic radius. Values vary slightly according to the type of crystal, but they are about 10–20 per cent higher than the corresponding covalent radii, suggesting that the bonding in a metallic crystal is weaker than covalent bonding. Compare, for example, the following metallic radii (in nm) with the covalent radii given in **b**.

Li 0.152	Na 0.186	K 0.227	Rb 0.248	Cs 0.263
Be 0.112	Mg 0.160	Ca 0.197	Sr 0.215	Ba 0.221

6 The ionic radii of the main group elements

The distance between the nuclei of two adjacent ions in an ionic crystal can be measured by X-ray analysis, and, if the ions are regarded as spheres, this distance is the sum of the two ionic radii.

Before any single ionic radius can be obtained from internuclear distances, it is necessary to decide the value of one radius. Various rather

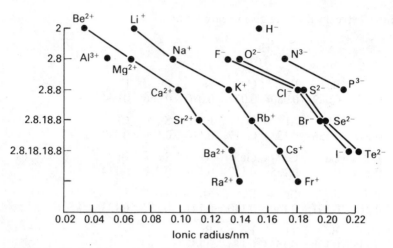

Fig. 66 *Ionic radii of common ions. The ions on each horizontal line all have the same arrangement of electrons.*

arbitrary methods have been used but once one value is decided, the others are easily obtainable from measurements on crystals. Typical figures are shown in Fig. 66, and the following points arise.

a The ions of elements in any one group of the periodic table increase in size as the relative atomic mass increases.

b For ions with the same arrangements of electrons, the size decreases as the atomic number increases. This is due to the increasing nuclear attraction for the electrons as the nuclear charge increases. Compare, for example, the ions O^{2-}, F^-, Na^+, Mg^{2+} and Al^{3+} (all with a 2.8 structure), and S^{2-}, Cl^-, K^+, Ca^{2+} and Sc^{3+} (all with a 2.8.8 structure).

c When two positively charged ions are formed by the same element, it is the one with higher charge that is smaller. This is because the more highly charged ion has fewer electrons, so that they are more tightly held. Compare, for example, Sn^{2+} and Sn^{4+}, Pb^{2+} and Pb^{4+}, Fe^{2+} and Fe^{3+}:

Sn^{2+} 0.112 nm Pb^{2+} 0.120 nm Fe^{2+} 0.076 nm
Sn^{4+} 0.071 nm Pb^{4+} 0.086 nm Fe^{3+} 0.064 nm

d The lowering of ionic size due to increasing ionic charge for positive ions is also shown by comparing the sizes of Na^+, Mg^{2+} and Al^{3+} (p. 155). For negative ions, increase in charge, i.e. addition of more and more electrons, leads to increase in size, e.g. F^-, O^{2-} and N^{3-}.

e A cation radius is smaller than the covalent radius for the same element, whereas an anion radius is larger. Anions are, in the main, larger than cations.

f Different methods of assessment give rather different values for ionic radii, and any values quoted tend to refer to only one type of crystal structure. Numerical values must, therefore, be treated with some caution.

7 Bonding by the elements lithium to chlorine

The gradation from metal to non-metal shown by these elements is also apparent in the general types of compound that they form. The metals lithium, sodium, beryllium, magnesium and aluminium can all form positive ions, and the compounds of lithium, sodium and magnesium are mainly ionic. But those of beryllium are mainly covalent because the small size and high charge on the Be^{2+} ion makes its formation difficult, and aluminium compounds show some ionic and some covalent characteristics.

The compounds of boron, carbon, silicon, nitrogen and phosphorus are almost exclusively covalent though N^{3-}, P^{3-} and C^{4-} ions do exist. Those of oxygen, sulphur, fluorine and chlorine may be covalent or ionic, the latter involving X^{2-} or X^- ions. In summary

Li	Be	B	C	N	O	F
Na	Mg	Al	Si	P	S	Cl
ionic compounds		covalent or partially covalent compounds	mainly covalent compounds		covalent or ionic compounds	

a Maximum covalency   The atoms lithium to fluorine only have electrons in the 1s, 2s and 2p orbitals. As there are no 2d orbitals available for bonding, and as the atoms are small so that they cannot pack many atoms around themselves, their maximum covalency is limited to 4.

The atoms sodium to chlorine are larger and have 3d orbitals available for bonding, even though these orbitals are not occupied in the ground states of the atoms. The maximum covalency rises, generally, to 6. This allows the existence of many compounds and ions of sodium to chlorine which have no counterparts for the elements lithium to fluorine. For example

$$[Mg(NH_3)_6]^{2-} \quad [AlF_6]^{3-} \quad [SiF_6]^{2-} \quad PCl_5 \quad SCl_4 \quad ClF_3$$
$$SF_6$$

The limited covalency of the elements lithium to fluorine also means that their compounds may have distinctly different properties from the corresponding compounds of sodium to chlorine. Compare, for example, the hydrolysis of tetrachloromethane with that of silicon tetrachloride (p. 317), the hydrolysis of nitrogen and phosphorus trichlorides (p. 317) and the stability of methane and silane (p. 217).

Elements beyond chlorine can have still higher maximum covalencies.

b Oxidation states The elements of Groups 1, 2 and 3, together with fluorine, generally have single oxidation states, but the other elements exhibit widely differing ranges:

Li	Be	B	C	N	O	F
+1	+2	+3	+4 to −4	+5 to −3	+2 to −2	−1

Na	Mg	Al	Si	P	S	Cl
+1	+2	+3	+4 to −4	+5 to −3	+6 to −2	+7 to −1

The highest positive oxidation state, equal to the group number, occurs when the elements are combined with the highly electronegative fluorine or oxygen. For example

+1	+2	+3	+4	+5	+6	+7
LiF	BeF_2	BF_3	CF_4			
NaF	MgF_2	AlF_3	SiF_4	PF_5	SF_6	
Li_2O	BeO	B_2O_3	CO_2	N_2O_5		
Na_2O	MgO	Al_2O_3	SiO_2	P_2O_5	SO_3	Cl_2O_7

NF_5 does not exist because of the low maximum covalency of nitrogen, and there is no ClF_7; IF_7 exists, because the iodine atom is larger than the chlorine atom.

The highest negative oxidation state occurs in compounds with the electropositive hydrogen. For example

−1	−2	−3	−4	−3	−2	−1
LiH	$(BeH_2)_x$	B_2H_6	CH_4	NH_3	H_2O	HF
NaH	$(MgH_2)_x$	$(AlH_3)_x$	SiH_4	PH_3	H_2S	HCl

The elements to the left are reducing agents; the reducing strength increasing, in general, on passing down a group. The elements on the right are oxidising agents, with the oxidising power increasing on going up a group.

c Formation of multiple bonds The elements sodium to chlorine have a much smaller tendency to form multiple bonds than the elements lithium to fluorine. Illustrative of this are the non-existence of S=S and P≡P molecules (except at high temperatures) as compared with O=O and N≡N; the lack of Si=Si, Si≡Si and Si=O bonds as compared with the prevalence of C=C, C≡C and C=O bonds; and the marked difference between the gaseous carbon dioxide, with its O=C=O molecules and solid silicon(IV) oxide which exists as a giant lattice because O=Si=O molecules are not formed.

d Catenation Catenation is the tendency of an atom to link with like atoms to form chains or rings (*catena* = chain). Carbon has unique powers of catenation; silicon forms only short chains and no rings (p. 217). In Group 4B then, the power of catenation falls in passing down the group. Contrariwise, it rises in passing down Groups 5B and 6B, with phosphorus and sulphur catenating more easily than nitrogen and oxygen.

e Hydrogen bond formation Hydrogen bonds, X—H$\cdots$X, are formed mainly in compounds where X is either fluorine, oxygen or nitrogen, i.e. has a high electronegativity. This affects the properties of compounds containing F—H, O—H and N—H bonds so that they may differ from the corresponding compounds with Cl—H, S—H and P—H bonds (p. 53).

8 Compounds of the elements lithium to chlorine

a Oxides All the elements form at least one oxide, typical ones being summarised below together with *i* their melting temperatures in °C and *ii* their boiling temperatures in °C.

	Li_2O(s)	BeO(s)	B_2O_3(s)	CO_2(g)	N_2O_5(s)	O_2(g)	F_2O(g)
i	> 1700	2550	460	− 56	32	− 219	− 224
ii		4120	1860	− 79 (sub)	41 (dec)	− 183	− 145

	Na_2O(s)	MgO(s)	Al_2O_3(s)	SiO_2(s)	P_4O_{10}(s)	SO_3(s)	Cl_2O(g)
i		2800	2145	1610		17	− 20
ii	1275	3600	2980	2230	300 (sub)	45	2

ionic crystals — covalent giant lattices — molecular crystals

In passing from left to right, the oxides change from the basic oxides of metals through amphoteric oxides to the acidic oxides of non-metals, as shown below.

Li_2O	BeO	B_2O_3	CO_2	N_2O_5		F_2O
Na_2O	MgO	Al_2O_3	SiO_2	P_4O_{10}	SO_3	Cl_2O
basic		amphoteric		acidic		

The hydroxides of the elements are similar. Those of lithium, sodium and magnesium are basic; those of aluminium and beryllium amphoteric; the remainder are acidic, generally existing as oxoacids after dehydration (p. 277).

There is a similar change in passing down those groups where the change from non-metal to metal is most marked, e.g. Group 4B (p. 198).

Further details of the types of oxides formed by the elements are given on p. 273.

b Chlorides　All the elements form at least one chloride, the main comparable ones being summarised below, together with *i* their melting temperatures in °C, *ii* their boiling temperatures in °C and *iii* their molar conductivities in the fused state in $S\,cm^2\,mol^{-1}$.

	LiCl(s)	BeCl$_2$(s)	BCl$_3$(g)	CCl$_4$(l)	NCl$_3$(l)	Cl$_2$O(g)	ClF(g)
i	606	404	-107	-23	-40	-20	-154
ii	1382	488	12.5	76	70	2	-101
iii	166	0.008	0	0	0	0	0

	NaCl(s)	MgCl$_2$(s)	AlCl$_3$(s)	SiCl$_4$(l)	PCl$_3$(l)	S$_2$Cl$_2$(l)	Cl$_2$(g)
i	800	715		-70	-112	-78	-102
ii	1440	1412	180 (sub)	58	75.5	59 (dec)	-35
iii	134	14.5	5×10^{-6}	0	0	0	0
		ionic	mainly covalent		covalent		

It is clear that the chlorides of lithium, sodium and magnesium are ionic. The remainder are essentially covalent, though there is some slight ionic character in beryllium and aluminium chlorides.

Further details of the halides of the elements are given on pp. 314–18.

c Hydrides　The major hydrides are summarised below, together with *i* their melting temperatures in °C and *ii* their boiling temperatures in °C.

	LiH(s)	(BeH$_2$)$_x$(s)	B$_2$H$_6$(g)	CH$_4$(g)	NH$_3$(g)	H$_2$O(l)	HF(g)
i	688		-165	-183	-78	0	-83
ii	dec		-93	-162	-33	100	20

	NaH(s)	(MgH$_2$)$_x$(s)	(AlH$_3$)$_x$(s)	SiH$_4$(g)	PH$_3$(g)	H$_2$S(g)	HCl(g)
i	dec			-185	-133	-89	-114
ii				-112	-88	-61	-85
	M$^+$H$^-$ crystals	see p. 151			molecular crystals		

As before, the physical properties indicate the general type of structure. The melting and boiling temperatures of HF, NH$_3$ and H$_2$O are higher than expected for hydrides in their groups, owing to hydrogen bonding (p. 53).

Further details of the hydrides of the elements are given on pp. 150–51.

11 Diagonal relationships

Certain diagonal relationships exist between elements in adjacent groups. These arise because the increase in electronegativity in elements in passing horizontally along a period is offset by the decrease in electronegativity in passing down a group. Therefore, diagonal pairs have very similar electronegativities (p. 40). The similarities are most significant in the following pairs.

a Lithium and magnesium The following summary shows the ways in which lithium resembles magnesium, and differs from the other alkali metals.

Li and Mg	Na, K, Rb and Cs
form ionic nitrides, $Li^+_3N^{3-}$ and $Mg^{2+}_3N^{3-}_2$, on heating in N_2	no reaction with N_2
form monoxides, Li_2O and MgO, on burning in air	form peroxides, e.g. Na_2O_2, or hyperoxides, e.g. KO_2
hydrogencarbonates known only in solution	solid hydrogencarbonates can be made
carbonates, hydroxides and peroxides decompose into oxides on heating	no similar decomposition
nitrates decompose to give oxides, NO_2 and O_2	nitrates decompose to give nitrites and O_2
phosphates, carbonates, fluorides and hydroxides only slightly soluble	corresponding compounds much more soluble
halides (except fluorides) are soluble in organic solvents	corresponding compounds much less soluble
compounds have covalent character	compounds predominantly ionic
sulphates do not form alums	sulphates are isomorphous and form alums

b Beryllium and aluminium The following survey shows how beryllium resembles aluminium, and differs from the other Group 2A metals.

Be and Al	Mg, Ca, Sr and Ba
compounds are mainly covalent	main compounds are ionic
metals react with concentrated solutions of alkalis to form H_2 and hydroxo-complexes	no reaction with alkalis
oxides and hydroxides are amphoteric	oxides and hydroxides are basic
fused chlorides have low boiling temperatures and conductivities; chlorides are readily hydrolysed and exist as dimers, Be_2Cl_4 and Al_2Cl_6 in the vapour state	chlorides have high conductivities and boiling temperatures; they are not hydrolysed and are ionic, e.g. $Ca^{2+}Cl^-_2$
form fluoro-complexes, e.g. $[BeF_4]^{2-}$ and $[AlF_6]^{3-}$	no fluoro-complexes formed

hydrolysed (handwritten note on left margin)

ionisated (handwritten note on right margin)

c Boron and silicon The ways in which boron resembles silicon and differs from aluminium are summarised below.

B and Si	Al
non-metals with very similar properties; non-conducting	metal with different physical properties; conducting
compounds are covalent	compounds are partially ionic
form acidic oxides, B_2O_3 and SiO_2, with giant lattice structures, which melt to give glassy solids	oxide is amphoteric, with high melting temperature
form series of covalent, volatile hydrides which readily ignite in air	solid, polymerised hydride, $(AlH_3)_x$
form covalent chlorides, BCl_3 and $SiCl_4$, which are readily hydrolysed	aluminium chloride less readily hydrolysed
form borides, e.g. MgB_2 and CaB_6, and silicides, e.g. Mg_2Si, with metals	no similar compounds

Questions on Chapter 9

1 The first member of a family (vertical group) of elements in the periodic table has properties which are not typical of the other members of the family. Discuss and explain this statement with respect to two non-metallic elements and one metal. (O)

2 What general methods are used for preparing the chlorides of the non-metallic elements? Describe the chief characteristics of this class of compound. Comment on the similarities and differences between the chlorides of the elements in each of the following pairs: (a) B and Al, (b) C and Si, (c) N and P, (d) O and S.

3 Compare and contrast the MO_2 oxides of carbon, silicon, sulphur, nitrogen and chlorine.

4 How and under what conditions can the typical hydrides of sulphur, phosphorus and calcium be made? What are their main differences in property, and how may these differences be explained?

5 What diagonal relationships can you discover between oxygen and chlorine?

6 (a) Why are F_2O and NF_3 not hydrolysed by water? (b) Predict the products of hydrolysis, if any, of the following: $BeCl_2$, $AlCl_3$, SF_6 and PCl_5.

7 Give a comparative account of the chemistry of potassium, calcium and aluminium, paying particular attention to the position of these elements (a) in the electrochemical series and (b) in the periodic table. (L)

8 Illustrate how the chemistry of elements and their compounds changes across a short period by writing a comparative account of the chemistry of the elements sodium to chlorine in the second short period, and of their oxides and hydrides. (O)

9 (a) Explain the trends in the atomic radii of the elements both along Period 3 (Na–Cl) and down Group 7 (F–I) of the periodic table. Show, with the aid of examples, how these trends help to explain the changes in chemical properties of the elements. (b) Explain how the effect of (i) heat and (ii) water on sodium chloride and aluminium chloride is influenced by the type of bonding present in each compound. (JMB)

10 Write an account of the more important oxides of the elements from sodium (Na; $Z = 11$) to chlorine (Cl; $Z = 17$), remembering that some of the elements have more than one well-characterised oxide. You may like to consider their composition, structure, thermal stability, and behaviour with water and/or acids and alkalis, but your account need not be concerned with all these points nor be confined to them. (L)

11 Discuss the statement that 'The chemistry of the head element is not typical of that of the remainder of the group'. (JMB)

12 Define the terms ionisation energy, electron affinity and electronegativity. Show how a consideration of the trends in the values of these terms may be used to explain the variation in the bond types and chemical properties of the chlorides, oxides and hydrides of the elements of the period sodium to chlorine. (JMB)

13 'Other factors being equal, the ionisation energy decreases in the order s > p > d > f.' Examine and explain this statement.

14 Comment on the changes in ionisation energy (a) on passing across the first short period and (b) on passing down Group 1A.

15 Compare the use of the ionisation energy of sodium with that of the enthalpy of formation of the $Na^+(aq)$ ion. Which do you think is the more important?

16 Choose any two metals and any two non-metals and show how they differ so far as any three physical properties and any three chemical properties are concerned.

17 'Semi-metals display properties mid-way between those of metals and those of non-metals.' Discuss this statement, with illustrative examples.

18 The elements silicon, phosphorus and sulphur occur in the second short period of the periodic table. With reference to the properties of the elements and their simple compounds, discuss the gradation in properties which they exhibit, and show how far these properties are consistent with the positions of the elements in the periodic table. (L)

19 What trends are discernible in (a) the oxides and (b) the chlorides of the elements lithium to chlorine?

20 A modern periodic table is divided into 'blocks' of elements – s-block, p-block, etc.

(a) Explain the basis on which elements are assigned to their respective blocks. (b) For three of the blocks, give an account of the chemical features which are peculiar to the elements in the block, and to their compounds. Metals can conveniently be classified in accordance with their position in the electrochemical series. Discuss the relation, if any, between the position of an element in that series and the 'block' it is in. (L)

21 Compare sodium, aluminium and sulphur with respect to their reactions with (a) oxygen and (b) chlorine. Your comparison should include the following: (i) the conditions of the reactions, and equations; (ii) the bonding in the products; (iii) the reactions, including equations, of the products with water to which a few drops of universal indicator have been added; (iv) how the characteristics (i), (ii) and (iii) show the change in the nature of an element on passing from left to right along a period in the periodic table. (AEB)

22* Comment on any general relationship you can discover between the enthalpy of atomisation and the bond energy in the covalently-bound elements.

23* Consider the factors that contribute to the enthalpy changes in reactions between hydrogen and (a) fluorine, (b) nitrogen and (c) carbon.

10
The noble gases

The discovery of the noble gases by Rayleigh, Ramsay and Travers, between 1894 and 1900, came some time after the establishment of the first periodic table classifications. As the discovery had not been foreseen, it was convenient that the new elements could be accommodated to the left of Group 1A in what was called Group 0, and that the group number fitted in very well by suggesting that the gases had a valency of zero, i.e. could not enter into chemical combination. Nowadays, the noble gases are usually written on the right-hand side of the long form of the periodic table.

The unusual stability of the noble gases also suggested that their atoms had particularly stable arrangements of electrons, now associated with fully occupied orbitals. Indeed, it was one of the foundation stones of early valency theory that atoms formed chemical bonds by gaining, losing or sharing electrons so that the atoms concerned could attain electronic arrangements like those in the noble gases. The stability of the electronic arrangement in the noble gas atoms is shown by the high values of their ionisation energies (p. 123).

1 Extraction and uses of the noble gases

All the non-radioactive noble gases occur in the atmosphere, with argon easily the most abundant, and xenon the least abundant:

	He	Ne	Ar	Kr	Xe
volume %	0.0005	0.0018	0.93	0.0001	0.00001

Helium is also found in some American sources of natural gas, in quantities up to about 5 per cent.

a Helium The gas is obtained from natural gas supplies, the other gases being liquefied to free the helium.

It is used as a substitute for hydrogen, which is flammable and explosive, for filling balloons; in air supplies to divers, because it is less soluble in blood than nitrogen; and in the welding of metals such as titanium, aluminium, magnesium and stainless steel, where an inert atmosphere is necessary to prevent oxidation.

Liquid helium is also used as a coolant when very low temperatures are required.

b Neon   A mixture of neon and helium can be obtained by the fractional distillation of liquid air, and neon is separated from the mixture by adsorption on charcoal at low temperature. It is used in the well-known neon lights.

c Argon   Argon, made by fractional distillation of liquid air, is the cheapest of the noble gases. It is most widely used to provide an inert atmosphere in welding operations, for mixing with neon in neon lights, and to fill inert-gas filled electric light bulbs.

d Krypton   A mixture of krypton and xenon is obtained by fractional distillation of liquid air, and the two gases are separated by low temperature adsorption on charcoal.

Krypton-filled bulbs are more efficient than those filled with argon but as they are more expensive they are only used for special purposes. Krypton is also finding increasing use in sign lighting.

e Xenon   Xenon is obtained as in **d**. It is used in chemical research into noble gas compounds, and as a filler in special lamps used in radar and in cinema projectors.

2 Melting and boiling temperatures

The noble gases are monatomic in the gaseous state, existing as single, discrete atoms. The relative lack of interaction between the atoms is shown by the fact that the noble gases approach ideal behaviour more closely than most other gases. In the solid state, the atoms are arranged in cubic close-packed structures (p. 50), each atom being surrounded by twelve equidistant neighbours, i.e. the coordination number is 12.

The van der Waals' forces (p. 55) holding the atoms together in the noble gas crystals are very weak, and this is shown by the very low melting temperatures, boiling temperatures and enthalpies of fusion and vaporisation; very little energy is required to break down the crystal structures or to boil the liquids:

	He	Ne	Ar	Kr	Xe	Rn
melting temperature/°C		−249	−189	−157	−112	−71
boiling temperature/°C	−269	−245	−186	−152	−109	−62
ΔH_{fus}/kJ mol^{-1}	0.02	0.33	1.2	1.6	2.3	2.9
ΔH_{vap}/kJ mol^{-1}	0.09	1.8	6.5	9.7	13.7	18

The trends suggest that the van der Waals' forces increase in strength as the atoms get bigger (p. 130). This is due to the greater polarisability (p. 16) as the number of electrons in the atoms increases. The same trends are apparent in other groups, such as the halogens, and in any homologous series of organic compounds.

Helium can only be solidified under pressure: at 26 atm it has a melting temperature of $-272.1\,°C$. Liquid helium is, also, very odd. When cooled to $-270.82\,°C$ it is converted into a unique liquid known as helium II. This has a thermal conductivity about 600 times that of copper, a very low viscosity, and it can flow *up* the surface of its container. There is, as yet, no fully satisfactory explanation of these unique properties.

how dense it is.

3 Noble gas compounds

The earlier ideas that the noble gases could not form chemical bonds were dispelled in 1962 when Bartlett made an orange–yellow solid by treating xenon with platinum(VI) fluoride at room temperature. He was led into doing this by a prior discovery that platinum(VI) fluoride reacted with oxygen to form an ionic compound, $[O_2]^+[PtF_6]^-$, and he argued that platinum(VI) fluoride might oxidise Xe to Xe^+ if it could oxidise O_2 to O_2^+, for the ionisation energies of oxygen ($117.9\,kJ\,mol^{-1}$) and xenon ($116.9\,kJ\,mol^{-1}$) are about equal. Bartlett's solid was first formulated as $Xe^+[PtF_6]^-$, but it is now known to have a more complicated structure.

This breakthrough soon led to the isolation of other compounds of xenon, e.g.

$$XeF_2 \quad XeO_3 \quad XeO_2F_2$$
$$XeF_4 \quad XeO_4 \quad XeOF_4$$
$$XeF$$

XeF_4, which is the easiest to make, forms quantitatively when a $1:5$ mixture of xenon and fluorine is heated for a few hours at $400\,°C$ and about 6 atm. On cooling, crystals of XeF_4 (melting temperature $117\,°C$) form.

KrF_2 has also been made, as have xenon and krypton compounds containing bonds between the noble gases and chlorine and nitrogen atoms. It seems likely that similar compounds of radon must exist, but the very short half-life of radon makes them difficult to investigate. No compounds of helium, neon or argon have yet been made.

It is the noble gases with the lowest ionisation energies that form compounds, but their nature and structure varies. The compounds of xenon and fluorine are stable and can be made by direct combination (as for XeF_4); they are exothermic compounds. The oxides of xenon are endothermic and explosive. The detailed nature of the bonding in all the compounds is not known, but the geometrical shapes of some of them agree with the ideas of electron pair repulsion (p. 27). Thus XeF_2 is linear, XeF_4 is square planar, XeO_3 is pyramidal, and XeO_4 is tetrahedral.

4 Clathrates

A clathrate is a substance in which small molecules or atoms are trapped within the crystalline framework of another substance. The name is derived from the Latin word, *clathratus*, which means 'enclosed behind bars'.

So-called hydrates of the noble gases can be formed by crystallising a solution of the gas in water at a high pressure and a low temperature. The 'hydrates' are only stable below 0 °C, and though their compositions can sometimes be expressed as a chemical formula, e.g. $Xe.6H_2O$, there is no real chemical bonding between the gas and the water. The noble gas atoms are trapped within an ice crystal. Similar substances can be formed with heavy water.

Phenol and benzene-1,4-diol (quinol), $C_6H_4(OH)_2$, can also form clathrates with some noble gases. When quinol is crystallised from an aqueous solution in the presence of argon at 40 atm, a clathrate with composition $[C_6H_4(OH)_2]_3Ar$ is formed. The argon atoms are trapped within a cage of hydrogen bonded (p. 52) quinol molecules. The argon can be liberated by melting or dissolving the crystalline clathrate. Xenon and krypton form similar clathrates.

Small molecules, such as O_2, SO_2, CH_4 and CH_3OH, can also form clathrates. It is a matter of size. Helium will not form a clathrate with quinol; its atom is, presumably, small enough to escape from the 'cage'. Methanol forms quinol clathrates, but the larger ethanol molecule does not.

Questions on Chapter 10

1 Write out the electronic configurations of the noble gases. What made scientists suspect that there was a gas in the atmosphere other than nitrogen, oxygen, carbon dioxide and water vapour, and how was the gas 'argon' first isolated? How was the relative atomic mass of argon determined?

The noble gases were once called the inert gases. State briefly how this concept of inertness contributed to the progress of valency theory. Why is the description 'inert gas' no longer considered appropriate? (JMB)

2 Discuss the use of the words 'mixture' and 'compound'.

3 Give an account of the discovery of the noble gases. Discuss the theoretical and experimental evidence showing that these gases are monatomic. What do you consider to be the most important practical and theoretical consequences of the discovery of the gases?

4 If the standard enthalpies of formation of XeF_2, XeF_4 and XeF_6 are -134, -262 and -381 kJ mol^{-1} respectively, calculate a probable value for the bond enthalpy of the Xe—F bond.

5 Predict the shapes of XeF_2, XeF_4 and XeF_6.

6 The noble gases are monatomic. Why is this? What atomicities are found in other elements?

7 Discuss the physical properties of the noble gases. In what ways do they resemble those of the halogens?

11
Hydrogen and hydrides

1 Sources of hydrogen

Hydrogen can be made in the laboratory by reaction of many electropositive metals with non-oxidising acids, by the action of reactive metals with water or steam, by the reaction of ionic hydrides with water, or by the action of aluminium or zinc with hot concentrated solutions of sodium hydroxide, e.g.

$$Zn(s)+2HCl(aq) \longrightarrow ZnCl_2(aq)+H_2(g)$$
$$2Na(s)+2H_2O(l) \longrightarrow 2NaOH(aq)+H_2(g)$$
$$NaH(s)+H_2O(l) \longrightarrow NaOH(aq)+H_2(g)$$
$$Zn(s)+2OH^-(aq)+2H_2O(l) \longrightarrow [Zn(OH)_4]^{2-}(aq)+H_2(g)$$

a Manufacture from petroleum products Hydrogen is mainly made from petroleum hydrocarbons, by a process known as steam reforming. In the primary stage, a mixture of carbon monoxide and hydrogen is obtained by treating either natural gas (mainly methane) or naphtha with steam at about 800 °C in the presence of a nickel catalyst:

$$CH_4(g)+H_2O(g) \longrightarrow CO(g)+3H_2(g)$$
$$\Delta H_m^{\ominus}(298\,K) = +206\,kJ\,mol^{-1}$$
$$C_xH_y+xH_2O(g) \longrightarrow xCO(g)+(x+y/2)H_2(g)$$

As the reactions are endothermic, the best yields are obtained at high temperatures. The resulting gas mixture is known as *synthesis gas.*

To produce still more hydrogen, synthesis gas is treated with more steam in the presence, first of a catalyst of iron(III) and chromium(III) oxides at 400 °C, and secondly, of copper-based catalysts at 250 °C:

$$CO(g)+H_2(g)+H_2O(g) \longrightarrow CO_2(g)+2H_2(g)$$
$$\Delta H_m^{\ominus}(298\,K) = -41\,kJ\,mol^{-1}$$

This reaction (the *shift reaction*) is exothermic and requires a low temperature for good yields. But too low a temperature makes the reaction very slow, so a compromise has to be struck.

b Dehydrogenation processes A number of catalytic dehydrogenation processes yield hydrogen as a by-product. It is obtained, for example, in the cracking of hydrocarbons, in the conversion of ethylbenzene into phenylethene (styrene), or in the conversion of propan-2-ol into propanone.

c Electrolytic methods Hydrogen is obtained as a by-product in the electrolysis of sodium chloride solution using a mercury cathode (p. 106) or in the electrolysis of hydrochloric acid (p. 313). Very pure hydrogen is also made by electrolysis of dilute solutions of sodium hydroxide or potassium hydroxide.

These electrolytic methods are, however, expensive unless the hydrogen is a by-product or unless a particularly pure supply is needed.

2 The bonding of hydrogen

a Formation of H^+   The H^+ ion, which is simply a proton, can exist in discharge tubes but it is always hydrated in aqueous solution. It is written as $H^+(aq)$ or as the *oxonium* ion, H_3O^+. There is, however, some evidence that the degree of hydration is higher, i.e. $H^+(H_2O)_x$. It is, generally, these hydrated ions that are meant when the simple term 'hydrogen ion' is used.

Isolated gaseous H^+ ions are not formed easily because the ionisation enthalpy of hydrogen is high:

$$H(g) \longrightarrow H^+(g) + e^- \qquad \Delta H_m^{\ominus}(298\,K) = 1317\,kJ\,mol^{-1}$$

An isolated proton would be about 50 000 times smaller than the small Li^+ ion, so it would have a very strong electrical field associated with it. This accounts for the ease of hydration of the H^+ ion and the high enthalpy of hydration (p. 18):

$$H^+(g) + H_2O(l) \longrightarrow H^+(aq) \qquad \Delta H_m^{\ominus}(298\,K) = -1091\,kJ\,mol^{-1}$$

Taken together with the enthalpy of atomisation ($218\,kJ\,mol^{-1}$), these figures give an absolute enthalpy of formation for the $H^+(aq)$ ion of $218 + 1317 - 1091 = +444\,kJ\,mol^{-1}$:

$$\tfrac{1}{2}H_2(g) + aq \longrightarrow H^+(aq) + e^- \qquad \Delta H_{f,m}^{\ominus}(298\,K) = +444\,kJ\,mol^{-1}$$

When relative values of the enthalpies of formation of hydrated ions are concerned, the value for $H^+(aq)$ is, arbitrarily, taken as zero (p. 17).

b Formation of H^-   This ion is present in the ionic or salt-like hydrides, e.g. Li^+H^-, formed by direct reaction between hydrogen and reactive metals (p. 150).

The electron affinity (ΔU) of hydrogen is $-72\,kJ\,(mol\,of\,H)^{-1}$, so that (p. 10)

$$H(g) + e^- \longrightarrow H^-(g) \qquad \Delta H_m^{\ominus}(298\,K) = -78\,kJ\,mol^{-1}$$

and it is only the high electropositivity of the metals forming ionic hydrides that makes the formation of H^- ions possible.

The ion cannot exist in aqueous solution for it reacts with water:

$$H^-(g) + H_2O(l) \longrightarrow H_2(g) + OH^-(aq)$$

c Formation of covalent bonds   Hydrogen can form fairly strong covalent bonds (p. 38) with most non-metals and will react directly with many of them, though the conditions of reaction vary enormously, from the explosive reactions with fluorine, chlorine and oxygen, through the reversible reactions with nitrogen, arsenic, bromine and iodine, to almost incomplete reaction with carbon even at very high temperatures.

The readiness with which hydrogen forms covalent bonds accounts for the fact that the element forms more compounds than any other.

d Hydrogen bonding   When hydrogen is covalently bonded to highly electronegative atoms, such as fluorine, oxygen and nitrogen, the resulting X—H bonds are highly polar. The positive charge on the small hydrogen atom enables it to form weak hydrogen bonds (p. 52) with other electronegative atoms.

3 Hydrogen as a reducing agent

Hydrogen can, theoretically, act as a reducing agent by taking oxygen out of a compound, by adding hydrogen, or by providing electrons, e.g.

$$PbO(s) + H_2(g) \longrightarrow Pb(s) + H_2O(l)$$
$$C_2H_2(g) + H_2(g) \longrightarrow C_2H_4(g)$$
$$\tfrac{1}{2}H_2(g) + H_2O(l) \longrightarrow H_3O^+ + e^-$$

a Reduction of metallic oxides   The temperature at which hydrogen will reduce metallic oxides can be obtained from the appropriate Ellingham diagrams (p. 113).

At low temperatures it is, theoretically, a stronger reducing agent than carbon, but this order reverses above about 700 °C. Carbon is, moreover, a more convenient chemical to use if it will fulfil the desired role. Hydrogen is used, however, in the production of high purity molybdenum and tungsten.

b Hydrogenation   Reduction in which hydrogen 'adds on' to multiple bonds is very common. The process usually requires a catalyst (p. 382) and is known as catalytic hydrogenation; high pressure and temperature may also be needed.

The best-known example is the conversion of inedible liquid oils into solid fats such as margarine. Other examples are the hydrogenation of benzene to cyclohexane, phenol to cyclohexanol and propanone to propan-2-ol.

c Reduction in solution   Hydrogen can function as a reducing agent in solution,

$$\tfrac{1}{2}H_2(g) \longrightarrow H^+(aq) + e^- \qquad E^{\ominus} = 0\,V$$

and it is this change that is taken as the arbitrary basis for comparing the strengths of other oxidising and reducing agents (p. 85).

Theoretically, hydrogen ought to reduce couples with positive electrode potentials under standard conditions, e.g.

$$
\begin{array}{ll}
H^+(aq) + e^- \longrightarrow \tfrac{1}{2}H_2(g) & E^\ominus = 0\,V \\
Fe^{3+}(aq) + e^- \longrightarrow Fe^{2+}(aq) & E^\ominus = +0.77\,V \\
\hline
\tfrac{1}{2}H_2(g) + Fe^{3+}(aq) \longrightarrow H^+(aq) + Fe^{2+}(aq) & E^\ominus = +0.77\,V
\end{array}
$$

The positive $E^\ominus$ value for the overall reaction suggests (p. 88) that it ought to take place, but like other similar reactions (p. 67) it is precluded on kinetic grounds, probably due to the high H—H bond energy.

4 Uses of hydrogen

The direct combination of hydrogen with nitrogen or chlorine is used in making ammonia or hydrochloric acid, and hydrogen is also used in many hydrogenation reactions (p. 147). In making methanol it is reacted with carbon monoxide at 250 °C and 50 atm pressure, with a catalyst of zinc oxide plus copper:

$$
CO(g) + 2H_2(g) \longrightarrow CH_3OH(l)
$$

Hydrogen is also used for filling balloons for various scientific purposes, in the atomic hydrogen torch (p. 149) and, in liquid form, as a rocket fuel.

5 Position in the periodic table

Hydrogen has the lowest relative atomic mass of all the elements and must, therefore, be placed first in the periodic table, but the best position for it is open to argument.

Its ability to lose an electron and form a positive ion suggests that it is similar to the alkali metals, so that it might be placed in Group 1A. But it is not a metal, and it forms many covalent compounds, which the alkali metals do not do.

The formation of H—X bonds, the diatomic nature of the hydrogen molecule, H—H, and the formation of the H⁻ ion, liken hydrogen to the halogens, so on these grounds, it might find a place in Group 7B.

Perhaps the unique nature of hydrogen is best represented by placing it above, and between, boron and carbon. This reflects its electronegativity value (p. 127) of 2.1, which is intermediate between those of boron (2.0) and carbon (2.5).

6 Forms of hydrogen

a Deuterium Deuterium, or heavy hydrogen, ^2_1H or D, is made from deuterium oxide, D_2O (p. 270), by electrolysis or by any of the reactions by which hydrogen can be made from water. Deuterium is slightly less reactive than hydrogen, because its bond enthalpy (439.3 kJ mol^{-1}) is slightly higher, but otherwise the two are chemically alike. What hydrogen will do, so will deuterium, so that it is easy to make deuterium compounds corresponding to well-known hydrogen compounds, e.g.

$$D_2O(l) + SO_3(s) \longrightarrow D_2SO_4(l)$$
$$Mg_3N_2(s) + 6D_2O(l) \longrightarrow 3Mg(OD)_2(aq) + 2ND_3(g)$$

The physical properties of deuterium and hydrogen are slightly different, and the difference in mass enables a particular hydrogen atom in a molecule to be 'labelled' by replacing it with a deuterium atom. This can be useful in investigations of reaction mechanisms or spectroscopic results.

Ordinary hydrogen contains about one part of deuterium in 6000.

b Tritium This isotope, ^3_1H or T, is made by bombarding ^6_3Li with neutrons,

$$^6_3Li + ^1_0n \longrightarrow ^3_1H + ^4_2He$$

and it occurs in minute quantities in the atmosphere, originating from the bombardment of nitrogen atoms by fast neutrons from cosmic rays.

Tritium is radioactive, so it is a better tracer than deuterium.

c Ortho- and para-hydrogen All diatomic molecules can exist in two forms depending on whether the nuclear spins are parallel or opposed. The smallness of the moment of inertia of the hydrogen molecule makes the effect particularly significant in this molecule.

In ortho-hydrogen the two nuclei spin in the same direction; in para-hydrogen the two spins are opposed:

ꙩ—ꙩ   ꙩ—ꙅ
ortho-hydrogen   para-hydrogen

At room temperature, the ratio of ortho- to para- is 3:1, but the para-form predominates near absolute zero. The interchange between the two forms is slow without catalysts such as charcoal. The two forms have been isolated by gas chromatography.

d Atomic hydrogen Hydrogen molecules can be dissociated into atoms by heat, radiation or electric discharge. If a stream of hydrogen is passed through an arc struck between tungsten electrodes, considerable dissociation takes place. The free atoms of hydrogen in the mixture recombine to form molecules on contact with a metal, which acts catalytically, and the release of energy produces a high temperature:

$$2H(g) \longrightarrow H_2(g) \qquad \Delta H = -436 \text{ kJ mol}^{-1}$$

This principle is put to use in the atomic hydrogen blow-pipe used for welding metals, particularly those such as aluminium, which may be oxidised by other flames.

7 Hydrides

Hydrogen will form some sort of 'compound' with almost any other element, except the noble gases, but the 'compound' is often ill-defined when the d-block metals are concerned. Simple binary hydrides fall reasonably clearly into three groups, as shown below.

Li	Be											B	C	N	O	F
Na	Mg											Al	Si	P	S	Cl
K	Ca	Sc	Ti	V	Cr	Mn	Fe	Co	Ni	Cu	Zn		Ge	As	Se	Br
Rb	Sr	Y	Zr	Nb	Mo	Tc	Ru	Rh	Pd	Ag	Cd		Sn	Sb	Te	I
Cs	Ba		Hf	Ta	W	Re	Os	Ir	Pt	Au	Hg		Pb	Bi	Po	At

ionic hydrides	d-block hydrides	covalent hydrides

Some complex hydrides, containing more than two elements, e.g. Li[AlH$_4$] (p. 167), are also important, as are many higher hydrides (p. 151).

a Ionic or salt-like hydrides These are formed by s-block metals, other than beryllium and magnesium, on heating the metal in a stream of hydrogen under pressure. They are white crystalline solids with high melting temperatures, made up of positive metallic ions and H$^-$ ions. The hydrides of the alkali metals have a crystal structure like that of sodium chloride; the structures of the hydrides of calcium, strontium and barium are less regular.

The existence of H$^-$ ions is shown by the fact that molten lithium hydride, or solutions of other hydrides in molten alkali halides, produce hydrogen at the anode on electrolysis, in quantities agreeing with Faraday's laws. The lattice enthalpies of the hydrides obtained from Born–Haber cycles are also in good agreement with the calculated values.

These hydrides, excepting that of lithium which is the most stable, decompose into the elements before melting. They all react with water,

$$H^- + H_2O(l) \longrightarrow H_2(g) + OH^-(aq)$$

and calcium hydride, which reacts at a convenient rate, is useful as a source of hydrogen.

b Covalent or volatile hydrides Most p-group elements form covalent hydrides which are generally gaseous. Typical examples are shown below, together with some standard enthalpies of formation in kJ mol^{-1}.

CH_4 -74.9	NH_3 -46.2	H_2O -242	HF -269
SiH_4 $+34.3$	PH_3 $+9.2$	H_2S -20.6	HCl -92.3
GeH_4 $+90.8$	AsH_3 $+66.4$	H_2Se $+29.7$	HBr -36.2
SnH_4 $+162.8$	SbH_3 $+145.1$	H_2Te $+99.6$	HI $+25.9$
PbH_4	BiH_3	H_2Po	

increase in stability ↑

────────→──────── general increase in stability ────────────→

Some of the elements, particularly carbon, silicon and boron, form series of hydrides (see section **d** below).

The thermal stability of the hydrides with respect to decomposition into the elements, generally increases in passing from left to right along a horizontal period, and decreases in passing down a group. Increased stability is, therefore, related to increase in electronegativity; it is also fairly closely related to the X—H bond enthalpy.

c d-block hydrides Many d-block metals can absorb large amounts of hydrogen. The gas can generally be liberated by pumping at a high enough temperature, and palladium is useful in removing hydrogen from gas mixtures.

During the absorption of the hydrogen, the crystal lattice of the metal is expanded, but not greatly distorted, and it appears that the small hydrogen atoms are situated between the metallic atoms; that is why these hydrides are sometimes called *interstitial hydrides*.

The amount of hydrogen that can be absorbed varies with the conditions, but hydrides with 'formulae' such as $TiH_{1.73}$, $TaH_{0.76}$, $PdH_{0.6}$ and $LaH_{2.8}$ have been reported. Such 'compounds' are very much like the metals from which they are made.

d Other hydrides Magnesium hydride, MgH_2, has a crystal structure like that of MgF_2, but it is not a genuine ionic hydride because it is less thermally stable and has a lower enthalpy of formation than other alkaline earth hydrides (p. 159).

Beryllium and aluminium hydrides have complex, polymeric structures that can be represented in an over-simplified way as $(BeH_2)_x$ and $(AlH_3)_x$. The aluminium hydride is particularly complex as it exists in a number of different forms.

e Higher hydrides A small group of elements clustered around carbon in the periodic table form a number of higher hydrides, some of them being compounds of great theoretical interest or practical use, e.g.

B_xH_{x+4} series, e.g. B_2H_6, $B_{10}H_{14}$	C_xH_{2x+2} series, e.g. CH_4, C_3H_8; plus unsaturated hydrocarbons	NH_3, ammonia N_2H_4, hydrazine HN_3, hydrogen azide	H_2O H_2O_2
B_xH_{x+6} series, e.g. B_4H_{10}, B_9H_{15}			
	Si_xH_{2x+2} series, $x = 1-6$	PH_3, phosphine P_2H_4, diphosphane	H_2S_x series, $x = 1-6$
	Ge_xH_{2x+2} series, $x = 1-3$		

These higher hydrides contain chains of X—X—X—X atoms; they are sometimes known as *catenated hydrides*. The number of hydrides formed tends to fall on passing down a periodic table group and, as a broad generalisation, the higher hydrides tend to decompose into the lower ones on heating.

Questions on Chapter 11

1 State five reactions of hydrogen, which illustrate its chief chemical properties, and give two industrial uses of the gas. Explain the different types of chemical bonding which occur in the following compounds: (a) sodium hydride, (b) hydrogen chloride, (c) potassium hydrogen fluoride. (OC)

2 What are the arguments for, and against, regarding hydrogen as a metal.

3 'Hydrogen is only acidic when bound to oxygen, sulphur or the halogens.' Is this statement true? Discuss it. (O)

4 Hydrogen could be placed in either Group 1, with the alkali metals, or in Group 7, with the halogens. Which position is the more appropriate? (QS)

5 Describe one method for the manufacture of hydrogen. Give an account of the reactions that can take place between hydrogen and (a) chlorine and (b) nitrogen, paying particular attention to the effect of varying the conditions of the reaction. (OC)

6 Discuss the ways in which hydrogen is bonded in inorganic compounds, giving one example in each case. Show, where possible, the outer electronic configurations in each example you select. Account for the shapes of the following molecules: (a) CH_4, (b) NH_3, (c) XeF_4. State and account for the relative strengths of hydrofluoric and hydrochloric acids in dilute and in concentrated solutions. (JMB)

7 Outline the methods available for the preparation of hydrides. In the molecules of methane, ammonia and water, the H–X–H bond angles are 109.5, 107 and 104.5 respectively. Account as fully as possible for these values. Compare the bonding in methane, sodium hydride and hydrogen fluoride in their solid state. (JMB)

8 To what extent can the hydrides of an element be regarded as the extreme end of a series of halides?

9 Discuss the shape of hydride molecules.

10 Compare and contrast the properties of the hydrides of the first short period elements with those of the second.

11 How would you try to discover whether lithium tetrahydridoaluminate(III) was a stronger reducing agent than sodium tetrahydridoborate(III)?

12 Outline the principal features in the chemistry of ionic and molecular hydrides. Starting with heavy water, D_2O, as the only source of deuterium, write reaction schemes for the preparation of the following: NaD, ND_3, $CH_2D.CH_2Br$, PD_4I, NaD_2PO_2. Suggest why the compound $LiAlH_4$ is considerably more soluble than $NaBH_4$ in ether. (OC)

13 Explain why lithium hydride is the most stable alkali metal hydride so far as decomposition into its elements is concerned.

14 Hydrogen resembles both the alkali metals and the halogens in its chemical behaviour. Survey the evidence which leads to this conclusion. Use the compounds which you have described to illustrate the changes in bond-type of the hydrides of elements in a period of the periodic table. (JMB)

15 What volume of hydrogen (at standard temperature and pressure) would be obtained by electrolysis of a solution of calcium hydride in a mixture of molten chlorides if a current of 10 A is passed for 161 min?

16* Use a Born–Haber cycle like that used for NaCl to calculate the lattice enthalpy of NaH.

17* What similarities are there between sodium hydride and sodium fluoride? Why is there such a big difference between their standard enthalpies of formation?

12
The s-block elements

The elements of Groups 1A and 2A are very much alike in kind though differing in degree. They all form mainly ionic compounds (which are colourless unless they contain a coloured anion) and have fixed oxidation states of $+1$ and $+2$ respectively.

The differences arise mainly from differences in ionic size and charge, and this results, for example, in some marked differences in solubility and thermal stability between Group 1A and Group 2A compounds.

Lithium, at the head of Group 1A, is in many ways unique, showing some diagonal relationships to magnesium (p. 137). Beryllium, likewise, is different from the other members of Group 2A and shows some resemblance to aluminium (p. 138).

Summaries of the main features of the compounds of Group 1A and Group 2A elements are given on pp. 166 and 179 respectively.

1 Ionic character of compounds

Lithium and beryllium, and to a lesser extent magnesium, form some covalent compounds (p. 162), but most of the other elements in Groups 1A and 2A form compounds which are predominantly ionic.

a Ionic radii   The ionic radii of the s-block metal ions are summarised in Fig. 67. As expected, there is a rise in passing down a group and the radii of the M^{2+} ions are smaller than those of the corresponding M^+ ion. The Al^{3+} ion, shown for comparison, is even smaller.

b Formation of ions by Group 1A metals   The ease of formation of $M^+(aq)$ ions from M in its standard state is measured by the standard electrode potential for the $M^+(aq)/M(s)$ couple or, less accurately,[*] by the standard enthalpy of formation of $M^+(aq)$. The order is $Li > K = Rb > Cs > Na$, as shown below:

	Li	K	Rb	Cs	Na
enthalpy of formation of $M^+(aq)/kJ\,mol^{-1}$	-278	-252	-251	-248	-240
electrode potential, $M^+(aq)/M(s)/V$	-3.045	-2.925	-2.925	-2.923	-2.714

[*]The standard free energy of formation must be used for more accurate results (p. 21).

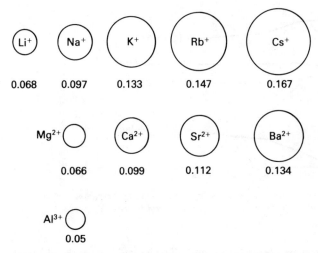

Fig. 67 *Ionic radii of s-block metal ions in nm. The Al^{3+} ion is shown for comparison.*

The figures quoted are on the arbitrary basis of a zero enthalpy of formation for $H^+(aq)$ and a zero electrode potential for the $H^+(aq)/\frac{1}{2}H_2(g)$ couple.

The irregularity of order comes about because three enthalpy terms are involved:

enthalpy of atomisation	ionisation enthalpy	enthalpy of hydration
$M(s) \longrightarrow M(g)$	$M(g) \longrightarrow M^+(g)$	$M^+(g) \longrightarrow M^+(aq)$

Each of these varies, as shown below and in Fig. 68.

	Li	K	Rb	Cs	Na
$M(s) \longrightarrow M(g)$	161	90	82	78	108
$M(g) \longrightarrow M^+(g)$	526	426	409	382	502
$M^+(g) \longrightarrow M^+(aq)$	-520	-320	-296	-264	-406
$M(s) \longrightarrow M^+(aq)$	167	196	195	196	204

The figures quoted here for the enthalpy of formation of the $M^+(aq)$ ions differ from those quoted above by approximately $444 \, kJ \, mol^{-1}$ (p. 20), as these figures are not on the arbitrary scale of $H^+(aq)$ having a zero enthalpy of formation.

The enthalpies of atomisation and hydration, and the ionisation enthalpies, all increase numerically in passing from caesium to lithium, as the ionic size decreases, but the variation is not regular (Fig. 68).

It is the particularly large value for the enthalpy of hydration of $Li^+(g)$ that is mainly responsible for the high negative electrode potential of the $Li^+(aq)/Li(s)$ couple, and the high ionisation enthalpy of Na(g) that causes the low value for the $Na^+(aq)/Na(s)$ couple.

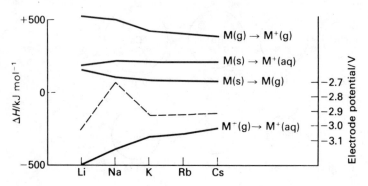

Fig. 68 *Showing (a) the energy terms contributing to the enthalpy of formation of M^+ (aq) ions and (b) the corresponding standard electrode potentials (broken line).*

c Formation of ions by Group 2A metals   The irregularity in the electrode potential values of M^+(aq) ions does not occur with M^{2+}(aq) ions, the order being Ba > Sr > Ca > Mg > Be as shown, though the values for barium, strontium and calcium are very close.

	Be	Mg	Ca	Sr	Ba
enthalpy of formation of M^{2+}(aq)/kJ mol^{-1}	-383	-467	-543	-546	-538
electrode potential, M^{2+}(aq)/M(s)/V	-1.85	-2.37	-2.87	-2.89	-2.90

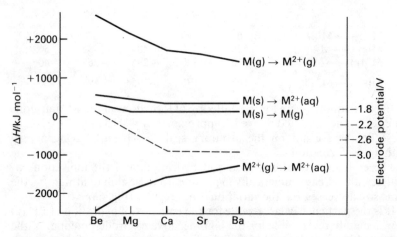

Fig. 69 *Showing (a) the energy terms contributing to the enthalpy of formation of M^{2+}(aq) ions and (b) the corresponding standard electrode potentials (broken line).*

The energy terms contributing are summarised as follows, on the same basis as in **b**, and plotted on Fig. 69.

	Be	Mg	Ca	Sr	Ba
$M(s) \longrightarrow M(g)$	324	148	177	164	180
$M(g) \longrightarrow M^{2+}(g)$	2669	2201	1747	1626	1480
$M^{2+}(g) \longrightarrow M^{2+}(aq)$	−2484	−1926	−1579	−1446	−1309
$M(s) \longrightarrow M^{2+}(aq)$	509	423	345	344	351

Because these are divalent ions, the values for the enthalpies of formation in the table above differ from those in the table on p. 156 by approximately $888 \, kJ \, mol^{-1}$.

d Lattice enthalpies of MX halides Alkali metal salts have high lattice enthalpies, and the ionic nature of the salts is shown by the good agreement between the calculated and measured values (p. 13). The measured values, from Born–Haber cycles, for the alkali metal halides are summarised below, in $kJ \, mol^{-1}$.

	fluoride	chloride	bromide	iodide
Li	−1035	−854	−813	−761
Na	−924	−788	−748	−704
K	−818	−718	−685	−649
Rb	−786	−695	−664	−631
Cs	−746	−675	−648	−618

For any given alkali metal, the lattice enthalpy of the halide becomes less negative in passing from the fluoride to the iodide. This is due to the increase in size in passing from F^- to I^-, which causes an increase in internuclear distances and a corresponding decrease in the attractive forces and, hence, in the lattice enthalpy.

For any given halide ion, the lattice enthalpy becomes less negative, for the same reasons, in passing from lithium to caesium.

When the internuclear distances are about equal as, for instance, in RbF and LiI, the lattice enthalpies are almost equal.

e Lattice enthalpies of MX$_2$ halides The values, from Born–Haber cycles, for the lattice enthalpies of halides of Group 2A metals are given below, in $kJ \, mol^{-1}$.

	fluoride	chloride	bromide	iodide
Be	−3525	−3039	−2919	−2817
Mg	−2931	−2525	−2423	−2321
Ca	−2619	−2254	−2156	−2072
Sr	−2484	−2152	−2062	−1969
Ba	−2341	−2054	−1971	−1874

The same trends are apparent as for the Group 1A metals and there is, again, good agreement between the measured and calculated values for many of the Group 2A halides. Discrepancies do occur, however, when the halides have layer lattice structures, e.g. $MgCl_2$, $MgBr_2$, MgI_2 and CaI_2, and in beryllium halides where there is considerable covalent character. In these cases the assumptions made in calculating the lattice enthalpies (p. 13) are not valid, for the crystal structures are not purely ionic.

For those compounds in which the structure is ionic, the lattice enthalpy of an $M^{2+}X^{2-}$ compound is considerably higher than that of an M^+X^- compound. This is because M^{2+} ions exert stronger fields than comparable M^+ ions; the former are both smaller and more highly charged. For comparable interionic distances, the lattice enthalpy is approximately four times higher when two doubly charged ions are concerned than for two singly charged ions, e.g.

	Li^+F^-	$Mg^{2+}O^{2-}$	K^+Cl^-	$Ba^{2+}S^{2-}$
interionic distance/nm	0.204	0.206	0.314	0.318
lattice enthalpy/kJ mol^{-1}	-1035	-3890	-718	-2703

2 Hydration of ions

A considerable input of energy is required to form $M^+(g)$ from an alkali metal ($610\,kJ\,mol^{-1}$ for sodium) and even more energy is required to form $M^{2+}(g)$ from an alkaline earth metal ($2349\,kJ\,mol^{-1}$ for magnesium). When these ions are formed in anhydrous crystals this energy is offset by the high values of the lattice enthalpies of the crystals concerned. When the ions are formed in solution, as $M^+(aq)$ and $M^{2+}(aq)$, the energy is offset by the high negative values for the enthalpies of hydration of the ions (p. 17) as summarised below (in $kJ\,mol^{-1}$) and in Figs. 68 and 69.

H^+	Li^+	Na^+	K^+	Rb^+	Cs^+
-1091	-520	-406	-320	-296	-264

Be^{2+}	Mg^{2+}	Ca^{2+}	Sr^{2+}	Ba^{2+}
-2484	-1926	-1579	-1446	-1309

Al^{3+}
-4680

The M^{2+} ions have a stronger electrical field than the M^+ ions and are more readily hydrated on that count. The Al^{3+} ion, included above for comparison, is even more easily hydrated. For both M^+ and M^{2+} ions, the smaller the size of the ion, the more readily hydrated it is.

The Li^+ ion is the most easily hydrated of the alkali metal ions, and this is illustrated by the high negative value for the enthalpy of hydration as given above, by the low ionic mobility (a measure of the velocity of an ion under a potential gradient) and by the number of salt hydrates formed.

The Li$^+$ ion is the smallest M$^+$ ion and ought, on grounds of size, to have the highest ionic mobility, the Cs$^+$ ion having the lowest. In fact, the order is reversed because Li$^+$ ions are more hydrated than Cs$^+$ ions so that the effective size of the former, in solution, is higher than that of the latter.

3 Enthalpy of formation of compounds

The standard enthalpy of formation of compounds gives a good measure of their relative stabilities so far as decomposition into their elements is concerned. The compounds of both Groups 1A and 2A elements have high negative values for their enthalpies of formation, indicating a high degree of thermal stability.

a Group 1A   Some standard enthalpies of formation for typical compounds are given below, in kJ mol^{-1}.

	MF	MCl	MBr	MI	MH	M$_2$O	MOH
Li	-610	-402	-350	-270	-90.6	-599	-485
Na	-574	-411	-360	-288	-56.4	-418	-426
K	-563	-436	-392	-328	-57.8	-363	-425
Rb	-551	-433	-391	-330	-47.7	-330	-418
Cs	-545	-447	-409	-351	-42.3	-318	-417

The only compounds that decompose into their elements on heating are the hydrides, with relatively low enthalpies of formation. Lithium hydride is the most stable; it melts without decomposition, but does decompose at about 1000 °C.

As a general rule, the enthalpy of formation falls in passing down Group 1A for compounds containing small anions, i.e. for the fluorides, hydrides and oxides. When the anion is large, e.g. I$^-$, the values rise in passing down the group. For intermediate anions, the change is random.

b Group 2A   Typical values, in kJ mol^{-1}, are as follows:

	MF$_2$	MCl$_2$	MBr$_2$	MI$_2$	MH$_2$	M$_3$N$_2$	MO	M(OH)$_2$
Mg	-1102	-642	-518	-360	-75	-461	-602	-925
Ca	-1214	-795	-675	-535	-186	-431	-635	-987
Sr	-1214	-828	-716	-567	-180	-391	-592	-960
Ba	-1200	-860	-755	-602	-179	-363	-553	-946

As for the alkali metal compounds, the values are highly negative, showing that the majority of the compounds are very stable so far as decomposition into their elements is concerned. It is only the hydrides that decompose, magnesium hydride being the least stable with a decomposition temperature of about 300 °C. The other hydrides decompose at about 600 °C but do not differ much in stability amongst themselves.

The order of stability on passing down the group varies from anion to anion, and is very irregular.

4 Thermal decomposition of compounds

Though only the hydrides of the alkali metals and the alkaline earths decompose into their elements on heating, many other compounds undergo some different type of thermal decomposition. This is particularly so for compounds of lithium and the alkaline earths, which decompose more readily than the corresponding compounds of the elements sodium to caesium. The following summary shows the main products on heating a number of compounds to a reasonably high temperature.

	Li	Na to Cs	Mg to Ba
hydroxide	$Li_2O + H_2O$	stable	$MO + H_2O$
carbonate	$Li_2O + CO_2$	stable	$MO + CO_2$
hydrogencarbonate	only known in solution	$M_2CO_3 + H_2O + CO_2$	only known in solution
nitrates	$Li_2O + NO_2 + O_2$	$MNO_2 + O_2$	$MO + NO_2 + O_2$

a General trends   In both Groups 1A and 2A, the decompositions take place less easily in passing down the group, i.e. it is the lithium and magnesium compounds that decompose most easily, as can be seen from the temperatures required to decompose the carbonates, given in section **b** below.

It is the high lattice enthalpies of Li_2O and MgO, as compared with those of the other oxides, that play a significant role, for they make the decomposition of the lithium and magnesium compounds more exothermic than similar decompositions for the other elements.

b Decomposition of carbonates   Lithium carbonate is the only alkali metal carbonate to decompose,

$$Li_2CO_3(s) \longrightarrow Li_2O(s) + CO_2(g)$$

but the alkaline earth carbonates all decompose, the temperature of decomposition under comparable conditions rising from that for magnesium carbonate to that for barium carbonate.

The decomposition of calcium carbonate can be considered in terms of the ΔH, ΔG and $T\Delta S$ values (p. 70) given below and plotted on Fig. 70.

temperature/K	298	500	700	900	1100	1300
$\Delta H/kJ\,mol^{-1}$	177.8	177.4	177.0	177.0	176.6	176.1
$\Delta G/kJ\,mol^{-1}$	130.1	97.5	65.3	33.9	2.1	-29.7
$T\Delta S/kJ\,mol^{-1}$	47.7	79.9	111.7	143.1	174.5	205.8

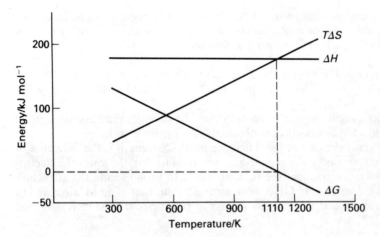

Fig. 70 *The temperature variation of ΔH, ΔG and $T\Delta S$ for the reaction*

$$CaCO_3(s) \longrightarrow CaO(s) + CO_2(g)$$

ΔG becomes negative above 1110 K.

Figure 70 shows that ΔG becomes negative at temperatures above approximately 1110 K. The ΔS change is positive as there is an entropy increase due to the conversion of a solid into a gas:

$$CaCO_3(s) \longrightarrow CaO(s) + CO_2(g) \qquad \Delta S^{\ominus}(298\,K) = 160.4\,J\,K^{-1}\,mol^{-1}$$
$$92.9 \qquad\qquad 39.7 \quad\; 213.6$$

ΔG only becomes negative when $T\Delta S$ becomes greater than ΔH. If it is assumed that ΔH and ΔS do not change with temperature, the temperature at which this happens is given by

$$T = \frac{\Delta H(298\,K)}{\Delta S(298\,K)} = \frac{177.8}{0.1604} = 1108\,K$$

Values obtained for other carbonates in the same way are (in °C):

Li_2CO_3	Na_2CO_3	$MgCO_3$	$CaCO_3$	$SrCO_3$	$BaCO_3$
1157	1863	404	826	1098	1370

5 Basic strength of oxides and hydroxides

The oxides and hydroxides of the alkali metals are significantly stronger than those of the alkaline earths, and, in each group, the basic strength increases in passing down the group.

A comparative measurement can be obtained from the free energy change for the reaction of the oxides or hydroxides with a chosen acidic oxide. For the reaction

$$CaO(s) + CO_2(g) \longrightarrow CaCO_3(s)$$

can be done

the free energy change is $-130.4\,kJ\,mol^{-1}$, the negative value indicating the feasibility of the reaction. The values for the reactions of other metallic oxides with carbon dioxide are as follows:

$$Li_2O\ -176\quad Na_2O\ -276\quad K_2O\ -356$$
$$MgO\ -66\quad CaO\ -130\quad SrO\ -144\quad BaO\ -216$$

The more highly negative the ΔG value, the stronger the basic strength of the oxide. Similar results are obtainable for the hydroxides.

The general cause of the difference in basic strength is the difference in the electronegativity of the metals concerned (p. 40); the lower the electronegativity, the stronger the basic strength. In broad terms, the attraction of an M^{2+} ion for an OH^- ion is stronger than that of an M^+ ion, and in both cases, the attraction gets weaker as the size of the ion increases. (See also p. 277.)

6 Flame coloration

The emission spectra of lithium, sodium and potassium have strong lines in the visible region. As a result, most of their compounds will colour a bunsen burner flame. Lithium compounds give a red colour, potassium compounds a lilac colour and sodium compounds a yellow colour. The chlorides give the strongest colours because they are volatile at the temperature of the bunsen burner flame. That is why flame tests are done by dipping the unknown substance in concentrated hydrochloric acid.

The lithium, sodium and potassium spectral lines which cause their flame colours are caused by 2p to 2s, 3p to 3s, and 4p to 4s electron shifts respectively.

Calcium, strontium and barium compounds give characteristic flame colours of red, crimson and apple-green.

7 Formation of covalent compounds

Polarisation considerations (p. 16) would predict that the smallest ion with the highest charge would tend to form covalent compounds the most easily. This is borne out by the fact that beryllium compounds are mainly covalent. Lithium and magnesium form some covalent compounds, but they are rare for the other metals.

a Covalent compounds of beryllium   Beryllium oxide, BeO, is a covalent, amphoteric oxide with chemical properties similar to those of the ionic oxides of aluminium and zinc. Beryllium chloride, $BeCl_2$, is also predominantly covalent; it is soluble in organic solvents and readily hydrolysed, but its low electrical conductivity in the fused state indicates some ionic character.

Beryllium complexes include $[BeF_4]^{2-}$, $[Be(NH_3)_4]^{2+}$ and basic beryllium ethanoate, $Be_4O(CH_3COO)_6$. In the basic ethanoate, a central oxygen atom is surrounded tetrahedrally by four beryllium atoms, and the six CH_3COO groups are arranged along the six edges of the tetrahedron.

b Covalent compounds of the alkaline earths The most important covalent compounds of the alkaline earths are the Grignard reagents formed from magnesium and alkyl or aryl halides in solution in an ether, e.g. CH_3MgI and C_6H_5MgBr. Dialkyl and diaryl compounds are also formed, e.g. $Mg(CH_3)_2$ and $Mg(C_6H_5)_2$.

The solubility of magnesium chloride, bromide and iodide in organic solvents, the discrepancy between the calculated and measured values of the lattice enthalpies (p. 158), and the fact that they have layer lattice structures, also suggest some measure of covalent bonding in these halides.

c Covalent compounds of the alkali metals Lithium forms alkyl and aryl compounds that are similar to Grignard reagents but more reactive. They are typical covalent compounds, and are used in organic synthesis and as polymerising agents. Sodium and potassium also form reactive alkyls and aryls, but these are essentially ionic and are much less useful than the lithium compounds.

Questions on Chapter 12

(See also pp. 176 and 186.)

1 The standard enthalpies of formation of $Ba(OH)_2$, BaO and $H_2O(g)$ are -946, -558 and $-242\,kJ\,mol^{-1}$ respectively, and the standard entropies are 112, 70, and $188\,J\,K^{-1}\,mol^{-1}$ respectively. Calculate the approximate temperature at which barium hydroxide might be expected to decompose into its oxide and water vapour.

2 The hydroxides of the alkaline earths decompose into the oxide and water more easily than the carbonates decompose into the oxide and carbon dioxide. This is because water is a weaker acid than carbon dioxide. Comment.

3 Examine the proposition that the lattice enthalpies of the alkali metal halides are inversely proportional to the interionic distance in the halide.

4 The enthalpy of atomisation of aluminium is $324\,kJ\,mol^{-1}$ and the enthalpy of hydration of $Al^{3+}(g)$ is $-4613\,kJ\,mol^{-1}$. The first three ionisation enthalpies are 578, 1817 and $2745\,kJ\,mol^{-1}$ respectively. Calculate the value for the enthalpy of formation of $Al^{3+}(aq)$ from Al(s) using these figures. Why does your answer differ from the value of $-525\,kJ\,mol^{-1}$ quoted in books of data?

5 Caesium is more reactive than lithium under anhydrous conditions. Why is this so?

6 What effect does polarisation have on the properties of compounds of the alkali metals and alkaline earths? Give illustrative examples.

7 The s-block of the periodic table contains Group 1A, the alkali metals, and Group 2A, the alkaline earths. Give an account of these two groups of elements and their compounds, paying particular attention to similarities and differences within the groups. (L)

8 'The properties of the first member of a group of elements in the periodic table are not typical of the group as a whole.' Discuss this with reference to the chemistry of the elements of Groups 1 (Li–Cs) and 2 (Be–Ba). You should include in your answer specific properties which differentiate lithium and beryllium from other members of their respective groups as well as the reasons for the differentiation. (JMB)

9 (a) Write the electronic structures of the Group 1 elements, lithium to caesium, in terms of s, p and d orbitals.

(b) For the elements lithium to caesium, state, and explain qualitatively, the general trends in (i) first ionisation energy, (ii) atomic radius.

(c) Give reasons for the following: (i) the large negative standard electrode potential of lithium; (ii) the differences in the structures of the crystal lattices of sodium chloride and caesium chloride; (iii) the fact that lithium salts are often hydrated but hydrated caesium salts are rare.

(d) Give *two* chemical properties of lithium or its compounds which are not typical of the other Group 1 elements and their compounds. By means of descriptions or equations show the ways in which these properties are atypical. (AEB)

13
The alkali metals (Group 1A)

$$M(s) + H_2O(l) \longrightarrow MOH(aq) + \tfrac{1}{2}H_2(g)$$

1 The elements

The alkali metals are soft and, for metals, have high electrical and thermal conductivities, but low melting and boiling temperatures and densities. The first three members of the group float on water, lithium being the lightest solid element. The metals all crystallise with a body-centred cubic lattice (p. 50).

The metals are extremely reactive and tarnish rapidly (*discoloured*) on exposure to air. They are, nevertheless, lustrous and silver-coloured when freshly made or cut. As they also react with cold water, and even with ice at very low temperatures, they must be stored under paraffin oil or naphtha, or (for rubidium and caesium) in air- and water-tight containers.

Because of the reactivity of the metals they do not occur native but they do occur in a wide variety of compounds that are so stable that the metal can only be extracted by electrolysis. This is generally done by using the fused chloride with added impurities to lower the melting temperature.

a Sodium Sodium is manufactured by the Downs process, in which sodium chloride (melting temperature 801 °C) is electrolysed in the molten state. This is done in a cylindrical cell with a central graphite anode and a surrounding steel cathode. An excess of calcium chloride is added so that the mixture melts at about 600 °C; it is this temperature lowering that makes the process feasible. The two electrodes are separated by a cylindrical steel gauze diaphragm so that the molten sodium, which floats to the top of the cathode compartment, is kept away from the gaseous chlorine formed at the anode.

Sodium reacts with oxygen and with cold water, with most non-metals (but not boron, carbon or nitrogen), and with alcohols and ammonia gas:

$$2Na(s) + 2ROH(l) \longrightarrow 2NaOR(s) + H_2(g)$$
$$2Na(s) + 2NH_3(g) \longrightarrow 2NaNH_2(s) + H_2(g)$$
$$\text{sodium amide}$$

(*sodium mixed with mercury*)

It forms a liquid or solid amalgam with mercury depending on the amount of sodium used; it also forms a liquid alloy with potassium and an alloy with lead.

Sodium is used in making sodium peroxide and sodium cyanide, and as a reagent in organic chemistry. The metal is a strong reducing agent and can be used in reducing titanium(IV) chloride in the manufacture of titanium. It is also used in sodium vapour lamps.

b Potassium Potassium occurs naturally as the chloride in carnallite,

$KCl.MgCl_2.6H_2O$, one of the major components of the Stassfurt deposits in Saxony, as sylvine, KCl, and as sylvinite, which is a mixture of KCl and $NaCl$. The metal is difficult to make by direct electrolysis as it vaporises very readily. It is generally obtained by reaction of molten potassium chloride with sodium vapour.

It is very much like sodium in appearance and properties, though slightly more reactive. Natural potassium is slightly radioactive because of a 0.012 per cent content of the radioactive isotope $^{40}_{19}K$.

2 Compounds of the alkali metals

The alkali metals have many common features, but lithium is in many ways unique.

a Ionic character   The alkali metals have a fixed oxidation state of $+1$ and their compounds are mainly stable ionic solids, that are colourless unless they contain a coloured anion. The ionic nature of the compounds is shown by the very good agreement between measured and calculated values of the lattice enthalpies.

The metals are strongly electropositive. The order of their ionisation enthalpies is $Li > Na > K > Rb > Cs$ (p. 154), in line with the increase in ionic radius (p. 155), but the order of the standard electrode potentials is $Li > K = Rb > Cs > Na$ (p. 154). *Shells are more away from nucleus ⇒ less attractive force*

b Hydration of ions   It is the high enthalpy of hydration of Li^+, due to its small size, that is mainly responsible for its high electrode potential. The ease of hydration falls on passing down the group to caesium.

Most lithium and sodium salts are hydrated, and many potassium salts are; but hydration is very rare for rubidium and caesium salts.

c Formation of covalent compounds   Lithium forms many important organometallic compounds which are covalent, e.g. ethyllithium, C_2H_5Li, but there are fewer similar compounds of sodium and potassium, and those that do exist are essentially ionic (p. 163).

d Solubility   Almost all the compounds of alkali metals are soluble in water, though the trends are varied. The solubility may rise, fall or reach a minimum in passing from lithium to caesium, depending on the anion. Of the relatively simple compounds, only LiF and Li_2CO_3 have low solubilities.

Na^+ ions give an almost quantitative precipitate with uranyl(VI) zinc ethanoate, and K^+ ions give precipitates with chlorate(VII) and hexanitrocobaltate(III) ions. These reagents are, therefore, useful in analysis.

e Thermal stability   The compounds have high negative enthalpies of *direct result of lattice enthalpy (directly proportional)* formation (p. 159) and only the hydrides will split up into the elements on heating, with lithium hydride being the most stable.

Other decompositions take place, particularly for lithium compounds. The stability of the compound generally increases in passing from lithium to caesium (p. 160).

the higher the metallic character the more basic.

f Basic strength of oxides and hydroxides   The oxides and hydroxides are strongly basic, the strength increasing in passing from lithium to caesium (p. 162). *decreases along a period.*

g Flame coloration   Compounds of the alkali metals give characteristic flame colorations (p. 162).

3 The hydrides

All five alkali metals form colourless hydrides, with the sodium chloride structure, by direct combination at temperatures varying from about $300\,°C$ for caesium to $600\,°C$ for lithium.

Lithium hydride is the most stable (p. 159), melting at about $700\,°C$ without decomposition, and not decomposing until about $1000\,°C$, whereas the hydrides of sodium and potassium decompose at about $400\,°C$:

$$2NaH(s) \longrightarrow 2Na(s) + H_2(g)$$

NaH

All the hydrides react with water to yield hydrogen:

$$H^- + H_2O(l) \longrightarrow H_2(g) + OH^-(aq)$$

a Sodium hydride   This is a good reducing agent, used for descaling iron and for making sodium tetrahydridoborate(III), $Na[BH_4]$, which is a selective reducing agent for reducing ketones to alcohols:

$$4NaH + (CH_3)_3BO_3 \longrightarrow Na[BH_4] + 3CH_3ONa$$

b Lithium tetrahydridoaluminate(III), $Li[AlH_4]$ This important re-agent is made by reaction between excess lithium hydride and an ethereal solution of aluminium chloride:

$$AlCl_3 + 4LiH \longrightarrow Li[AlH_4] + 3LiCl$$

It is a white crystalline solid used as a selective reducing agent. It reduces carboxylic acids, esters, aldehydes and ketones to alcohols, but will not reduce C=C bonds. In inorganic chemistry it will reduce chlorides to hydrides, carbon dioxide to carbon, and sulphates to sulphides.

4 The oxides

All the alkali metals form oxides, M_2O, which contain the O^{2-} ion and are strongly basic:

$$O^{2-} + H_2O(l) \longrightarrow 2OH^-(aq)$$

Lithium oxide is not affected by heating (p. 159) but the others, or the metals alone, form peroxides or hyperoxides on heating in excess oxygen. The peroxides, M_2O_2, are ionic and contain the $[O-O]^{2-}$ ion. They react with water to liberate hydrogen peroxide,

$$O_2{}^{2-} + 2H_2O(l) \longrightarrow 2OH^-(aq) + H_2O_2(l)$$

and with carbon dioxide to form oxygen:

$$2M_2O_2(s) + 2CO_2(g) \longrightarrow 2M_2CO_3(s) + O_2(g)$$

Hyperoxides, MO_2, contain $O_2{}^-$ ions and liberate hydrogen peroxide and oxygen when treated with water:

$$2O_2{}^- + 2H_2O(l) \longrightarrow 2OH^-(aq) + H_2O_2(l) + O_2(g)$$

They also, like peroxides, convert carbon dioxide into oxygen:

$$4MO_2(s) + 2CO_2(g) \longrightarrow 2M_2CO_3(s) + 3O_2(g)$$

5 The hydroxides

The hydroxides, MOH, are formed by the action of water on the metals or the dimetal oxides, and by the electrolysis of aqueous solutions of the chlorides.

The hydroxides are white solids and, with the exception of lithium hydroxide, they all fuse without decomposition. They are strong bases containing the OH^- ion. The basic strength increases in passing from lithium to caesium.

a Sodium hydroxide, NaOH (caustic soda)   This important alkali, which must be handled with care because it is caustic, is mainly manufactured by the electrolysis of brine (p. 106).

Sodium hydroxide is white, but slightly translucent; it is also deliquescent. It melts at 322 °C without decomposing, and it is very soluble in water, with the evolution of much heat, and in ethanol.

In the laboratory it is used in neutralisation reactions, to liberate ammonia from ammonium salts, and to precipitate insoluble hydroxides from salt solutions. Amphoteric hydroxides will dissolve in excess sodium hydroxide, and acidic oxides also react with it. In preparative chemistry it is used in making phosphine, sodium chlorate(I) and chlorate(V) and sodium methanoate.

It is also used in the purification of bauxite (p. 188), in making soap, as a cleansing agent, and in the manufacture of paper, rayon, dyes and petroleum products.

b Potassium hydroxide, KOH (caustic potash)   This is manufactured by the electrolysis of potassium chloride solution. It is a white translucent alkali, soluble in both water and ethanol. Both aqueous and alcoholic solutions of it are used in organic chemistry. Potassium hydroxide is also used in making soap; it gives a softer product than sodium hydroxide.

Potassium hydroxide is more hygroscopic and more soluble than sodium hydroxide. This accounts for its use in organic chemistry, and for its use as a drying agent and for absorbing acidic gases.

6 The halides

All the alkali metal halides are ionic compounds and they generally crystallise with the sodium chloride structure. They are all very thermally stable (p. 159) and soluble in water. The chloride, bromide and iodide of lithium are also soluble in alcohol and ethoxyethane.

a Sodium chloride, NaCl   Common salt occurs naturally and extensively as rock salt and it is found in sea water to the extent of about 3 per cent. It can be mined as a solid or pumped from underground deposits as a saturated solution known as *brine*; in hot parts of the world it can also be obtained by solar evaporation of sea water. The rock salt or brine is purified by treatment with sodium carbonate, sodium hydroxide and a soluble barium salt to remove Ca^{2+}, Mg^{2+} and SO_4^{2-} ions. The product is suitable for industrial use, but table salt must be purified further.

Sodium chloride is an essential part of the diet and it is also used as a food preservative. In chemical industry it is used to manufacture sodium, sodium hydroxide, chlorine, sodium carbonate, sodium chlorate(I) and sodium chlorate(V). Other uses are in glazing earthenware, in regenerating water softeners, in the salting out of soap and in clearing ice and snow from roads.

b Potassium chloride, KCl   This is obtained from the Stassfurt deposits, which contain carnallite (p. 178), by fractional crystallisation. Alternative supplies come from eastern Europe and from deposits of sylvinite (p. 166) near Whitby. It is the most important potassium compound, and is used as a fertiliser and in making most other potassium compounds.

7 The carbonates and hydrogencarbonates

The carbonates are ionic salts containing the CO_3^{2-} ion. They are all thermally stable apart from lithium carbonate (p. 160), whereas other metallic carbonates decompose on heating.

The alkali metal carbonates are also unique because they are soluble in water, though lithium carbonate is only slightly soluble. All the carbonates dissolve, too, in carbonic acid to form hydrogencarbonates, $MHCO_3$, in solution. The only hydrogencarbonates that exist in the solid state are those of the Group 1A metals (again, excepting lithium).

Hydrated sodium carbonate is efflorescent, whereas potassium carbonate is deliquescent, and potassium hydrogencarbonate is much more soluble than the sodium compound.

* a process in which a crystalline hydrate looses water forming a powdery deposit on the crystals.

a Sodium carbonate, Na_2CO_3 This is made in the Solvay process (p. 109) and it exists in the anhydrous state (*soda ash*), as a *monohydrate*, $Na_2CO_3.H_2O$, and more commonly as the decahydrate, $Na_2CO_3.10H_2O$ (*washing soda*). The decahydrate effloresces in air to the monohydrate. The anhydrous salt is made by heating the hydrates or by decomposing sodium hydrogencarbonate. Sodium carbonate melts at 852 °C without decomposition, but there may be some evolution of carbon dioxide if it is heated in silica containing vessels, due to reaction between the silica and the hot carbonate.

As sodium carbonate is a salt of a strong base and a weak acid, it is hydrolysed (p. 209) to give an alkaline solution,

$$CO_3^{2-}(aq) + H_2O(l) \longrightarrow HCO_3^-(aq) + OH^-(aq)$$

which is used in volumetric analysis. Sodium carbonate is used industrially in making glass, as a water softener, in detergents, and in the manufacture of sodium hydroxide, soap and paper.

b Sodium hydrogencarbonate, $NaHCO_3$ This is a white crystalline solid made by the Solvay process (p. 109). It decomposes on heating,

$$2NaHCO_3(s) \longrightarrow Na_2CO_3(s) + H_2O(l) + CO_2(g)$$

and though it is an acid salt, gives an alkaline aqueous solution because of hydrolysis.

Sodium hydrogencarbonate is used as an ingredient of baking powder and, medicinally, as an antacid. — *used for acidic stomach.*

c Potassium carbonate, K_2CO_3 (potash or pearl ash) This salt cannot be made by the Solvay process in the same way as sodium carbonate because potassium hydrogencarbonate is much more soluble in water than sodium hydrogencarbonate. The process can, however, be adapted to make potassium carbonate by using alcoholic solutions. Alternatively, potassium hydroxide can be neutralised by carbon dioxide and the resulting potassium hydrogencarbonate heated.

Potassium carbonate is used in making potassium cyanide and potassium chromate(VI).

8 The nitrates

The nitrates of the alkali metals show marked thermal stability as compared with other metallic nitrates. They melt without decomposition, but, excepting lithium nitrate, they decompose to the nitrite and oxygen on stronger heating. *Their thermal stability increases on moving down a group.*

a Sodium nitrate, $NaNO_3$ (Chile saltpetre) This salt occurs as the main component of caliche in Chile, and it is extracted by hot water from that source. It is also formed when nitrogen dioxide is passed into sodium hydroxide solution: *ore material of $CaCO_3$*

Na^+OH^-

$$2NaOH(aq) + NO_2(g) \longrightarrow NaNO_2(aq) + NaNO_3(aq) + H_2O(l)$$

It melts at 310 °C and decomposes above that temperature:

$$2NaNO_3(s) \longrightarrow 2NaNO_2(s) + O_2(g)$$

It is used as a nitrogenous fertiliser and in making potassium nitrate.

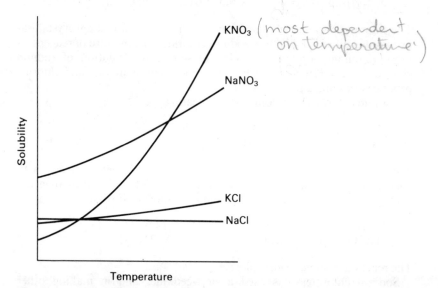

KNO₃ *(most dependent on temperature)*

Solubility

NaNO₃

KCl

NaCl

Temperature

Fig. 71 *Preparation of potassium nitrate.*

b Potassium nitrate, KNO$_3$ (nitre) Hot, saturated solutions of sodium nitrate and potassium chloride react together to give a mixture containing K$^+$, Na$^+$, Cl$^-$ and NO$_3^-$ ions. At the temperature used, sodium chloride is the least soluble salt present, so it crystallises out. On cooling, potassium nitrate becomes the least soluble product, so it crystallises out. The solubilities are shown in Fig. 71.

Potassium nitrate is used in making black powder (gunpowder), which is a mixture of potassium nitrate, sulphur and carbon, and in other explosive mixtures. It is better for this purpose than sodium nitrate because it is not deliquescent. and as a fertilizer.

9 Other compounds of sodium

Stopped here

a Disodiumtetraborate-10-water, Na$_2$B$_4$O$_7$.10H$_2$O (borax) This occurs naturally and is mainly used in making glasses, as a mild antiseptic and as a volumetric standard. It reacts with acids to form boric acid, e.g.

$$Na_2B_4O_7(s) + 2HCl(aq) + 5H_2O(l) \longrightarrow 4H_3BO_3(aq) + 2NaCl(aq)$$

and is used in making other boron compounds (p. 194).

b Sodium chlorate(V), NaClO$_3$ This is formed, together with sodium chloride, by treating hot, concentrated sodium hydroxide solution with chlorine:

$$3Cl_2(g) + 6NaOH(aq) \longrightarrow NaClO_3(aq) + 5NaCl(aq) + 3H_2O(l)$$

The chlorate(V) can be obtained from the mixture by fractional crystallisation as it is more soluble than sodium chloride. The mixture of reagents is most commonly obtained by electrolysing a hot solution of sodium chloride in a cell in such a way that the sodium hydroxide and chlorine produced can mix and react together.

Sodium chlorate(V) decomposes on heating,

$$4NaClO_3(s) \longrightarrow 3NaClO_4(s) + NaCl(s)$$
$$3NaClO_4(s) \longrightarrow 3NaCl(s) + 6O_2(g)$$

and, in the molten state, it acts as a strong oxidising agent. It reacts with concentrated sulphuric acid to liberate chloric(V) acid, which then decomposes further into chloric(VII) acid and chlorine dioxide:

$$NaClO_3(s) + H_2SO_4(aq) \longrightarrow HClO_3(aq) + NaHSO_4(aq)$$
$$3HClO_3(aq) \longrightarrow HClO_4(aq) + 2ClO_2(g) + H_2O(l)$$

The reaction is liable to be explosive.

Sodium chlorate(V) is used as a weedkiller and in making other chlorates(V) and chlorates(VII).

c Sodium cyanide, NaCN Sodium cyanide is made by reaction between ammonia, sodium and carbon at 350 °C, or by neutralisation of hydrogen cyanide by sodium hydroxide.

It is a white crystalline solid with the sodium chloride structure, and is used in the extraction of gold and silver in electroplating industries, in case hardening of steels, and as a poison.

d Sodium nitrite, NaNO₂ The absorption of nitrogen dioxide by sodium hydroxide solution, which produces sodium nitrate, also serves as the main source of sodium nitrite. It is used in diazotisation in the dyestuffs industry, and as a corrosion inhibitor.

e Sodium phosphate(V) Reaction of phosphoric(V) acid, H_3PO_4, with the theoretical amount of sodium carbonate or sodium hydroxide produces disodium hydrogenphosphate(V), $Na_2HPO_4.12H_2O$, and this is commonly known as sodium phosphate(V). Sodium dihydrogenphosphate(V), NaH_2PO_4, and trisodium phosphate(V), Na_3PO_4, can be made by further treatment with phosphoric(V) acid or sodium hydroxide respectively.

Disodium hydrogenphosphate(V) loses water on heating, and forms sodium heptaoxodiphosphate(V):

$$2Na_2HPO_4(s) \longrightarrow Na_4P_2O_7(s) + H_2O(l)$$

On melting sodium dihydrogenphosphate(V) and then rapidly cooling it, sodium polytrioxophosphate(V), $(NaPO_3)_n$, is formed.

Phosphates(V) are used in making detergents, as water softeners, and in baking powders, self raising flour, fertilisers and tonic medicines.

f Sodium potassium 2, 3-dihydroxybutanedioate (tartrate), $NaKC_4H_4O_6.4H_2O$ (Rochelle salt) A colourless crystalline solid used medicinally, in making Fehling's solution and in silvering mirrors.

g Sodium silicates Fusion of sodium carbonate and silica at 1300 °C gives a variety of products in which the Na_2O/SiO_2 ratio varies considerably, depending on the original mixture used. If the Na_2O/SiO_2 ratio is 1:1, the product is Na_2SiO_3 which is called sodium silicate:

$$Na_2CO_3(s) + SiO_2(s) \longrightarrow Na_2SiO_3(s) + CO_2(g)$$

Silicate mixtures with a Na_2O/SiO_2 ratio between about 2 and 4 are known as water glass; they are glass-like mixtures that are soluble in water.

Silicate mixtures are used in making silica gel, adhesives and detergents, and in treating stonework.

h Sodium sulphate, Na_2SO_4 This is made by heating sodium chloride and concentrated sulphuric acid at a high temperature; sodium hydrogensulphate is first formed:

$$NaCl(s) + H_2SO_4(aq) \longrightarrow NaHSO_4(aq) + HCl(g)$$
$$NaHSO_4(aq) + NaCl(aq) \longrightarrow Na_2SO_4(aq) + HCl(g)$$

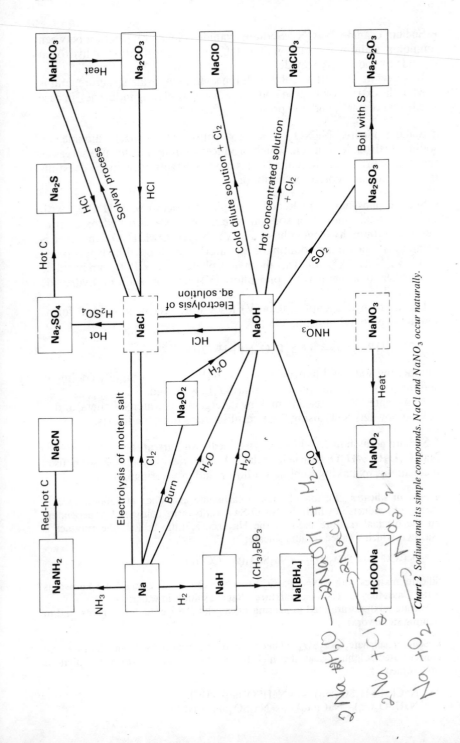

Chart 2 Sodium and its simple compounds. NaCl and NaNO₃ occur naturally.

It is known in the anhydrous form (salt-cake), as a heptahydrate, and as the decahydrate, $Na_2SO_4.10H_2O$ (Glauber's salt).

It can be reduced to sodium sulphide (used as a depilatory) by heating with carbon, and it is used as a purgative and in the manufacture of wood pulp, glass and detergents.

i Sodium sulphite, Na_2SO_3 This is mainly obtained by reaction between sodium carbonate and sulphur dioxide:

$$Na_2CO_3(aq) + SO_2(g) + H_2O(l) \longrightarrow Na_2SO_3(aq) + H_2CO_3(aq)$$

It is a colourless crystalline compound used in the paper, brewing and textile industries, and in making sodium thiosulphate.

j Sodium thiosulphate, $Na_2S_2O_3$ The commonest form of this salt is the pentahydrate, $Na_2S_2O_3.5H_2O$ (*hypo*). It is made by heating sodium sulphite solution with sulphur under slight pressure:

$$Na_2SO_3(aq) + S(s) \longrightarrow Na_2S_2O_3(aq)$$

It is widely used in photography because it will dissolve silver halides, and in volumetric analysis (p. 91) because it will reduce iodine:

$$AgX(aq) + 2S_2O_3{}^{2-}(aq) \longrightarrow [Ag(S_2O_3)_2]^{3-}(aq) + X^-(aq)$$
$$2Na_2S_2O_3(aq) + I_2(aq) \longrightarrow 2NaI(aq) + Na_2S_4O_6(aq)$$

10 Other compounds of potassium

a Potassium chlorate(V), $KClO_3$ This is prepared by treating calcium or sodium chlorates(V) with potassium chloride. It decomposes on heating in the same way as sodium chlorate(V).

b Potassium cyanide, KCN Made by heating a mixture of potassium carbonate and carbon with ammonia, or by reaction between potassium hydroxide and hydrogen cyanide. It is a very poisonous substance that is very soluble in water. It liberates hydrogen cyanide with acids and, slowly, with water.

c Potassium dichromate(VI) and potassium chromate(VI) See p. 361.

d Potassium ethanedioates (oxalates) Neutralisation of ethanedioic (oxalic) acid by potassium hydroxide or potassium carbonate produces either the normal salt potassium ethanedioate, $K_2C_2O_4.H_2O$, or the acid salt, potassium hydrogen ethanedioate, KHC_2O_4. The latter decomposes in cold solution to deposit crystals of potassium quadroxalate or tetroxalate, $KHC_2O_4.H_2C_2O_4.2H_2O$ (*salts of sorrel* or *salts of lemon*). It is used to remove iron and ink stains, and as a reducing agent in volumetric analysis.

e Potassium manganate(VII) See p. 366.

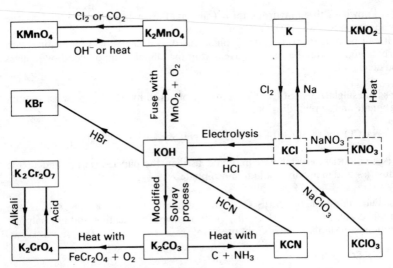

Chart 3 *Potassium and its simple compounds. KCl and KNO₃ occur naturally.*

Questions on Chapter 13

1 Outline the properties you would expect francium to have.

2 List (a) the first ionisation energies and (b) the standard electrode potentials of the alkali metals. Comment.

3 How do sodium and potassium and their compounds differ?

4 In what ways is lithium unlike the other alkali metals?

5 Taking common salt as the electrolyte, illustrate the statement that 'different products can be obtained from the same electrolyte by carrying out the electrolytic process under different conditions'.

6 Na_2CO_3 and Na_2SO_3 are very similar formulae. In what ways do sodium carbonate and sodium sulphite (a) resemble, and (b) differ from, each other?

7 Compare the properties of an alkali metal with a halogen from the point of view of (a) the appearance, (b) the reducing or oxidising power, (c) the nature of the hydrides, (d) the nature of the hydroxides, (e) the nature of the oxides, (f) the existence of salts of oxoacids, (g) the electrical conductivity, (h) the boiling point.

8 How, and under what conditions, does sodium hydroxide react with other elements?

9 'The electrolysis of aqueous sodium chloride involves oxidation, reduction, and the displacement of ionic equilibrium.' Carefully explain this statement. Outline how this process under varying conditions has been adapted for the manufacture of (a) sodium hydroxide, (b) sodium chlorate(I), (c) sodium chlorate(V). (L)

10 (a) How are the existence of families and periods in the classification of the elements explained in terms of atomic structure? (b) Rubidium is next to potassium in Group 1 of the periodic table. From your knowledge of the chemistry of potassium and sodium and their compounds, discuss the properties you would expect to find in rubidium, and in its hydride, hydroxide, carbonate and chloride. (O)

11 (a) Explain why lithium differs from the other alkali metals in some respects and give *two* examples of the differences. (b) What type of chemical compounds constitutes the major source of alkali metals? Suggest a reason why the alkali metals occur in these forms. (c) Indicate by means of diagrams and equations how sodium carbonate is manufactured from a naturally occurring sodium compound. State *two* large scale uses of sodium carbonate. (OC)

12 How does sodium resemble and how does it differ from a metal such as iron?

13 To what extent would you expect (a) the NH_4^+ ion and (b) the Tl^+ ion to be like the K^+ ion?

14 Compare and contrast the reactions of the different alkali metals with (a) oxygen, (b) hydrogen, (c) chlorine, (d) water.

15* Examine the proposition that, for salts of strong acids, the lithium salt is usually the most soluble of the alkali metal salts, whereas it is the least soluble for salts of weak acids.

16* Examine the solubilities of some Group 1A compounds and draw what conclusions you can about the trends in passing from lithium to caesium.

14
The alkaline earths (Group 2A)

1 The elements

The six elements in Group 2A resemble each other in much the same way as the alkali metals of Group 1A. Beryllium is the odd one out, and radium is radioactive.

Magnesium, calcium, strontium and barium are known as the alkaline earths. All four metals are reactive (though less so than the alkali metals), with barium being the most and magnesium the least reactive. Thus, calcium, strontium and barium react with cold water, whereas magnesium will only react with hot water or steam. The metals combine directly with oxygen, sulphur, nitrogen and all the halogens, and all but magnesium also combine directly with hydrogen. The metals also react readily with dilute acids to form salts, but they will not react with alkalis.

The metals are widely distributed in nature, particularly as the carbonates and sulphates, and the high stability of the compounds means that electrolytic methods of extraction are necessary.

a Magnesium Magnesium occurs as carnallite, $MgCl_2.KCl.6H_2O$, magnesite, $MgCO_3$, and dolomite, $MgCO_3.CaCO_3$. It is extracted in a variety of ways. Fused carnallite can be electrolysed, or a mixture of chlorides for electrolysis can be obtained by first heating the carbonate ores and then heating the resulting oxides with carbon and chlorine, e.g.

$$MgCO_3(s) \longrightarrow MgO(s) + CO_2(g)$$
$$MgO(s) + C(s) + Cl_2(g) \longrightarrow MgCl_2(s) + CO(g)$$

Alternatively, the magnesium chloride in sea water can be precipitated as magnesium hydroxide by adding slaked lime:

$$MgCl_2(aq) + Ca(OH)_2(aq) \longrightarrow Mg(OH)_2(s) + CaCl_2(aq)$$

Heating the magnesium hydroxide gives magnesium oxide from which magnesium chloride can be obtained as above.

Magnesium is a silver–white metal of low density mainly used in making light alloys, e.g. magnalium, Duralumin and Electron. It is also used, mixed with potassium chlorate(V) or barium peroxide, in making flash powders.

b Calcium Many calcium compounds, e.g. the carbonate, sulphate, phosphate(V), silicate and halides, occur naturally. The metal is obtained by electrolysis of a fused mixture of calcium chloride and calcium fluoride, the latter lowering the melting point.

Calcium is a soft, silvery, malleable metal. It is very reactive but is not used extensively.

c Barium Occurs as heavy spar or barytes, $BaSO_4$, and as witherite, $BaCO_3$. The carbonate is converted into the chloride by reaction with hydrochloric acid, and barium is obtained by electrolysis of the fused chloride.

It is a silvery-white, soft metal.

2 Compounds of the alkaline earths

The main features of the chemistry of the alkaline earths are similar in kind, but different in degree, to those of the alkali metals (p. 166).

a Ionic character The alkaline earth metals have a fixed oxidation state of $+2$, and their compounds are mainly stable ionic solids, which are colourless unless a coloured anion is present. The ionic nature of the compounds is shown by the generally good agreement between calculated and measured lattice enthalpies, but there are some discrepancies when the compounds concerned have layer lattice structures, e.g. $MgCl_2$, $MgBr_2$, MgI_2 and CaI_2 (p. 57).

The metals are electropositive, readily forming M^{2+} ions. The ionisation energies to form such ions (Be > Mg > Ca > Sr > Ba) are in the expected order of increasing ionic size (p. 155). The order of standard electrode potential is reversed (Ba > Sr > Ca > Mg > Be), though barium, strontium and calcium have almost equal values.

b Hydration of ions The smallest M^{2+} ion is the most easily hydrated, and many magnesium, but few barium, salts are hydrated.

c Formation of covalent compounds The compounds of beryllium (p. 162) are mainly covalent, and magnesium has the greatest tendency towards covalency of the alkaline earths; barium has the least. The Grignard reagents, e.g. CH_3MgI and C_6H_5MgBr, are typical covalent compounds of magnesium.

Compounds of Magnesium with

d Solubility The compounds of the alkaline earths have much lower general solubilities than those of the alkali metals. The fluorides, sulphates, formula hydroxides and carbonates, for example, are all insoluble apart from BeF_2, $BeSO_4$, $MgSO_4$ and $Ba(OH)_2$.

R Mg X

organometalic

e Thermal stability The compounds have high negative enthalpies of formation, and only the hydrides split up into the elements on heating, with magnesium hydride being the least stable.

Partial thermal decomposition is common; much commoner than for the alkali metal compounds. The carbonates, hydrogencarbonates, hydroxides, nitrates and peroxides all decompose on heating. The barium compounds require the highest temperatures; magnesium compounds the lowest.

f Basic strength of oxides and hydroxides   The oxides and hydroxides are weaker bases than those of the alkali metals, the strength increasing in passing down the group.

g Flame coloration   Calcium (red), strontium (crimson) and barium (green) give distinctive flame colorations (p. 162).

3 Beryllium

Beryllium differs from magnesium, calcium, strontium and barium by forming compounds with greater covalent character (p. 162) and by forming more complexes. In the main, too, the compounds of beryllium are less thermally stable; they are also toxic. Beryllium is, in fact, very like aluminium (p. 138), and similar to zinc.

Beryllium oxide, BeO, is a covalent, amphoteric substance with properties like aluminium and zinc oxides; the peroxide, BeO_2, does not exist. Beryllium chloride, $BeCl_2$, has a low conductivity when molten but is predominantly covalent; it is soluble in organic solvents and is readily hydrolysed. Typical examples of beryllium complexes are $[BeF_4]^{2-}$ and $[Be(NH_3)_4]^{2+}$.

4 The oxides

All the metals form both an oxide and a peroxide. The oxides, formed on heating the hydroxides, nitrates or carbonates, are white powders with very high melting points; they are not easily reduced and are strongly basic. Like the dimetal oxides of the alkali metals, they are ionic compounds which crystallise with the sodium chloride structure.

The peroxides are formed either by the direct heating of the oxide (for barium) or by adding hydrogen peroxide to a solution of a salt (for strontium, calcium and magnesium). The peroxides all form hydrates (magnesium peroxide is, in fact, only known as a hydrate), all decompose on heating into the oxide and oxygen, and all react with dilute acids to give hydrogen peroxide. They are ionic compounds containing the peroxide ion, $(O—O)^{2-}$, and with structures like that of calcium dicarbide (p. 184) but with the dicarbide ions replaced by peroxide ions.

Hyperoxides (p. 274) are not formed.
superoxides

a Uses of the oxides   Magnesium oxide is used as a refractory for furnace linings and as an antacid. Calcium oxide (quicklime) is used as a drying agent, in building and agriculture, and in making refractory bricks. Barium peroxide is used as a bleaching agent.

5 The hydroxides

The hydroxides, which are ionic compounds containing the OH^- ion, are formed by reaction between the oxides and water. Barium oxide reacts most readily, with the evolution of much heat; magnesium oxide reacts least readily. Alternatively, precipitation methods can be used, because the alkaline earth hydroxides are white, sparingly soluble solids. They decompose into the oxide and water on heating.

Barium hydroxide is the strongest base, but it is weak compared with the strongest alkali metal hydroxides.

a Calcium hydroxide, $Ca(OH)_2$ (slaked lime)   This is made from the theoretical amounts of calcium oxide and water:

$$CaO(s) + H_2O(l) \longrightarrow Ca(OH)_2(s)$$

It is used in removing acid gases in industrial processes, in treating acid (sour) soils, in water softening, and in making bleaching powder (p. 322). It is also used as a suspension in water (milk of lime) or as a solution (limewater).

6 The halides

All the dihalides of magnesium, calcium, strontium and barium are known. They are white, soluble compounds with the exception of the sparingly soluble fluorides and they almost all form well defined hydrates. The halides are predominantly ionic, with magnesium halides showing the greatest tendency towards covalency. Thus, magnesium chloride is hydrolysed to a basic salt on boiling with water, and the chloride, bromide and iodide of magnesium all dissolve in organic solvents, form complexes with ethers, aldehydes and ketones, and have layer lattice structures.

a Magnesium chloride, $MgCl_2$ This is obtained from carnallite or by reaction between magnesium oxide and hydrochloric acid. The hexahydrate, which is very deliquescent, is the best known form, but there are other hydrates. Heating any of the hydrates produces a basic chloride. Anhydrous magnesium chloride has, therefore, to be obtained by heating the hydrate in a stream of hydrogen chloride, or by heating the oxide with carbon and chlorine.

b Calcium chloride, $CaCl_2$ This is obtained as a cheap by-product of the Solvay process, or by reaction between calcium carbonate or calcium oxide and hydrochloric acid. The hydrate crystallising from solution is $CaCl_2.6H_2O$ but this gives a dihydrate on heating at $200\,°C$ and the anhydrous salt on stronger heating. Anhydrous calcium chloride is a useful and cheap drying agent, but it cannot be used to dry alcohols or ammonia as it forms compounds, e.g. $CaCl_2.4CH_3OH$ and $CaCl_2.8NH_3$, with them.

c Barium chloride, $BaCl_2$ This is made by treating either barium carbonate or barium sulphide with hydrochloric acid. It is a white, crystalline dihydrate, $BaCl_2.2H_2O$, used in the analysis of sulphates and for making precipitated barium sulphate.

d Calcium fluoride, CaF_2 (fluorspar)   This occurs naturally as fluorspar or fluorite. It is used to lower the melting point of metallic halides in metallurgical processes, as a flux, and as a source of fluorine and other fluorides.

7 The carbonates

These occur naturally, and calcium carbonate is particularly widely distributed. They decompose into the oxide and carbon dioxide on heating and are insoluble in water. They dissolve, however, in carbonic acid to form hydrogencarbonates which are known only in solution and which decompose on heating into the carbonate, carbon dioxide and water. Magnesium and calcium hydrogencarbonates occur in some hard waters causing temporary hardness.

a Magnesium carbonate, $MgCO_3$ (magnesite)   A white, insoluble solid obtainable by reaction between sodium hydrogencarbonate solution and a solution of a soluble magnesium salt, i.e.

$$Mg^{2+}(aq) + 2HCO_3^-(aq) \longrightarrow MgCO_3(s) + CO_2(g) + H_2O(l)$$

The product obtained on adding sodium carbonate to a similar Mg^{2+} solution is a basic carbonate of variable composition but approximating to the formula $MgCO_3.Mg(OH)_2.4H_2O$. These basic carbonates are used as antacids and mild laxatives, e.g. milk of magnesia.

b Calcium carbonate, $CaCO_3$ (chalk, marble and limestone)   A very widespread chemical; other ores include calcite, aragonite, Iceland spar and dolomite ($MgCO_3.CaCO_3$).

The compound is dimorphic, with a hexagonal calcite structure as the stable form at ordinary temperatures and pressures, and a rarer rhombic, aragonite structure. Calcite is anisotropic (p. 57), having different thermal conductivities and refractive indices in different directions.

Limestone is used in making sodium carbonate (in the Solvay process), glass, steel, cement (p. 186) and quicklime; when finely ground it is also used for treating heavy acid soils. Chalk is used in making dentifrices, putty, pigments, glass and paper. Blackboard chalk is not calcium carbonate, but mainly calcium sulphate; French chalk is magnesium silicate. Marble and limestone are used for building.

8 The sulphates

The sulphates, like the carbonates, occur naturally. They are thermally stable, ionic solids, showing a marked decline in solubility in passing from magnesium to barium. Barium sulphate is very insoluble ($2.4\,\text{mg\,dm}^{-3}$ at $20\,°C$).

a Magnesium sulphate, $MgSO_4$ This occurs naturally as kieserite, $MgSO_4.H_2O$, or can be made by reaction between magnesium or its oxide, hydroxide or carbonate, and sulphuric acid. The heptahydrate, $MgSO_4.7H_2O$ (Epsom salt), is probably the best known form; it is used as a mild purgative.

b Calcium sulphate, $CaSO_4$ (anhydrite) This occurs naturally in the anhydrous form as anhydrite, or the dihydrate form as gypsum, selenite or alabaster. It is a sparingly soluble solid, causing permanent hardness in water.

On heating the dihydrate at about $100\,°C$, some water of crystallisation is lost and plaster of Paris, calcium sulphate-$\frac{1}{2}$-water, is formed:

$$2CaSO_4.2H_2O(s) \longrightarrow (CaSO_4)_2.H_2O(s) + 3H_2O(l)$$

At a higher temperature, all the water of crystallisation is driven off and the anhydrous salt remains.

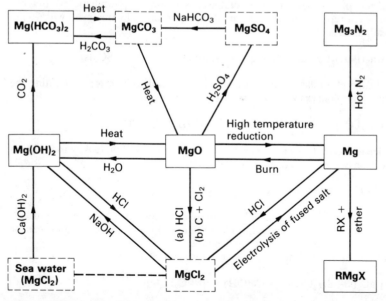

Chart 4 Magnesium and its simple compounds. $MgCl_2$, $MgCO_3$ and $MgSO_4$ occur naturally.

Plaster of Paris sets solid, with slight expansion, on mixing with water and standing. It is widely used for making plaster casts and moulds. Plaster board, which is used in building, is also made from plaster of Paris.

c Barium sulphate, $BaSO_4$ (heavy spar and barytes)   Occurs naturally or is manufactured by adding sulphuric acid to a solution of barium chloride. The natural ores are used in making most barium compounds, generally after an initial reduction of the sulphate to sulphide by heating with carbon:

$$BaSO_4(s) + 4C(s) \longrightarrow BaS(s) + 4CO(g)$$

This initial reduction is necessary because barium sulphate is so very insoluble; hence its use in qualitative and quantitative analysis for barium and sulphate ions.

Precipitated barium sulphate is used as a filler for paper and rubber, and in making paints.

9 Other compounds of calcium

a Calcium phosphates(V)   Calcium phosphate(V), $Ca_3(PO_4)_2$, occurs as rock phosphate or fluoroapatite, $CaF_2.3Ca_3(PO_4)_2$, and is the main source of phosphorus. The phosphate(V) can be made by precipitation on mixing solutions of disodium hydrogen phosphate(V) and calcium chloride. It is a white, insoluble solid.

Calcium phosphate(V) is too insoluble to be of use as a phosphatic fertiliser, but it can be converted into the widely used superphosphate by treatment with 70 per cent sulphuric acid:

$$Ca_3(PO_4)_2(s) + 2H_2SO_4(aq) \longrightarrow Ca(H_2PO_4)_2(s) + 2CaSO_4(s)$$

Triple superphosphate, containing no calcium sulphate is made by treating the calcium phosphate(V) with phosphoric(V) acid:

$$Ca_3(PO_4)_2(aq) + 4H_3PO_4(aq) \longrightarrow 3Ca(H_2PO_4)_2(s)$$

b Calcium dicarbide, CaC_2   This is made by heating carbon (coke) and calcium oxide in an electric furnace at about 2000 °C:

$$CaO(s) + 3C(s) \longrightarrow CaC_2(s) + CO(g)$$

The process produces greyish-black lumps of impure dicarbide, but colourless, pure crystals, with a sodium chloride structure of Ca^{2+} and C_2^{2-} ions, can be made by heating calcium hydride and ethyne.

Calcium dicarbide reacts with water to form ethyne,

$$CaC_2(s) + 2H_2O(l) \longrightarrow Ca(OH)_2(aq) + C_2H_2(g)$$

and with hot nitrogen, at 1000 °C, to form calcium cyanamide.

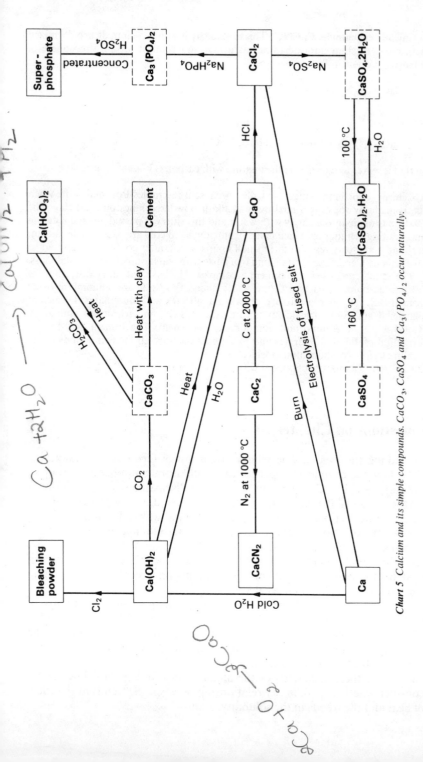

Chart 5 Calcium and its simple compounds. CaCO₃, CaSO₄ and Ca₃(PO₄)₂ occur naturally.

$Ca + 2H_2O \longrightarrow Ca(OH)_2 + H_2$

$2CaO + O_2 \rightarrow 2CaO$

c Calcium cyanamide, CaCN$_2$ This is made by heating calcium dicarbide, mixed with some calcium fluoride to act as a flux and catalyst, in an atmosphere of nitrogen at about 1000 °C:

$$CaC_2(s) + N_2(g) \longrightarrow Ca^{2+}NCN^{2-}(s) + C(s)$$

It is slowly hydrolysed by water,

$$CaCN_2(s) + 3H_2O(l) \longrightarrow 2NH_3(g) + CaCO_3(s)$$

so that it serves as a fertiliser; the mixture with carbon is known as nitrolime.

d Cement   Portland cement, so called because it resembles Portland stone when set, is made by roasting a mixture of chalk or powdered limestone and clay. The mixture is made into a slurry with water and this slurry is passed down a rotating and slanting cylinder up which flames from powdered coal are passed. The calcium carbonate and clay react to form small lumps which, after cooling, are ground to give cement. Cement sets, on mixing to a paste with water, into a tough, solid mass.

The actual composition of cement is complex. The basic calcium oxide, from the calcium carbonate, reacts with the acidic silicon(IV) oxide and aluminium oxide from the clay to give a mixture of calcium silicates and aluminates. Dicalcium silicate, $(CaO)_2SiO_2$, tricalcium silicate, $(CaO)_3SiO_2$, and tricalcium aluminate, $(CaO)_3Al_2O_3$, are important components of cement. The cement sets by the formation of different hydrates and by interaction between the compounds in the mixture to form calcium aluminosilicates.

Concrete is a mixture of cement with sand and gravel.

Questions on Chapter 14

1 Taking the elements beryllium to barium of Group 2A as examples, show how the chemical properties vary with atomic number in a group of the periodic table. Confine your answer to a discussion of the properties of (a) the elements, (b) the oxides, (c) the chlorides, (d) the sulphates. (OC)

2 What deductions can you make concerning (a) beryllium hydride, (b) beryllium oxide, (c) radium sulphate and (d) radium hydroxide from your knowledge of the alkaline earths?

3 What similarities and what differences are there between the effect of heat on the compounds of the alkali metals and on the corresponding compounds of the alkaline earths?

4 Discuss the part played by (a) lattice enthalpy and (b) enthalpy of hydration in solubility considerations.

5 Compare and contrast the values of the enthalpies of formation of the fluorides, hydrides and oxides of the alkali metals with those of the same compounds of the alkaline earths (refer to p. 159).

6 Taking the compounds of the alkali metals and the alkaline earths, consider whether there is any relationship between the trends in the effect of heat and the trends in the solubility.

7 What is meant by (a) a period, (b) a group, in the periodic table? The number of elements in a period can be 2, 8, 18 or 32; explain briefly why this is so. From a consideration of its properties and those of its compounds, justify the position of barium in the periodic table. (C)

8 Compare and contrast the properties of the alkali and alkaline earth metals and their more important compounds.

9 Compare the nature of the naturally occurring compounds of the Group 1A and Group 2A metals.

10 What part do compounds of calcium and magnesium play in the hardness of water?

11 Explain the meaning of the terms 'first electron affinity', 'first ionisation energy' and 'electronegativity'. Discuss the trends in first ionisation energy and electronegativity in Groups 1A and 2A of the periodic table. In each case, relate your answer to the chemical properties of the elements. Mention and explain three other trends which occur in one or both of these periodic groups. Discuss briefly the relationships which exist between the two groups of elements. (L)

12 (a) How is calcium metal obtained? (Technical details are not required.) Why is this the preferred method? (b) Describe the chemical properties of (i) calcium oxide, (ii) calcium hydride, in each case giving reasons for the stated properties. (c) Some calcium compounds are used industrially on a large scale. Select any *one* such compound and describe in outline how and why it is used in industry. (d) Calcium does not form any stable compounds in which it has an oxidation state of $+3$. Briefly discuss the reasons for this. (OC)

13* Draw up a table of some of the important physical properties of the Group 1A and Group 2A metals and comment on any points of significance.

15
Boron and aluminium

1 Extraction of aluminium

Aluminium is extracted from bauxite, which is often formulated as $Al_2O_3.2H_2O$ but which is really a mixture of the mono- and trihydrates, $Al_2O_3.H_2O$ and $Al_2O_3.3H_2O$. The bauxite is treated to give anhydrous aluminium oxide and a solution of this in molten cryolite, $Na_3[AlF_6]$, together with aluminium fluoride, AlF_3, is electrolysed to give aluminium at the cathode and oxygen at the anode (p 116).

The bauxite ore is first powdered, and then heated with sodium hydroxide solution under pressure; the aluminium oxide reacts, because of its amphoteric nature, to form a complex salt which is often called, in an over-simplified way, an aluminate (p 116):

$$Al_2O_3(s) + 2NaOH(aq) + 7H_2O(l) \longrightarrow 2Na[Al(OH)_4(H_2O)_2] (aq)$$

Impurities in the ore, notably iron(III) oxide, can be filtered off as they do not react with sodium hydroxide. The hot solution is agitated by passing compressed air in, and 'seed crystals' of very finely divided $Al_2O_3.3H_2O$ are added. After some time, and on cooling, most of the aluminium present precipitates as the hydrated oxide. That hydrate is filtered off and heated, at around 1000 °C, to form anhydrous aluminium oxide.

Naturally occurring cryolite was originally used in the process but the natural supply is now inadequate, so that cryolite has to be synthesised by reaction between aluminium oxide, hydrofluoric acid and sodium hydroxide.

A typical electrolytic cell or pot (Fig. 72) is made of steel lined with a thermal insulation and a carbon cathode made from baked metallurgical

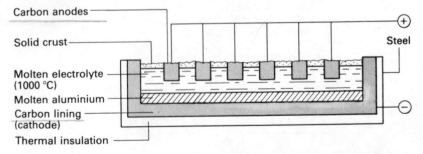

Fig. 72 *An electrolytic cell or pot for making aluminium. The cell is about 2 m × 3 m × 6 m in size and the anodes are long blocks about 0.5 m square in cross-section. To achieve the very high current required, the gap between the anode and cathode is much smaller (about 5 cm) than indicated in the diagram. A modern smelter has well over a thousand cells in operation.*

coke, pitch and tar; the anodes are made from petroleum coke and pitch. The oxygen produced during electrolysis 'burns' the anodes away so that they have to be lowered into the cell as necessary and replaced periodically. The aluminium is molten at the temperature of the cell; it collects at the bottom and is syphoned off from time to time. Fresh aluminium oxide and fluoride are added through the solid crust on top of the molten electrolyte.

The process uses a lot of electricity, each kilogram of aluminium requiring about 13 kW h in a modern cell. It is necessary to pass very high currents, and this is achieved by keeping the gap between the anodes and the cathode very small. Aluminium smelters can only be operated economically where cheap electricity is available, so most of the world's plants are fed by hydroelectric schemes.

2 Properties and uses of aluminium

The metal is reactive, but it is usually coated with a tenacious layer of oxide that greatly lowers its chemical reactivity. Thus 'ordinary' aluminium will not react with air or oxygen or steam. But if the oxide layer is broken down by amalgamation or by sodium chloride, the aluminium reacts vigorously with air and with steam to form aluminium oxide.

Many acids and alkalis also break down the oxide layer and react with the aluminium to liberate hydrogen, e.g.

$$2Al(s) + 6HCl(aq) \longrightarrow 2AlCl_3(aq) + 3H_2(g)$$

$$2Al(s) + 2NaOH(aq) + 10H_2O(l)$$
$$\longrightarrow 2Na[Al(OH)_4(H_2O)_2](aq) + 3H_2(g)$$

but hot oxidising acids, particularly nitric acid, build up the oxide layer and render aluminium passive. Hot aluminium also reacts directly with chlorine, sulphur, nitrogen, phosphorus and carbon.

The facility with which aluminium will combine with oxygen to form an oxide,

$$4Al(s) + 3O_2(g) \longrightarrow 2Al_2O_3(s) \qquad \Delta H_m^{\ominus}(298\,K) = -3340\,kJ\,mol^{-1}$$

means that aluminium can be used to reduce other metallic oxides such as Fe_2O_3, Cr_2O_3, Mn_3O_4 and Co_3O_4 (p. 114).

Aluminium is used in making electric cables, and pots and pans. The oxide layer can be thickened by making aluminium the anode in a bath of sulphuric acid. The oxygen produced at the anode builds up the layer in a process known as *anodising*. The oxide layer can be coloured by many dyes, and anodised aluminium is used in making many small articles and for decorative work in building. Most aluminium is used, however, for making light alloys, e.g. magnalium and Duralumin (p. 178).

3 Extraction and properties of boron

Boron is made, in an impure amorphous form, by reducing boron oxide with hot magnesium:

$$B_2O_3(s) + 3Mg(s) \longrightarrow 3MgO(s) + 2B(s)$$

In the amorphous form, it is a dark powder. As the pure element, which is difficult to make, it is a hard, black solid that melts at about 2300 °C and has a structure made up of B_{12} icosahedra.

It is less reactive than aluminium, but forms B_2O_3, B_2S_3 and boron trihalides on heating with oxygen, sulphur and the halogens, respectively. It also reacts at high temperatures with carbon and with nitrogen or ammonia. With carbon, a carbide, BC, that is similar to silicon carbide, is formed; with nitrogen or ammonia, the product is a white, slippery compound, BN, which has a layer structure like that of graphite, but with alternating boron and nitrogen atoms within the hexagonal rings.

4 Compounds of boron and aluminium

Boron is a non-metal: it does not form B^{3+} ions and all its binary compounds are covalent. It is, in some ways, like silicon (p. 138). On the other hand, aluminium is a metal which does form trivalent ions; but many of its compounds are predominantly covalent, and it resembles beryllium in some ways (p. 138).

The non-metallic nature of boron shows in its ability to form borides, e.g. TiB_2, with metals; in the formation of two series of hydrides, B_nH_{n+4} and B_nH_{n+6}; in the formation of an acidic oxide, B_2O_3, and a hydroxide, $B(OH)_3$ which is really boric acid, H_3BO_3; and in the formation of easily hydrolysed halides.

Aluminium forms no compounds with metals and only one simple hydride; its oxide and hydroxide are amphoteric; and its halides are either resistant to hydrolysis (AlF_3) or only partially hydrolysed.

a Covalent compounds   Both boron and aluminium exhibit an oxidation state of $+3$ and those compounds which are covalent do show some likenesses.

As the elements only have three valency electrons, the formation of three single covalent bonds provides only six electrons around the central atom, and the resulting compounds, e.g. BF_3 and $AlCl_3$, are electron deficient. As a result, they are very good electron acceptors, function as Lewis acids (see section **5b** below) and form complexes, e.g. $Li[AlH_4]$, $Na[BH_4]$, $[BF_4]^-$ and $[AlF_6]^{3-}$, the last two illustrating the fact that the maximum covalency of boron is 4 whilst that of aluminium is 6.

Alternatively, the compounds may exist as polymers, e.g. $(AlH_3)_x$ and Al_2Cl_6.

b Ionic compounds Anhydrous Al^{3+} ions occur in anhydrous AlF_3 or anhydrous $Al(ClO_4)_3$, and the hydrated ion, $[Al(H_2O)_6]^{3+}$, is found in many hydrates, e.g. $Al_2(SO_4)_3.18H_2O$ and $Al(NO_3)_3.9H_2O$, and in aqueous solution. Neither B^{3+} nor any hydrated ions are known.

Ionisation energy values show that an input of $6887\,kJ\,mol^{-1}$ is required to convert $B(g)$ into $B^{3+}(g)$. The corresponding value for Al is still high $(5140\,kJ\,mol^{-1})$, but the $Al^{3+}(g)$ ion is small enough to have a high polarising effect (p. 16) so that it is readily hydrated with a high enthalpy of hydration, and a consequent release of energy:

$$Al^{3+}(g)+6H_2O(l)\longrightarrow Al(H_2O)_6{}^{3+}(aq) \quad \Delta H_m^{\ominus} = -4680\,kJ\,mol^{-1}$$

It is this high enthalpy of hydration which accounts (p. 17) for the high negative electrode potential of aluminium $(-1.67\,V)$, which is close to that of beryllium $(-1.70\,V)$.

The small size of the Al^{3+} ion also means that ionic compounds of aluminium have high lattice enthalpies; as would be expected, it is the fluoride that is the most ionic.

c The $[Al(H_2O)_6]^{3+}$ ion The high polarising effect of the Al^{3+} ion and its tendency to form a strong bond with oxygen weakens the O—H bond of the water molecules in the $[Al(H_2O)_6]^{3+}$ ion. So much so, in fact, that the hydrated ion dissociates and acts as an acid:

$$[Al(H_2O)_6]^{3+} + H_2O \longrightarrow [Al(OH)(H_2O)_5]^{2+} + H_3O^+$$
$$[Al(OH)(H_2O)_5]^{2+} + H_2O \longrightarrow [Al(OH)_2(H_2O)_4]^+ + H_3O^+$$
$$[Al(OH)_2(H_2O)_4]^+ + H_2O \longrightarrow Al(OH)_3(H_2O)_3 + H_3O^+$$

The pK_a value for $[Al(H_2O)_6]^{3+}$ is 4.9, making it comparable in strength with ethanoic acid ($pK_a = 4.76$).

The final dissociation, resulting in the formation of hydrated aluminium hydroxide will only take place if basic anions are present to lower the H_3O^+ ion concentration; OH^-, $CO_3{}^{2-}$ and S^{2-} ions all serve this purpose. As a result, addition of $CO_3{}^{2-}$ or S^{2-} ions to a solution of aluminium salts precipitates the hydrated hydroxide and not the carbonate or sulphide.

Other hydrated trivalent ions, e.g. $[Fe(H_2O)_6]^{3+}$, act in the same way as $[Al(H_2O)_6]^{3+}$, so that they are also acidic.

5 Halides of boron

The chloride, bromide and iodide of boron can all be made by direct combination between the elements at a high temperature. They have trigonal planar MX_3 molecules, and they fume in moist air because of rapid hydrolysis (p. 316) to form boric and hydrochloric acids, e.g.

$$BCl_3(g)+3H_2O(l)\longrightarrow H_3BO_3(aq)+3HCl(aq)$$

a Boron trifluoride, BF₃ This is the most important halide of boron for it is one of the strongest Lewis acids known and, as such, it is widely used as a catalyst in organic reactions. It is a pungent colourless gas manufactured by heating boron oxide with calcium fluoride and concentrated sulphuric acid:

$$B_2O_3(s) + 3CaF_2(s) + 3H_2SO_4(aq)$$
$$\longrightarrow 3CaSO_4(s) + 3H_2O(l) + 2BF_3(g)$$

Unlike the other halides, it is only partially hydrolysed; the products are boric acid and a strong acid known as tetrafluoroboric(III) acid, H[BF₄].

$$4BF_3(g) + 3H_2O(l) \longrightarrow 3H[BF_4](aq) + H_3BO_3(aq)$$

b Lewis acids and bases Lewis defined an acid as a substance that can accept a pair of electrons, and a base as a substance that can donate a pair of electrons. This extended the idea of acids and bases, e.g.

acid	base	

$$\text{H}_3\text{O}^+ \ + \ \text{OH}^- \ \longrightarrow \ 2\text{H}_2\text{O}$$

$$\begin{array}{ccccc} & \text{F} & & \text{H} & & & \text{F} & \text{H} \\ & | & & | & & & | & | \\ \text{F}-\text{B} & + & {}^\times_\times\text{N}-\text{H} & \longrightarrow & \text{F}-\text{B}\leftarrow\text{N}-\text{H} \\ & | & & | & & & | & | \\ & \text{F} & & \text{H} & & & \text{F} & \text{H} \end{array}$$

$$\text{AlCl}_3 \ + \ 4\text{Cl}^- \ \longrightarrow \ [\text{AlCl}_4]^-$$

$$\text{Fe}^{3+} \ + \ 6\text{CN}^- \ \longrightarrow \ [\text{Fe(CN)}_6]^{3-}$$

When used in organic chemistry, strong Lewis acids, such as BF₃ and AlCl₃, facilitate the formation of electrophilic carbonium ions, e.g.

$$\text{RCl} + \text{AlCl}_3 \longrightarrow \ \underset{\substack{\text{carbonium} \\ \text{ion}}}{\text{R}^+} \ + [\text{AlCl}_4]^-$$

6 Halides of aluminium

Hot aluminium combines with all the halogens to form trihalides. The fluoride, AlF₃, is ionic and does not react with water or dissolve in organic solvents. The other halides are mainly covalent, and are partially hydrolysed by water and soluble in organic solvents. Aluminium chloride is the most important; it is used in organic chemistry in much the same way as BF₃.

a Aluminium chloride, AlCl₃ Treatment of aluminium with hydrochloric acid gives a solution which deposits crystals of AlCl₃.6H₂O, but these are hydrolysed on heating and cannot be used for making the anhydrous salt. This is made by passing dry chlorine over hot or molten aluminium, or industrially, by heating aluminium oxide with carbon and chlorine:

$$2Al(s) + 3Cl_2(g) \longrightarrow 2AlCl_3(s)$$
$$Al_2O_3(s) + 3C(s) + 3Cl_2(g) \longrightarrow 2AlCl_3(s) + 3CO(g)$$

The anhydrous salt is a white crystalline, deliquescent solid with a structure containing layers of $AlCl_6$ units, the layers being linked together by van der Waals' forces. The solid sublimes at 180 °C but density measurements show that it exists as a dimer, Al_2Cl_6 (p. 25), in the vapour state up to about 400 °C. Above this temperature the dimer begins to dissociate into $AlCl_3$ molecules, the process being complete at about 800 °C (Fig. 73).

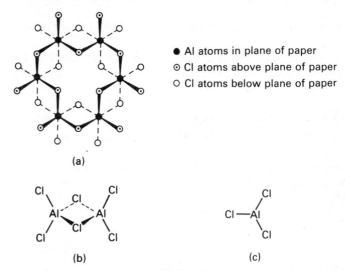

● Al atoms in plane of paper
⊙ Cl atoms above plane of paper
○ Cl atoms below plane of paper

(a)

(b)   (c)

*Fig. 73 The structure of aluminium chloride. (**a**) The general arrangement of $AlCl_6$ units in a layer of the layer lattice structure of anhydrous solid $AlCl_3$; each aluminium atom is surrounded by six chlorine atoms and each chlorine atom is linked to two aluminium atoms. (**b**) At about 180 °C the solid sublimes, giving a vapour containing Al_2Cl_6 molecules; each atom of aluminium is surrounded almost tetrahedrally by four chlorine atoms. (**c**) Above about 400 °C, the Al_2Cl_6 dimer splits up into trigonal planar $AlCl_3$ molecules.*

The Al_2Cl_6 dimer is also formed when aluminium chloride is dissolved in some non-polar solvents, e.g. benzene. But when a non-polar solvent has lone pairs and can act as an electron donor, the aluminium chloride acts as a Lewis acid (p. 192), forming a complex with the solvent by accepting the lone pair. In ether solution for example, aluminium chloride exists as a tetrahedral $R_2O \longrightarrow AlCl_3$ complex.

Anhydrous aluminium chloride reacts with water,

$$AlCl_3(s) + 6H_2O(l) \longrightarrow [Al(H_2O)_6]^{3+} + 3Cl^-(aq)$$

the energy necessary to break the Al—Cl bonds being provided by the high enthalpy of hydration of the Al^{3+} ion (p. 18). The hydrated ions then react further with water, making the solution acidic.

7 Oxoacids and oxides of boron

a Oxoacids   Boron forms a number of oxoacids, with boric acid, H_3BO_3, being the commonest; it can be regarded as the trihydroxide of boron, $B(OH)_3$. It is made by treating a hot solution of the naturally occurring *borax*, disodiumtetraborate-10-water, $Na_2B_4O_7.10H_2O$, with acids, e.g.

$$Na_2B_4O_7(aq) + 2HCl(aq) + 5H_2O(l) \longrightarrow 4H_3BO_3(aq) + 2NaCl(aq)$$

White crystals of H_3BO_3, with a hydrogen bonded structure (p. 55), form on cooling.

It is a weak acid, used as a mild antiseptic in eye lotions and lint. On heating, dehydration takes place, either to boron oxide or to tetraboric acid, $H_2B_4O_7$ (of which borax is a salt):

$$2H_3BO_3(s) \longrightarrow B_2O_3(s) + 3H_2O(l)$$
$$4H_3BO_3(s) \longrightarrow H_2B_4O_7(s) + 5H_2O(l)$$

b Boron oxide, B_2O_3   This is a weakly acidic oxide made by strong heating of boric acid; it forms as a glass-like, white solid. It reacts with metallic oxides to form borates, and these are present, together with silicates (p. 218), in hard glasses such as Pyrex.

8 Aluminium oxide, Al_2O_3 (alumina)

This occurs naturally as bauxite, corundum and white sapphire; with other metallic oxides as impurities it also occurs as ruby, amethyst, blue sapphire and emerald. It can be made by heating aluminium hydroxide or many other aluminium compounds. It is a white refractory material that is almost insoluble in water but soluble in acids and alkalis because it is amphoteric. The reaction with acids and alkalis is affected, however, by the initial heat treatment and by the fact that the oxide exists in more than one form.

Aluminium oxide is used in making other aluminium compounds, furnace linings, and as a catalyst or catalyst support and a packing in chromatographic columns. Corundum and impure corundum (emery) are also used as abrasives, and fusion of aluminium oxide with other metallic oxides can be used to make synthetic gemstones.

9 Aluminium hydroxide, $Al(OH)_3$

This is precipitated in a hydrated form as a white, gelatinous mass on adding ammonia solution or a little sodium hydroxide solution to a solution of an aluminium salt. Its composition is variable and its properties change on standing or heating. In one form it is insoluble in both acids and alkalis, but it exists more commonly as an amphoteric hydroxide with a layer lattice structure, similar to that of $AlCl_3$, in which $Al(OH)_6$ units in the layers are linked by hydrogen bonds. In its reactions with acids it acts as a base, i.e.

$$Al(OH)_3(s) + 3H^+(aq) \longrightarrow Al^{3+}(aq) + 3H_2O(l)$$

forming aluminium salts. In its reactions with bases it acts as an acid, forming 'aluminate' ions:

$$Al(OH)_3(s) + OH^-(aq) + 2H_2O(l) \longrightarrow [Al(OH)_4(H_2O)_2]^-(aq)$$

The systematic name for the ion is diaquatetrahydroxoaluminate(III). The ion would exist as $[AlO_2]^-$ when fully dehydrated; this latter ion is the real aluminate ion, and it exists in fused aluminates.

10 Aluminium sulphate, $Al_2(SO_4)_3$

This occurs naturally but is usually manufactured by reacting aluminium oxide with hot concentrated sulphuric acid.

It is a white crystalline solid that is reasonably soluble in water and which decomposes into aluminium oxide, sulphur(VI) oxide and sulphur dioxide on heating. It forms double salts (alums) with other sulphates, and is used as a source of Al^{3+} ions in water and sewage treatment, and in paper making.

11 Alums

Alums are double sulphates, with a general formula $M^I M^{III}(SO_4)_2.12H_2O$ or $M^I_2SO_4 M^{III}_2(SO_4)_3.24H_2O$, where M^I may be Na^+, K^+, Rb^+, Cs^+, Tl^+ or NH_4^+ and M^{III} includes Al^{3+}, Cr^{3+}, Fe^{3+}, Mn^{3+} and Co^{3+}. They can all be made by mixing equimolar quantities of solutions of the two sulphates concerned. The alums crystallise out, each cation being associated with six molecules of water of crystallisation (p. 17).

Potash alum or aluminium potassium sulphate-12-water, and ammonium alum or aluminium ammonium sulphate-12-water, are the commonest alums:

$KAl(SO_4)_2.12H_2O$ $NH_4Al(SO_4)_2.12H_2O$

potash alum   ammonium alum

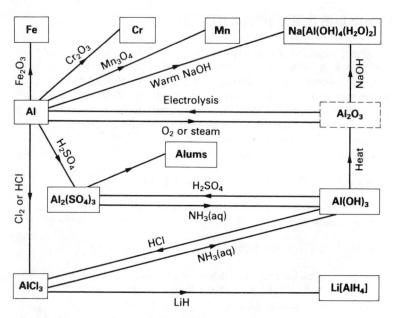

Chart 6 *Aluminium and its simple compounds. Al_2O_3 occurs naturally.*

Potash alum is used in dyeing, as a mordant. A fabric to be dyed is dipped into a solution of alum and then steam heated. Hydrolysis of the potash alum produces aluminium hydroxide within the fibres and certain dyestuffs are absorbed by the hydroxide to form coloured lakes.

12 Aluminosilicates

Silicates (p. 218) contain tetrahedral SiO_4^{4-} groups arranged in chains, layers or three-dimensional structures. In aluminosilicates, some of the SiO_4^{4-} groups are replaced by tetrahedral AlO_4^{5-} groups, the cation groups being changed to maintain electrical neutrality. Such aluminosilicates are found in micas, feldspars, zeolites and some clays.

Questions on Chapter 15

1 Give an account of the sources and extraction of aluminium and describe the principal chemical and physical features of its oxide, hydroxide, chloride and sulphate. (OC)

2 Explain a method by which the relative molecular mass of aluminium chloride could be measured at 450 °C.

3 Iron(III) chloride and aluminium chloride have superficially similar formulae, $FeCl_3$ and $AlCl_3$. Describe and explain how they differ.

4 Describe the extraction of aluminium metal from purified bauxite.

How and under what conditions does aluminium react with (a) hydrochloric acid, (b) nitric acid, (c) sodium hydroxide, (d) iodine and (e) iron(III) oxide? (JMB)

5 What predictions would you make about the chemistry of gallium and its simple compounds?

6 Compare and contrast the properties of magnesium chloride, aluminium chloride and phosphorus trichloride.

7 Give an account of the complexes of aluminium.

8 'The chemistry of boron is dictated by its small size, its electron configuration and the limitation of its coordination number to a maximum value of four.' Discuss this statement in the light of the chemistry of boron and in comparison with that of aluminium. (JMB)

9 Outline the process whereby pure aluminium oxide is obtained from bauxite, and the principles of the extraction of the metal by electrolysis. Give an account of the uses of the metal, emphasising its advantages and disadvantages in each case. Give the structure of aluminium chloride. Describe, giving essential experimental details, a method for obtaining a pure specimen of anhydrous aluminium chloride from aluminium. (L)

10 Comment on *four* of the following statements.

(a) Anhydrous aluminium chloride cannot be prepared by the evaporation of an aqueous solution of aluminium chloride. (b) Anhydrous aluminium chloride is used as a catalyst in certain organic reactions. (c) Aluminium metal is prepared by an electrolytic method and not by heating aluminium oxide with coke in a blast furnace. (d) The compound borazine, $B_3N_3H_6$, possesses physical properties which closely resemble those of benzene, C_6H_6. (e) Boric acid, H_3BO_3, ionises very slightly in water, to give the ions H_3O^+ and $B(OH)_4^-$. (O)

11 Describe some important uses of aluminium, and relate them to the physical and/or chemical properties of the metal.

12 Compare and contrast (a) $[BH_4]^-$ and $[AlH_4]^-$, (b) B_2O_3 and Al_2O_3, (c) AlF_3 and BF_3.

13 Why is it that boron has some likenesses to silicon, and aluminium to beryllium? List any four similarities in each case.

14 The standard molar enthalpies of formation of the oxides of aluminium, iron(III) and chromium(III) are -1669, -822 and $-1128\,kJ\,mol^{-1}$ respectively. Calculate the enthalpy changes in the following reactions.

$$2Al(s) + Fe_2O_3(s) \longrightarrow Al_2O_3(s) + 2Fe(s)$$
$$2Al(s) + Cr_2O_3(s) \longrightarrow Al_2O_3(s) + 2Cr(s)$$

Comment on the values obtained.

15* Use enthalpies of formation to compare the enthalpies of combustion of $C_2H_6(g)$ and $B_2H_6(g)$. Would diborane be suitable for use as a fuel?

16* Use standard enthalpies of formation to find the enthalpy changes in the reactions between aluminium and sodium oxide and between magnesium and aluminium oxide. Comment on the results.

*In all compounds C forms
the strongest bonds. that is
Thermal stability decreases
down a group.

16
Group 4B and carbon

1 General trends

The major, simple trend in the Group 4B elements is the marked change from the non-metallic carbon and silicon, through the semi-metal germanium, to metallic tin and lead. This accounts for changes in the structures and physical properties of the elements; changes in the thermal stability of the hydrides (from the stable methane to the very unstable PbH_4); changes from the acidic oxides, CO_2 and SiO_2, to the amphoteric oxides of tin and lead; and a general increase in the stability of ionic compounds in passing from carbon to lead. But, as summarised below, many other factors have to be taken into account.

a Structure of the elements   Carbon (as diamond), tin (as grey tin), and germanium and silicon (which do not have allotropes) all exist as atomic or covalent crystals (p. 43), in which each atom is tetrahedrally surrounded by four others, as in diamond. The C—C bond in diamond is strong ($346 \, kJ \, mol^{-1}$) but the bond becomes less and less strong in passing down the group, until it is only about $152 \, kJ \, mol^{-1}$ in grey tin. It is one of the main features of carbon that it forms strong bonds, both with itself and with many other elements.

C also has a graphite form, with a layer structure (p. 202); the other allotropes of tin have metallic crystals; and lead, with only one form, also has metallic crystals.

b Oxidation states   All the Group 4B elements exhibit oxidation states of -4 and $+4$, and some of them have oxidation states of $+2$. They form many compounds and ions with similar formulae, as shown in the summary on p. 199. But such a summary masks the marked differences between many of the compounds or ions.

c Formation of ions   The compounds of the Group 4B elements are predominantly covalent, particularly those in the -4 or $+4$ oxidation states. But ionic character enters in more and more on passing from carbon to lead, as would be expected from the increase in metallic character and the decrease in electronegativity.

C^{4-} ions occur in some carbides (p. 211), but there are no M^{4+} or M^{2+} ions of carbon or silicon. M^{4+} ions occur in the MO_2 oxides of germanium, tin and lead, and perhaps in some M^{4+} salts. Simple Ge^{2+} and Sn^{2+} ions may occur in some salts; and Pb^{2+} ions are found in the MX_2 halides (excepting the iodide) and in some Pb^{2+} salts. Both $M^{4+}(aq)$ and $M^{2+}(aq)$ ions are so extensively hydrolysed that they have no real existence, except for $Pb^{2+}(aq)$ (p. 228).

Some oxidation states of Group 4B elements (uncertain or unstable species are shown in brackets)

-4	$+2$		$+4$	
CH_4	CO		CO_2	$CO_3{}^{2-}$
			CS_2	
			CCl_4	
SiH_4	(SiO)		SiO_2	$SiO_4{}^{4-}$
			SiS_2	
	(SiF_2)		$SiCl_4$	$[SiF_6]^{2-}$
GeH_4	(Ge^{2+})		Ge^{4+}	$[Ge(OH)_6]^{2-}$
	GeO		GeO_2	
	GeS		GeS_2	
	$GeCl_2$		$GeCl_4$	$[GeCl_6]^{2-}$
SnH_4	(Sn^{2+})	$[Sn(OH)_6]^{4-}$	Sn^{4+}	$[Sn(OH)_6]^{2-}$
	SnO		SnO_2	
	SnS		SnS_2	
	$SnCl_2$	$[SnCl_4]^{2-}$	$SnCl_4$	$[SnCl_6]^{2-}$
(PbH_4)	Pb^{2+}	$[Pb(OH)_6]^{4-}$	Pb^{4+}	$[Pb(OH)_6]^{2-}$
	PbO		PbO_2	
	PbS			
	$PbCl_2$	$[PbCl_4]^{2-}$	$PbCl_4$	$[PbCl_6]^{2-}$

Silicon forms complex ions in the $+4$ oxidation state, and germanium, tin and lead form complexes in both the $+4$ and $+2$ oxidation states.

d Formation of single covalent bonds   All the elements form covalent MX_4 compounds, where X is a halogen or a hydrogen atom; the molecules are tetrahedral in shape.

In the hydrides, there is a marked drop in bond energy from $412\,kJ\,mol^{-1}$ for the C—H bond to an estimated $253\,kJ\,mol^{-1}$ for Sn—H. This is reflected in a marked decrease in thermal stability in passing from CH_4 to SnH_4; and PbH_4 is so unstable that its existence is open to doubt.

In the chlorides there is, again, a general decrease in bond energy in passing down the group, but the C—Cl bond is weaker than would be expected. CCl_4 is, however, resistant to hydrolysis because of the low covalency maximum for carbon, and $GeCl_4$ is surprisingly resistant, too. The other MCl_4 compounds are readily hydrolysed.

The SiO_4 grouping, containing four strong Si—O bonds, plays a large part in the chemistry of silicon, but there is no similar CO_4 grouping.

e Maximum covalency   The maximum covalency (p. 133) for carbon is 4 but for silicon, with available 3d orbitals, it is 6. This means that silicon compounds are more 'open to attack' than carbon compounds. $SiCl_4$ for example, is so readily hydrolysed that it fumes in moist air, whereas CCl_4

because they have available d orbitals.

will not react with sodium hydroxide solution. It is for the same reason that the silanes, Si_xH_{2x+2}, unlike the alkanes, are spontaneously flammable in air, and readily hydrolysed by sodium hydroxide solution.

Moreover, the higher maximum covalency for elements beyond carbon allows the formation of ions such as $[SiF_6]^{2-}$, $[GeCl_6]^{2-}$ and $[Pb(OH)_6]^{2-}$ (p. 225).

f Formation of chains and rings   Carbon atoms can link together in extremely long chains, i.e. the element has a high degree of *catenation*; hence the very large number of aliphatic homologous series of organic compounds. The chain lengths in silicon and germanium compounds are limited to about 6 and 9 respectively; and catenation of tin and lead occurs only in some organo-compounds.

The ability of carbon to form such long, stable chains is accounted for by the strength of the C—C bond, and by the low maximum covalency of carbon, which means that the carbon chains are not 'open to attack'. *more stable*

Carbon atoms can also link together in rings containing C—C bonds, so that there are large numbers of alicyclic organic compounds. There are no similar compounds for the other elements.

g Formation of multiple bonds   Carbon can form a number of strong multiple bonds, e.g. C=C, C≡C, C=O, C=S and C≡N, but the other elements do not bond in these ways. That is why, for example, carbon dioxide, with molecules containing C=O bonds is a gas, whereas silicon dioxide, containing Si—O bonds, is a solid. It is also why there is such an extensive carbocyclic and heterocyclic chemistry of carbon, with no counterpart for silicon or any of the other Group 4B elements.

h Oxidation state of +2   There is a very marked increase in the stability of the +2 oxidation state, compared with the +4 state, in passing from carbon to lead.

It is, indeed, doubtful whether carbon ever really exists in the +2 state. CO appears to show carbon in the +2 state, but it is only formally so, as the carbon is not simply forming two bonds, i.e. it is not divalent. Better examples of carbon in the +2 state are found in the *carbenes*, :CF_2 and :CR_2, but these only exist as transient intermediates. Likewise, SiO and SiF_2 are very unstable.

The +2 state is, however, much more common in the chemistry of germanium, tin and lead. For germanium and tin, the +4 state is the more stable, but for lead it is the +2 state. The reason for the existence of these +2 compounds is not fully understood.

i Summary   Carbon differs from the other elements in having a maximum covalency of 4, in its ability to form strong covalent bonds with a number of different elements, in its degree of catenation, and in its ability to form ring compounds and multiple bonds. Other general trends within the group are summarised below, with M standing for the elements.

C	Si	Ge	Sn	Pb

increase in   metallic character
                basic nature of oxides *and hydroxides*
                ionic character of bonding
                stability of +2 oxidation state

decrease in   thermal stability of hydrides
                M—M bond energy
                M—H bond energy
                M—Cl bond energy (except C—Cl)
                degree of catenation

Both carbon and silicon are in many ways unique. To keep the discussions manageable, the detailed chemistry of carbon is dealt with in this chapter; silicon is dealt with in Chapter 17 (p. 213); and germanium, tin and lead are dealt with together, in Chapter 18 (p. 222).

2 Diamond and graphite

a Occurrence and manufacture   Diamonds are mined mainly in South Africa, but graphite is distributed much more widely; in the United Kingdom it occurs in Cumberland.

The natural supply of graphite cannot meet the full demand, and it is manufactured by the Acheson process in which an electric current is passed through a mixture of petroleum coke and silica for 24–30 hours. The process requires a lot of electric power and is carried out at Niagara Falls. Graphite is, thermodynamically, slightly stabler than diamond and, theoretically, it might be expected that diamonds would change into graphite:

$$C(\text{graphite}) \longrightarrow C(\text{diamond}) \qquad \Delta H_m^{\ominus}(298\,\text{K}) = 1.883\,\text{kJ}\,\text{mol}^{-1}$$

The change is, however, so slow under normal conditions as to be negligible. The reverse change can, however, be made by subjecting graphite to very high pressures and temperatures, with or without a transition metal as a catalyst. Small industrial diamonds are made commercially in this way, but the synthetic gemstones cannot compete with the natural product yet.

b Structure   Diamond has covalently linked carbon atoms arranged tetrahedrally (Fig. 26, p. 42). The arrangement is regular, and the atoms are strongly bound by sp^3 bonds; it is sometimes described as a giant molecule or giant lattice. The C—C bond distance is 0.154 nm, which is the same as that found in aliphatic hydrocarbons.

In the stablest form of graphite, layers of hexagonally-linked carbon atoms are arranged above and below each other in an ABABA... layer

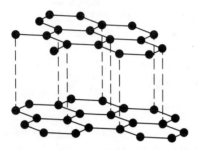

Fig. 74 Graphite.

lattice (Fig. 74). The bond distance between carbon atoms in the hexagons is 0.142 nm, which is almost equal to the C—C bond distance in benzene. The bonding within the hexagons is sp^2 bonding, and the remaining electrons on each carbon atom pair up into a delocalised system of π bonds. The distance between the layers is 0.34 nm, which is approximately twice the van der Waals' radius of carbon. A second form of graphite, with an ABCABC... arrangement of the layers, as compared with the normal ABAB... arrangement, also occurs.

c Physical properties Because of the compactness and strong binding in the structure, diamond is a very hard, non-volatile solid; it is transparent. Graphite is also non-volatile, but it is black and opaque; it is also soft, because the layers can slip over each other relatively easily. The graphite structure is much more 'open' than that of diamond, so that the density of graphite is lower. The delocalised system of π bonds in graphite is associated with the crystal as a whole, and it is these 'mobile' electrons that make graphite an electrical and thermal conductor.

Graphite is also anisotropic, i.e. its physical properties differ depending on the direction in which they are measured. Thus the electrical and thermal conductivities are greatest parallel to the layers whilst the coefficient of expansion is greatest normal to the plane of the layers.

Both diamond and graphite are insoluble in all known solvents.

d Pyrographite If it was possible to make a piece of graphite in which all the layers of the various crystallites were orientated in the same direction, the anisotropy would be particularly marked. This has recently been achieved by the slow decomposition of hydrocarbons at temperatures above 2000 °C. The product, known as pyrographite, has a thermal conductivity 80 times greater, and an electrical conductivity 1200 times greater, in the plane of the layers than across them.

The differential thermal conductivity and high melting temperature make the material very useful for lining rocket nozzles or re-entry shields. By placing the pyrographite in the right direction, it is possible to conduct the heat away from, and not into, the main body.

e Carbon fibres Carbon fibres can be made by carefully heating fibres of materials such as polypropenenitrile until they 'char' to carbon. If such carbon fibres are incorporated into plastics, a very stiff and strong material with a low density can be made.

f Graphitic compounds Graphite forms a number of 'compounds' in which atoms, ions or molecules position themselves between the layers. Liquid potassium, for example, reacts to form a series of compounds with formulae KC_8, KC_{16}, KC_{24} and KC_{40}, depending on the number of spaces between the layers that are occupied by potassium atoms. The electrical conductivity of these compounds is higher than that of graphite. Other species, e.g. Cl_2, HSO_4^- and FeS_2, can also be introduced between the layers.

With strong oxidising agents such as fuming nitric acid and potassium chlorate(V), graphitic oxides, with formulae varying around C_2O, can be made, and graphite reacts with fluorine at about 600 °C to form fluorides, $(CF)_x$. These compounds are non-conducting.

g Uses of diamond and graphite Diamonds are used as jewels but far more are used industrially in oil-well drills, cutting tools, glass cutters, watch bearings, dies and abrasives.

Graphite is used in making pencils; it is mixed with various amounts of clay to give hard or soft pencils. It is also used in making electrodes, graphite crucibles, high temperature dies and plungers, furnace linings, brushes for electric motors, lubricants, e.g. oildag (a colloidal suspension of graphite in oil), and as a neutron moderator in atomic piles.

absorbs neutrons.

3 'Amorphous' carbons

There are a number of forms of so-called amorphous carbon which may vary considerably in purity; they contain microcrystalline forms of graphite.

a Charcoals Charcoal can be made by heating wood, nut shells, bones, sugar or even blood. Wood charcoal is the commonest; animal charcoal, from bones, contains calcium phosphate(V); sugar charcoal is the purest variety.

Charcoal has a porous structure made up of minute graphite crystals. It is a good absorbent, particularly when activated by heating in steam; this cleans and develops the pores. Charcoal finds use in decolorising sugar solutions, in gas masks, in the production of high vacua, and as a support for metal catalysts.

b Carbon blacks These finely divided forms of carbon are made by incomplete combustion of carbonaceous materials. Carbon black itself is made from natural gas or oils. It is mainly used in making rubber tyres, and provides a convenient source of a relatively cheap black powder for making black paints, inks and polishes, e.g. shoe polish.

Lamp black is similar but is obtained from lamp or vegetable oils.

4 Carbon dioxide, CO_2

a Preparation   Carbon dioxide is formed when carbon is burnt in excess air or oxygen; graphite needs a temperature of about 700 °C, and diamond a slightly higher temperature. The gas can be made in the laboratory by reaction of most carbonates with most strong acids, or by heating most carbonates or hydrogencarbonates. Industrially it is obtained as a by-product in heating limestone to make quicklime, in the brewing industry by fermentation of sugars, and in the manufacture of hydrogen (p. 145).

b Properties   Carbon dioxide is a colourless, odourless gas, which is soluble in water and is easily liquefied and solidified by cooling.

Some carbon dioxide is present as the free gas in aqueous solution, but a little of the gas reacts with water to form a solution of carbonic acid, H_2CO_3:

$$CO_2(g) + H_2O(l) \rightleftharpoons H_2CO_3(aq)$$

Carbonic acid is a very weak, dibasic acid,

$$H_2CO_3(aq) \rightleftharpoons H^+(aq) + HCO_3^-(aq) \qquad (pK_a = 6.37)$$
$$HCO_3^-(aq) \rightleftharpoons H^+(aq) + CO_3^{2-}(aq) \qquad (pK_a = 10.33)$$

which is readily decomposed into carbon dioxide and water on heating. Carbon dioxide is absorbed by potassium or barium hydroxide solutions and can be estimated in this way.

Limewater (a solution of calcium hydroxide) is turned milky by carbon dioxide and this serves as a test, though it is not very sensitive. The milkiness is caused by the formation of a precipitate of calcium carbonate. Further carbon dioxide reacts with this insoluble calcium carbonate and converts it into soluble calcium hydrogencarbonate so that the milkiness disappears.

$$Ca(OH)_2(aq) + CO_2(g) \longrightarrow CaCO_3(s) + H_2O(l)$$
$$CaCO_3(s) + H_2O(l) + CO_2(g) \longrightarrow Ca(HCO_3)_2(aq)$$

As the calcium hydrogencarbonate decomposes on heating, to re-form calcium carbonate, the milkiness reappears on heating.

c Uses of carbon dioxide   Carbon dioxide is dissolved in water under pressure to make soda water and other carbonated beverages, and it is used as a solid ('dry ice') as a convenient refrigerant in, for example, ice cream containers. It sublimes into a gas at −78 °C under 1 atm pressure.

As carbon dioxide will not support combustion it is also widely used in fire extinguishers, though very strongly burning substances such as magnesium decompose it and continue to burn in the oxygen released.

In chemical manufacturing it is used in making ammonium sulphate (p. 236) and urea (carbamide).

d Structure of carbon dioxide   The arrangement of electrons in the CO_2 molecule is described on p. 41, and the arrangement of the molecules in the molecular crystal on p. 56.

e Comparison with silicon dioxide   See p. 216.

5 Carbon monoxide, CO

a Preparation   Carbon monoxide is made by reducing carbon dioxide by passing it through red-hot carbon, or by the dehydration of methanoic acid using concentrated sulphuric acid:

$$HCOOH(l) \longrightarrow CO(g) + H_2O(l)$$

The gas can be collected over water.

b Properties   Carbon monoxide is a colourless gas. It is insoluble in water but can be absorbed by an ammoniacal solution of copper(I) chloride (p. 387). It is dangerously poisonous, particularly as it has no smell. It kills by reacting with the haemoglobin in blood to form carboxy-haemoglobin; this prevents the formation of oxyhaemoglobin which normally transfers oxygen to the tissues.

Carbon monoxide burns with a characteristic blue flame to form carbon dioxide.

$$2CO(g) + O_2(g) \longrightarrow 2CO_2(g)$$

In an analogous reaction, carbonyl sulphide, COS, is formed by passing carbon monoxide and sulphur vapour through a hot tube:

$$CO(g) + S(g) \longrightarrow COS(g)$$

Carbon monoxide reacts with chlorine in sunlight or when a mixture of the gases is passed over hot charcoal as a catalyst; the product is carbonyl chloride, $COCl_2$ (phosgene), which is a very poisonous gas:

$$CO(g) + Cl_2(g) \longrightarrow COCl_2(g)$$

Although carbon monoxide is obtained by dehydration of methanoic acid, it will not react with water to re-form the acid, so it is not a true acid anhydride. It will, however, react with a hot solution of sodium hydroxide, particularly under pressure, to form sodium methanoate:

$$CO(g) + NaOH(aq) \longrightarrow HCOONa(s)$$

Carbon monoxide is a strong reducing agent, reducing many metallic oxides to metals. It plays an important part in many metallurgical processes (p. 114). Carbon monoxide also reduces iodine(V) oxide to iodine (p. 318), and this reaction can be used to estimate carbon monoxide by titrating the liberated iodine with standard solutions of sodium thiosulphate (p. 91).

c Metallic carbonyls Carbon monoxide reacts with many transitional metals of Groups 6, 7 and 8 to form metallic carbonyls, e.g. $[Ni(CO)_4]$, $[Fe(CO)_5]$ and $[Cr(CO)_6]$. They are covalent compounds which are volatile and soluble in organic solvents. The arrangement of the ligands is tetrahedral in $[Ni(CO)_4]$, trigonal bipyramidal in $[Fe(CO)_5]$ and octahedral in $[Cr(CO)_6]$. As the central atom does not provide any electrons for the bonding, its oxidation state is zero.

Tetracarbonylnickel(0) is the most useful carbonyl as it plays an important part in the purification of nickel by the Mond process (p. 382). Decomposition of pentacarbonyliron(0) is also used for making pure, finely divided iron.

Higher carbonyls, e.g. $[Fe_2(CO)_9]$, $[Fe_3(CO)_{12}]$ and $[Co_4(CO)_{12}]$, are also known.

d Uses of carbon monoxide The main uses of carbon monoxide, often mixed with hydrogen in the form of synthesis gas (p. 145), are in industrial organic chemistry. Typical reactions are with hydrogen to form methanol (p. 148) and with alkenes to form aldehydes (the oxo process):

$$CO(g) + 2H_2(g) \longrightarrow CH_3OH(l)$$
$$C_2H_4(g) + H_2(g) + CO(g) \longrightarrow C_2H_5CHO(l)$$

e Bonding in carbon monoxide CO and N_2 have the same number of electrons available for bonding; they are said to be *isoelectronic*. It is not surprising, therefore, that the bonding in the two molecules is similar. An over-simplified representation of the CO molecule, in (a) below, shows it containing two covalent and one dative bonds. The better representation in (b) denotes a σ bond and two π bonds between the carbon and oxygen atoms.

(a) C≝O   (b) C⫶O

6 Halides of carbon

Tetrachloromethane, CCl_4 (carbon tetrachloride), is the most important halide of carbon. It is made by the chlorination of methane or by passing chlorine into hot carbon disulphide in the presence of iron(III) chloride as a catalyst:

$$CH_4(g) + 4Cl_2(g) \longrightarrow CCl_4(l) + 4HCl(g)$$
$$CS_2(l) + 2Cl_2(g) \longrightarrow CCl_4(l) + 2S(s)$$

It is a colourless, toxic liquid with a characteristic smell, and is used in making chlorofluorohydrocarbons (p. 310).

Tetrafluoromethane, CF_4, can also be made by fluorinating methane, or by direct reaction between carbon and fluorine. It is a gas, whereas CBr_4 and CI_4 are solids, which are yellow and red respectively. The melting temperatures rise from $-185\,°C$ for CF_4 to $171\,°C$ for CI_4, in line with the increasing relative molecular mass.

The thermal stability of the compounds increases as the C—X bond gets shorter and stronger, so that CI_4 is readily decomposed but CF_4 is extremely stable. $_\wedge-ve\ values$

None of the halides reacts with water or sodium hydroxide solution, even though the expected reactions are thermodynamically feasible. This is attributed to the non-availability of d orbitals on the carbon atom, which cannot, therefore, act as an acceptor for either water molecules or hydroxide ions (p. 216). and it can't donate e's .

7 Hydrides of carbon

The very high degree of catenation of carbon, together with its ability to form single and multiple bonds, and ring compounds, shows up most noticeably in the limitless number of hydrocarbons that exist.

a Methane, CH_4 This occurs very widely as natural gas, but it is also formed when aluminium carbide or beryllium carbide react with water, e.g.

$$Al_4C_3(s) + 12H_2O(l) \longrightarrow 4Al(OH)_3(s) + 3CH_4(g)$$

These two carbides contain slightly distorted C^{4-} ions. Other carbides (p. 211) give other hydrocarbons when they react with water.

Methane, like all other hydrocarbons, will burn in excess oxygen to form carbon dioxide and water; with a limited supply of oxygen, carbon and carbon monoxide may also be formed. The ignition temperature is about $500\,°C$.

Methane is an exothermic compound. It cannot be made by direct combination of carbon and hydrogen, but it will decompose into the elements at about $800\,°C$:

$$CH_4(g) \longrightarrow C(s) + 2H_2(g) \qquad \Delta H^\ominus = 74.9\,kJ\,mol^{-1}$$

Like tetrachloromethane, and for the same reasons (p. 216), it will not react with water or sodium hydroxide solution.

A summary of the properties of CH_4, as compared with SiH_4, is given on p. 218.

8 Carbon disulphide, CS_2

This can be made by heating carbon and sulphur in an electric furnace, but in the modern method, sulphur vapour and methane are heated to a high temperature in the presence of a catalyst:

$$C(s) + 2S(s) \longrightarrow CS_2(l)$$
$$CH_4(g) + 4S(g) \longrightarrow CS_2(l) + 2H_2S(g)$$

The commercial product has a nasty smell due to the presence of impurities; pure carbon disulphide is a colourless liquid with a pleasant smell. The compound is very toxic and highly flammable, and its structure is like that of carbon dioxide (p. 41).

Just as carbon dioxide forms carbonates with bases, so carbon disulphide forms thiocarbonates with sulphides, e.g.

$$CO_2(g) + Na_2O(s) \longrightarrow Na_2CO_3(s)$$
$$CS_2(l) + Na_2S(s) \longrightarrow Na_2CS_3(s)$$

Carbon disulphide is a very good solvent for sulphur, phosphorus, rubber and waxes. Its main uses are in making viscose rayon, cellophane, tetrachloromethane (p. 206), ammonium thiocyanate, thiocarbamide (urea), xanthates and other sulphur containing organic compounds.

9 Carbonates

a Preparation The carbonates of the alkali metals (except lithium) and of NH_4^+ are soluble, but others are insoluble. Many of the insoluble carbonates occur naturally. The soluble ones can be made by saturating a solution of the required alkali with carbon dioxide and then adding a further, equimolar amount of alkali:

$$OH^-(aq) + CO_2(g) \longrightarrow HCO_3^-(aq)$$
$$HCO_3^-(aq) + OH^-(aq) \longrightarrow CO_3^{2-}(aq) + H_2O(l)$$

Carbonates of the alkaline earth metals (except $MgCO_3$) can be made by precipitation reactions between carbonate ions and the required metallic ion, e.g.

$$Ca^{2+}(aq) + CO_3^{2-}(aq) \longrightarrow CaCO_3(s)$$

But $MgCO_3$ and normal carbonates of other divalent or trivalent ions cannot be made in the same way. This is because solutions of carbonates are alkaline (see below), whereas solutions containing ions such as Cu^{2+} and Zn^{2+} are acidic, and those containing Al^{3+} and Fe^{3+} are even more so (p. 191). Addition of a solution of carbonate ions to solutions containing Cu^{2+} or Zn^{2+} gives a precipitate of a *basic* carbonate which is an individual substance containing metal ions together with CO_3^{2-} ions and OH^- ions, but which functions like a mixture of a normal carbonate with

CO$_3^-$ ⟶ OH$^-$ + CO$_2$

Group 4B and carbon [Al(H$_2$O)$_6$]$^{3+}$ ⟶ [Al(H$_2$O)$_5$(OH)]$^{2+}$ 209 4H$^+$
⟶ H$^+$ + Al(H$_2$O)$_4$(OH)$_2$]$^+$

a hydroxide, e.g. $CuCO_3.Cu(OH)_2$. With Al^{3+} or Fe^{3+}, the precipitate is a hydroxide (p. 191).

b Properties Carbonates are ionic, crystalline solids that contain CO_3^{2-} ions. These ions are planar, with bond angles of 120° (p. 42). The detailed crystal structure depends mainly on the size of the cation.

The soluble carbonates give alkaline solutions, owing to hydrolysis:

$$CO_3^{2-}(aq) + H_2O(l) \rightleftharpoons HCO_3^-(aq) + OH^-(aq)$$

Many carbonates react with most strong acids to form carbon dioxide.

Many carbonates will also decompose on heating to give carbon dioxide, their thermal stability depending on the charge and size of the cation (p. 160). Cations of low size and high charge give the most unstable carbonates. So much so, that iron(III) carbonate and aluminium carbonate cannot be made.

Sodium carbonate (p. 170) and calcium carbonate (p. 182) are particularly useful.

10 Hydrogencarbonates

The only hydrogencarbonates that can be obtained as solids are those of the alkali metals (excepting lithium); within the crystals, the HCO_3^- ions are hydrogen bonded (p. 55). They crystallise out from the solutions obtained by saturating a solution of the hydroxide or carbonate of the required cation with carbon dioxide, e.g.

$$NaOH(aq) + CO_2(g) \longrightarrow NaHCO_3(aq)$$
$$Na_2CO_3(aq) + H_2O(l) + CO_2(g) \longrightarrow 2NaHCO_3(aq)$$

Potassium hydrogencarbonate is more soluble than sodium hydrogencarbonate, so its crystallisation from solution is more difficult; that is why potassium carbonate cannot be made by the Solvay process (p. 110).

The hydrogencarbonates decompose at temperatures below 100 °C, so that decomposition takes place on boiling their aqueous solutions, e.g.

$$2NaHCO_3(aq) \longrightarrow Na_2CO_3(aq) + CO_2(g) + H_2O(l)$$

Aqueous solutions of the hydrogencarbonates are less alkaline than those of the normal carbonates, i.e. they contain a lower concentration of OH^- ions. They can, therefore, be used to precipitate normal carbonates of ions such as Cu^{2+}, e.g.

$$Cu^{2+}(aq) + 2HCO_3^-(aq) \longrightarrow CuCO_3(s) + H_2O(l) + CO_2(g)$$

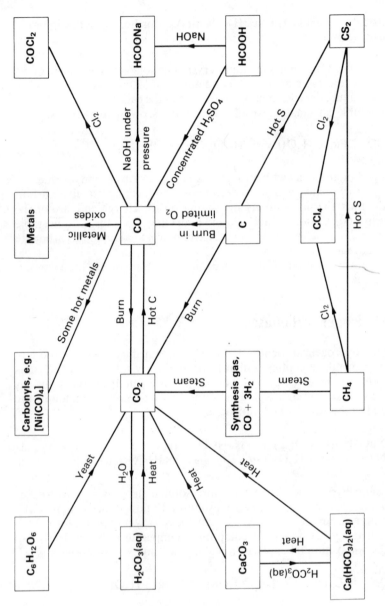

Chart 7 Carbon and its simple compounds.

Hydrogencarbonates of the alkaline earths, e.g. $Ca(HCO_3)_2$ and $Mg(HCO_3)_2$, occur only in solution. They decompose into the carbonates even at room temperature, and play a large part in the hardness of water and the formation of stalactites and stalagmites (p. 268).

11 Carbides

Binary compounds of carbon with elements of lower electronegativity (other than hydrogen) are known as carbides. They are hard solids with high melting points, generally made by heating the element or its oxide with carbon to a very high temperature.

a Ionic carbides The carbides of the alkali metals and calcium, strontium and barium contain dicarbide ions, $[C{\equiv}C]^{2-}$, in sodium chloride type crystals. They give ethyne on hydrolysis. Calcium dicarbide, CaC_2 (p. 184) is typical. The corresponding copper(I) and silver(I) compounds, Cu_2C_2 and Ag_2C_2, can be made by passing ethyne into ammoniacal solutions of Cu^+ and Ag^+ salts. They are coloured solids (reddish and white respectively), which are liable to explode when subjected to heat or mechanical shock.

Magnesium carbide, Mg_2C_3, yields propyne, C_3H_4, amongst its hydrolysis products, suggesting that it contains the C_3^{4-} ion. Aluminium and beryllium carbides, Al_4C_3 and Be_2C, give methane on hydrolysis; they contain distorted C^{4-} ions, but are not fully ionic.

b Covalent carbides Silicon and boron carbides, SiC and B_4C, are typical covalent carbides, and are made by heating the oxides at high temperatures with carbon.

Silicon carbide (carborundum) is used as a refractory, only decomposing above about 2000 °C, and as an abrasive. Boron carbide is even harder than carborundum; it is used in making heat radiation shields.

c Interstitial carbides Close-packed metallic crystals with atoms having radii larger than 0.13 nm have spaces between the atoms into which carbon atoms can fit. The resulting interstitial carbides, made by heating the powdered metal or its oxide with carbon at a high temperature, are metallic in nature but are harder and more brittle than the pure metal because the interstitial carbon atoms lower the possibilities of atoms slipping within the crystal. Tungsten carbide, WC, is typical; it is used for making tools and dies.

Questions on Chapter 16

(See also pp. 220 and 228.)

1 Compare the occurrence, manufacture, structure and uses of graphite and diamond.

2 Give an account of the preparation, properties, structure and uses of carbon dioxide.

3 Compare and contrast the dioxides of carbon, sulphur and nitrogen.

4 Compare and contrast carbon dioxide and carbon disulphide.

5 Comment on the structures of carbon dioxide and carbon monoxide.

6 Describe briefly *one* laboratory method and *one* industrial method for the preparation of carbon monoxide. Why is carbon monoxide poisonous? How and under what conditions does carbon monoxide react with (a) sodium hydroxide, (b) iron(III) chloride, (c) chlorine, (d) nickel, (e) water? (JMB)

7 Give three physical properties which distinguish diamond from graphite. How are these differences related to their crystalline structures? Describe how you could prepare a pure specimen of carbon monoxide in the laboratory. How is this compound used for the purification of nickel? (C)

8 List some important uses of carbonates and hydrogencarbonates.

9 Starting from barium carbonate labelled with ^{14}C, $Ba^{14}CO_3$, outline methods for making (a) $^{14}CH_3OH$, (b) $^{14}CH_4$ and (c) ^{14}CO.

10 Write short notes on (a) carbides, (b) carbonyls, (c) organometallic compounds, (d) carbanions, (e) carbonium ions and (f) halides of carbon.

11* Compare the enthalpy change, the free energy change and the entropy change in the two reactions

$$C(s) + O_2(g) \longrightarrow CO_2(g)$$
$$C(s) + \tfrac{1}{2}O_2(g) \longrightarrow CO(g)$$

Comment on the figures.

12* Use the standard enthalpies of formation to calculate the enthalpy change in the formation of carbon monoxide by reduction of carbon dioxide by carbon. Taking the bond enthalpy of the $C{=}O$ bond in carbon dioxide as $803.3\,kJ\,mol^{-1}$, and the enthalpy of atomisation of graphite as $715\,kJ\,mol^{-1}$, calculate the bond enthalpy in carbon monoxide. What value do you obtain by applying the same method of calculation to the combustion of carbon monoxide into carbon dioxide?

13* Discuss the conditions under which the following reactions will take place: (a) $2C + O_2 \longrightarrow 2CO$, (b) $C + O_2 \longrightarrow CO_2$, (c) $2CO + O_2 \longrightarrow 2CO_2$.

17
Silicon*

[handwritten margin note: more reactive down a group because IE decreases]

1 The element

Silicon is the second most abundant element in the earth's crust (28 per cent). It does not occur as the element itself, but is very common in rocks, which are mainly made up of silicon(IV) oxide, SiO_2 (silica), and silicates (p. 218).

Silicon exists as an amorphous brown powder or in a less reactive crystalline, lustrous form. Both varieties have the diamond structure, the amorphous form being microcrystalline.

a Manufacture Crystalline silicon is manufactured by heating crushed coke and sand in an electric furnace:

$$SiO_2(s) + 2C(s) \longrightarrow Si(s) + 2CO(g)$$

The product is not very pure, but is suitable for use in alloys such as ferro-silicon, which is made by incorporating iron(III) oxide with the coke and sand being heated.

Very pure silicon is required for making silicon chips, and this is manufactured by heating silicon tetrachloride with zinc in an inert atmosphere, or by reducing trichlorosilane with hydrogen at a high temperature:

$$SiCl_4(l) + 2Zn(s) \longrightarrow Si(s) + 2ZnCl_2(s)$$
$$SiHCl_3(l) + H_2(g) \longrightarrow Si(s) + 3HCl(g)$$

The amorphous silicon produced can be purified by zone refining. In this process, heat is applied at one end of a rod of silicon to give a thin band of molten silicon. The rod is then passed through a furnace so that the molten zone moves from one end to the other. As each molten zone re-solidifies, it deposits pure crystals so that the impurities are swept along the rod in the molten zone. The process can also be used for purifying many metals.

Single large crystals of silicon, which are required for silicon chips, are made by withdrawal of a seed crystal, as it grows, from a bath of molten silicon in an inert atmosphere.

b Reactions of silicon Silicon, though not very reactive, is more so than carbon. It reacts with oxygen, sulphur and metals at lower temperatures

---

*The reader is advised to read the summary of the trends in Group 4B (pp. 198–201) before reading this chapter.

than carbon does, and it also reacts with chlorine, nitrogen, sodium hydroxide solution and hydrofluoric acid, which will not react directly with carbon.

c Uses of silicon   Silicon is used in bronze and steel alloys, particularly in spring steels. It is also used in making silicones (p. 220) and, with small amounts of boron or phosphorus incorporated, as a semiconductor.

2 Silicon (IV) oxide, SiO_2 (silica)

This very important constituent of the earth's crust occurs mainly as quartz, but it can be converted into another crystalline form – cristobalite – at high temperatures.

Cristobalite has a structure similar to that of diamond, but with an oxygen atom midway between each pair of silicon atoms. The structure can also be regarded as made up of SiO_4 tetrahedra joined together by the oxygen atoms at the corners of the tetrahedra (Fig. 75). In quartz, the structure still contains SiO_4 tetrahedra, but they are linked in spiral arrangements around an axis. The spiral may be right- or left-handed, giving optically active dextro- and laevo-forms.

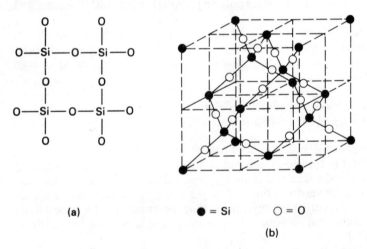

(a)

● = Si   ○ = O

(b)

Fig. 75 *The structure of silicon (IV) oxide. (**a**) A two-dimensional representation of the linkage of SiO_4 units. (**b**) The arrangement in the cristobalite form. The structure is like that of diamond (p. 42) but with silicon instead of carbon atoms, and with oxygen atoms between each pair of silicon atoms.*

Many of the naturally occurring forms of silica are impure. Sand contains iron(III) oxide and other metallic oxides; amethyst is a violet form of quartz; agate is quartz with purple and brown bands; diatomite or kieselguhr is the remains of minute diatoms.

Silica can be made as a finely divided white powder by roasting the precipitate formed on adding sulphuric acid to a solution of sodium silicate.

a Physical properties   All the various forms of silicon(IV) oxide begin to soften at about 1600 °C and are completely molten at 1710 °C. The liquid is viscous but can be shaped in the same way as molten glass. It does not form crystalline silica on cooling; instead it solidifies into silica glass (quartz glass), which has a very low coefficient of expansion and is very transparent to infrared and ultraviolet light.

b Chemical properties   Silicon(IV) oxide is not chemically reactive, but it is attacked by hydrofluoric acid,

$$SiO_2(s) + 4HF(aq) \longrightarrow SiF_4(g) + 2H_2O(l)$$
$$SiF_4(g) + 2HF(aq) \longrightarrow H_2[SiF_6](aq)$$

and by fused alkalis or carbonates. The latter reaction gives a mixture of silicates with different $SiO_2 : Na_2O$ ratios (p. 173), e.g.

$$SiO_2(s) + 2NaOH(l) \longrightarrow Na_2SiO_3(s) + H_2O(l)$$
$$SiO_2(s) + 4NaOH(l) \longrightarrow Na_4SiO_4(s) + 2H_2O(l)$$

c Silica gel   The sodium silicates with Na_2O/SiO_2 ratios between 1.5 and 4 are known as water glasses; they are clear, glass-like compounds that dissolve in water to give viscous solutions. On acidification, a solution of water glass sets into a gel, $SiO_2.xH_2O$, which contains varying amounts of water. The gel is almost fully dehydrated on heating and the product, containing about 5 per cent of water, is still called silica gel even though it is a hard, granular material. It has a large surface area and is a good absorbent.

d Uses of silicon(IV) oxide   Sand is widely used in making mortar, cement, concrete, glass and refractory silica bricks. Fused silica is used in making optical lenses and prisms and other scientific apparatus; it can be drawn out into quartz threads. Articles that must be able to stand extreme thermal shocks are also made of fused silica.

Silica gel is used as a drying and dehydrating agent; it can be regenerated by heating. The cobalt(II) salts that are incorporated in self-indicating silica gel show up pink when the gel has absorbed moisture, and become blue again on heating. Silica gel also serves as a catalyst and as a catalyst support; it is also used in making silicon carbide and, when powdered, as a filler for rubber.

Kieselguhr is used as a medical absorbent both externally and internally, as a filtration medium and as a thermal insulator.

e Comparison with carbon dioxide   The main similarities and differences

[handwritten annotations at top: "C max covalency +4. Si can expand to use their d orbitals."]

between carbon dioxide and silicon(IV) oxide are summarised below.

CO_2	SiO_2
colourless gas	hard, brittle, solid *(due to bonding)*
melts at $-56\,°C$	melts at $1610\,°C$
occurs naturally	occurs naturally *(covalent bonding)*
$\Delta H_{f,m}^{\ominus} = -394\,kJ\,mol^{-1}$	$\Delta H_{f,m}^{\ominus} = -859\,kJ\,mol^{-1}$
molecular crystal	atomic crystal
$C{=}O$ bonds	$Si{-}O$ bonds
acidic oxide	acidic oxide
$+H_2O \longrightarrow H_2CO_3$	no reaction with water
$+NaOH \longrightarrow$ carbonates	$+NaOH \longrightarrow$ silicates
no reaction with HF	$+HF \longrightarrow SiF_4$ or $H_2[SiF_6]$
$+$ Hot C $\longrightarrow$ CO	$+$ Hot C $\longrightarrow$ SiC or Si

[handwritten annotation: "because A'yes B not, Si'yes C not"]

3 Silicon tetrahalides

Silicon forms all four tetrahalides, SiX_4, by direct combination between silicon and the appropriate halogen. They are all readily hydrolysed by water to silicon(IV) oxide, e.g.

$$SiCl_4(l) + 2H_2O(l) \longrightarrow SiO_2(s) + 4HCl(aq)$$
$$\Delta H_m^{\ominus}(298\,K) = -317\,kJ\,mol^{-1}$$

Thermodynamically, the corresponding reaction with tetrachloromethane is equally feasible,

$$CCl_4(l) + 2H_2O(l) \longrightarrow CO_2(g) + 4HCl(aq)$$
$$\Delta H_m^{\ominus}(298\,K) = -353\,kJ\,mol^{-1}$$

but it does not take place. This must be due to kinetic factors, and is probably associated with the fact that the maximum covalency of carbon is 4, whereas that of silicon is 6 (p. 133).

The higher maximum covalency of silicon also allows the formation of $[SiF_6]^{2-}$ ions when silicon tetrafluoride reacts with excess hydrogen fluoride or with F^- ions (p. 314):

$$SiF_4 + 2HF \longrightarrow H_2[SiF_6]$$
$$SiF_4 + 2F^- \longrightarrow [SiF_6]^{2-}$$

[handwritten annotations: 2 factors — thermodynamics (energy wise) — Kinetically feasible because C doe .]

The hexafluorosilicic(IV) acid, $H_2[SiF_6]$, and its salts resist hydrolysis, for the silicon atom bonded to six fluorine atoms can act neither as donor nor as acceptor.

4 Hydrides of silicon

Silicon forms a series of hydrides similar to the alkanes with a general formula Si_xH_{2x+2}, but the chain length is limited to 6 for silicon, whereas it seems to be limitless for carbon. Because silicon forms no ring compounds or multiple bonds, as carbon does, there are no silicon analogues of the alkenes, alkynes or aromatic hydrocarbons. *you need a spark.*

Silane, SiH_4, is formed, along with other members of the series, when magnesium silicide is hydrolysed:

$$Mg_2Si(s) + 4HCl(aq) \longrightarrow SiH_4(g) + 2MgCl_2(aq)$$

The hydrolysis of aluminium carbide to give methane is a similar reaction.

Pure silane can be obtained by fractional distillation of the mixture from the hydrolysis of magnesium silicide, but it is best made by reaction between silicon tetrachloride or silicon(IV) oxide and lithium tetrahydridoaluminate(III) in solution in an ether:

$$SiCl_4(l) + Li[AlH_4] \longrightarrow SiH_4(g) + LiCl(s) + AlCl_3(s)$$

Silane is a gas, but higher members of the series are liquids. The silanes, which are endothermic compounds, are less stable than the exothermic alkanes. The difference is due to the fact that the Si—Si bond is weaker than the C—C bond, and to the higher enthalpy of atomisation of carbon. Silane decomposes into its elements at about 450 °C, whereas methane requires a temperature of about 800 °C:

$$SiH_4(g) \longrightarrow Si(s) + 2H_2(g) \qquad \Delta H_m^{\ominus}(298\,K) = -34.3\,kJ\,mol^{-1}$$
$$CH_4(g) \longrightarrow C(s) + 2H_2(g) \qquad \Delta H_m^{\ominus}(298\,K) = 74.9\,kJ\,mol^{-1}$$

Silanes are also far more reactive than the corresponding alkanes, partly because the Si—Si bond is weak, but also because silicon has a maximum covalency of 6 (p. 133). Thus, silanes catch fire spontaneously in air, whereas methane has an ignition temperature of about 500 °C; silane is hydrolysed in alkaline solution,

$$SiH_4(g) + 2NaOH(aq) + H_2O(l) \longrightarrow Na_2SiO_3(s) + 4H_2(g)$$

whereas methane is not; and silane is a strong reducing agent, unlike methane.

a Comparison of methane and silane   The main similarities and differences between methane and silane are summarised below.

CH$_4$	SiH$_4$
colourless gas	colourless gas
boils at $-161\,°C$	boils at $-112\,°C$
C—H bond enthalpy = 413 kJ mol^{-1}	Si—H bond enthalpy = 323 kJ mol^{-1}
bond more polar *due to electronegativity diffrence*	bond less polar
$\Delta H_{f,m}^{\ominus} = -74.9$ kJ mol^{-1}	$\Delta H_{f,m}^{\ominus} = 34.3$ kJ mol^{-1}
decomposes into elements at about 800 °C	decomposes into elements at about 450 °C
weak reducing agent	strong reducing agent *due to covalen*
catches fire in air at 500 °C	catches fire spontaneously
no reaction with NaOH(aq)	hydrolysed by NaOH(aq)

E.N ↓ down the group (handwritten annotation)

5 Silicon disulphide, SiS$_2$

Sulphur reacts with silicon at a lower temperature than with carbon, to form silicon disulphide. It is a white solid, whereas carbon disulphide is a volatile liquid. The latter contains CS$_2$ molecules, containing double bonds, but the needle-shaped crystals of silicon disulphide are made up of SiS$_4$ tetrahedra linked into long chains by their edges (Fig. 76) and held together by van der Waals' forces.

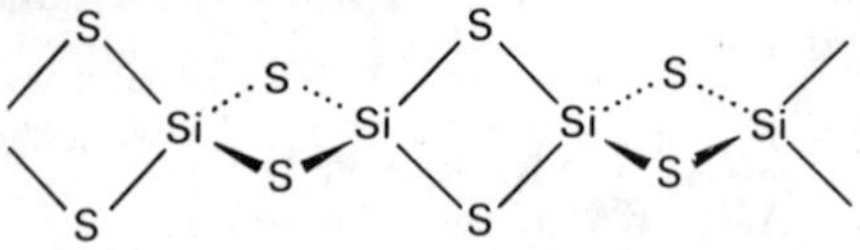

Fig. 76 SiS$_4$ units in a chain in SiS$_2$.

Silicon disulphide is rapidly hydrolysed by cold water:

$$SiS_2(s) + 2H_2O(l) \longrightarrow SiO_2(s) + 2H_2S(g)$$

6 Silicates

Silicates occur naturally in great abundance and are manufactured to some extent (p. 220). They are all made up of tetrahedral SiO$_4$ units but differ in the way these units are linked together. Alkali metal silicates are soluble in water, but others are insoluble. They are all thermally stable.

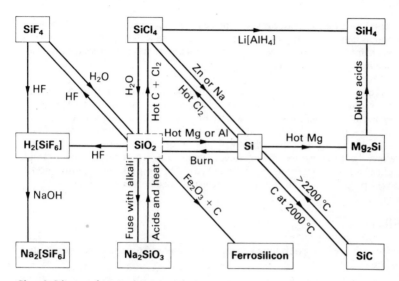

Chart 8 *Silicon and its simple compounds. SiO_2 and the silicates occur naturally.*

Simple silicates, e.g. olivine, Mg_2SiO_4, and zircon, $ZrSiO_4$, contain SiO_4^{4-} ions with metallic cations distributed in the spaces between the anions. In other silicates, the SiO_4 units are linked together through the oxygen atoms at the corners of the tetrahedra. The linking can be in pairs, threes, single chains, double chains, cyclic sheets and frameworks. The SiO_4 unit has a silicon:oxygen ratio of 1:4. As units share oxygen atoms, this ratio falls until it is 1:2 (SiO_2), when all four oxygen atoms in the SiO_4 tetrahedra are shared. Similarly, the charge on the SiO_4 unit falls from -4 on the single unit to zero for SiO_2.

The arrangement of the anions in a crystal is an open one, so that cations of the right size can fit into the spaces. Many cations have a suitable size but Ca^{2+}, Mg^{2+}, K^+, Na^+ and Fe^{2+} occur most commonly. OH^- and F^- ions, with similar sizes, may also occur.

In the SiO_4 unit, it frequently happens that some of the silicon atoms are replaced by aluminium, the Al^{3+} ions being of similar size to the Si^{4+} ion; the products are known as aluminosilicates (p. 196). The replacement of SiO_4^{4-} units by AlO_4^{5-} units means that extra monovalent cations have to be introduced, or that monovalent cations have to be replaced by divalent ones.

a Manufacture of glass Common, or soda lime, glass is made by fusing white sand with sodium and calcium carbonates; it is a supercooled solution of sodium and calcium silicates. Other glasses contain other metallic cations, e.g. potash glass is harder than soda lime glass; crown glass contains K^+ and Ba^{2+} silicates; flint glass contains lead silicates; and many transition metals give coloured glasses. Green or brown glass, which is cheaper than white glass, is made from sand containing iron oxide impurities.

The silicate content of glass can also be replaced to some extent by borates and phosphates(V). Borates give a glass with a higher softening point and lower coefficient of expansion, e.g. Pyrex; calcium phosphate gives opalescence in glass.

b Ion exchangers The cations in some aluminosilicates, particularly the zeolites, are very loosely held and can be replaced by other cations, so that they can function as cation exchangers. At one time they were widely used for this purpose as, for example, in water softening, but they have been largely replaced by more effective materials made from resins (p. 269).

c **Molecular sieves**   When a zeolite containing water is heated, it loses the water and a structure containing many small channels remains. This can be used as a drying agent for it will re-absorb water. But some dehydrated zeolites will also absorb gases and liquids whose molecules are of the right size and shape to fit into the available channels whilst they will not absorb other molecules. They can, therefore, be used as molecular sieves, being able to separate, for example, straight-chain hydrocarbons from branched-chain or ring ones. Many modern catalysts, particularly in the petroleum industry, are also based on zeolites.

7 Silicones

Silicones are a recently developed group of polymerised organosilicon compounds based on

$$-O-\overset{|}{\underset{|}{Si}}-O-\overset{|}{\underset{|}{Si}}-O-$$

chains with alkyl and aryl groups attached to the silicon atoms. The compounds may be rubber-like solids or liquids; they are colourless, odourless, non-volatile, insoluble in water and chemically unreactive. They are also good electrical insulators, are strongly water-repellent, and the liquids have a low surface tension so that they are good wetting agents.

Silicone oils, with chain lengths of about ten silicon atoms, are good lubricants and good hydraulic fluids, and they have the particular advantage that they maintain their viscosity over temperature ranges from about $-40\,°C$ to $200\,°C$. Silicone greases and rubbers, with longer chain lengths, also maintain their properties at different temperatures. Silicone grease is used on burette taps.

Liquid silicones are very useful release agents because of their non-stick properties. The water repellent and spreading properties make the silicones useful in waterproofing textiles and in furniture, leather, car and other polishes. Silicones also improve the spreading qualities and the water and heat resistance of paints and varnishes.

Questions on Chapter 17

(See also pp. 211 and 228.)

1 Compare the physical and chemical properties of crystalline silicon and graphite.
2 Compare the nature of the bonding in compounds of carbon and silicon.
3 Why is it that silicon(IV) oxide and carbon dioxide are so different? Why do they have any similarities?

4 Suggest reasons why the degree of catenation in silicon compounds is so much lower than in carbon compounds.

5 Compare the properties and reactions of silicon with those of (a) carbon, (b) boron, showing how far the first comparison illustrates the resemblance between the elements of the same group in the periodic table, and how far the second illustrates the so-called 'diagonal relationship'. (L)

6 Compare and contrast (a) CCl_4 and $SiCl_4$, and (b) CF_4 and CI_4.

7 Write short notes on (a) zone refining, (b) silicones, (c) silicosis, (d) silica gel, (e) ion exchangers, (f) silicates and (g) silanes.

8* By considering the enthalpies of atomisation of M(s) and hydrogen, and the bond energies, M—H, calculate the enthalpy changes in the reaction

$$M(s) + 2H_2(g) \longrightarrow MH_4(g)$$

when M is graphite and when it is silicon. Comment on the result you obtain.

9* What factors make methane exothermic whereas silane is endothermic?

10* Compare the enthalpy changes for the reactions of methane and silane with (a) oxygen and (b) water.

11* Use standard enthalpies of formation and standard entropies to estimate the temperature at which carbon will reduce silica to silicon.

18
Germanium, tin and lead*

1 Extraction of the elements

a Germanium   The main ores of germanium are sulphide ores, but they are neither rich nor widespread. The element is, therefore, generally obtained as a by-product of other processes. The final stage is the high temperature reduction of germanium(IV) oxide, GeO_2, by hydrogen or carbon.

For use as a semiconductor, the germanium must be exceptionally pure. For this purpose, hydrogen is used as the reducing agent, and the final purification is by zone refining, as for silicon (p. 213).

b Tin   Cassiterite or tinstone, SnO_2, is the main source of tin. The powdered and washed ore is first roasted to remove arsenic and sulphur impurities as the volatile oxides. The concentrated ore is then reduced by heating with anthracite:

$$SnO_2(s) + 2C(s) \longrightarrow Sn(s) + 2CO(g)$$

The tin is refined by first melting it and pouring it away from any infusible residues. Air is then passed through the molten tin and metallic impurities are oxidised to form a surface scum which can be skimmed off. Tin is not a cheap metal, so it is commercially viable to recover it from scrap tin plate.

c Lead   Lead is made from galena, PbS. The crushed and concentrated ore is first roasted to convert it into the oxide (p. 115):

$$2PbS(s) + 3O_2(g) \longrightarrow 2PbO(s) + 2SO_2(g)$$

It is then fed into a small blast furnace together with coke and limestone and some scrap iron. The limestone forms a slag with impurities, and the lead(II) oxide is reduced to lead by the carbon. Any lead(II) sulphide which might be present is reduced to lead by the iron:

$$PbO(s) + C(s) \longrightarrow Pb(s) + CO(g)$$
$$PbS(s) + Fe(s) \longrightarrow Pb(s) + FeS(s)$$

Further purification processes are necessary to remove other metals present in the original ore.

---

*The reader is advised to read the summary of the trends in Group 4B (pp. 198–201) before reading this chapter.

2 Structure, properties and uses

a Germanium Germanium has a diamond-like structure (p. 42). It is a
hard, brittle material with a melting temperature of 937 °C; it is used as a
semiconductor.

b Tin Tin has three allotropic forms,

$$\text{grey } (\alpha) \text{ tin} \xrightleftharpoons{13.2\,°C} \text{white } (\beta) \text{ tin} \xrightarrow{161\,°C} \text{rhombic } (\gamma) \text{ tin}$$

with white tin being the normal form at room temperature. Tin is unique
in having a non-metallic, diamond-like structure in one allotrope (grey
tin), and metallic structures in the other two allotropes.

Tin is used in plating steel. The steel, after being cleaned with sulphuric
acid, washed and dried, is dipped into molten tin; alternatively, the plating
is carried out electrolytically. Tin is also used as a component of many
alloys, e.g. pewter (75Sn : 25Pb), tinsmith's solder (60Sn : 39Pb : 1Sb) and
bronze (90Cu : 10Sn).

c Lead Lead exists in only one form, which has a metallic structure. It
has a low melting temperature (327 °C) and a high density. It is used as a
shield for X-rays and radiation, in lead–acid accumulators and in alloys,
e.g. plumber's solder (67Pb : 32Sn : 1Sb) and type metal (75Pb : 15Sb : 10Sn).
It is also used, because of its resistance to corrosion and its malleability, in
roofing work.

3 Compounds of the elements

The main chemical trends in passing from germanium to lead (p. 201) are
an increase in the basic character of the oxides, an increase in the stability
of the $+2$ oxidation state compared with the $+4$ state, and a decrease in
the number and the stability of the hydrides.

The elements are not very reactive, as indicated by the standard
electrode potentials for tin and lead,

$$Sn^{2+}(aq)/Sn(s) \qquad E^{\ominus} = -0.14 \text{ V}$$
$$Pb^{2+}(aq)/Pb(s) \qquad E^{\ominus} = -0.13 \text{ V}$$

They will, however, all react on heating with oxygen, chlorine or sulphur
to form oxides, chlorides and sulphides respectively. For germanium and
tin, the oxidation state in the product is generally $+4$; for lead it is
generally $+2$. The elements will also react with concentrated acids and
with alkalis, but the reactions of lead may be impeded by the insolubility
of the products, as many lead compounds have low solubility.

All three elements form covalent hydrides, MH_4, in which the oxidation state is -4. PbH_4 and SnH_4 are both very unstable, but there are nine hydrides of germanium, Ge_xH_{2x+2}, that correspond to the alkanes and the silanes. These are, surprisingly, less reactive than the silanes.

The compounds with oxidation states $+4$ or $+2$ are predominantly covalent, but they may be ionic or have significant ionic character. The M^{4+} and M^{2+} ions do exist, particularly in the solid fluorides and oxides, and Pb^{2+} ions can exist in aqueous solution even though they are hydrolysed to some extent (p. 228). Because the $+2$ state becomes more stable in passing from germanium to lead, the $+4$ compounds of lead are oxidising agents, whereas the $+2$ compounds of germanium and tin are reducing agents. Some typical changes which reflect this, and which are easy to bring about, are summarised below.

GeO $\longrightarrow$ GeO$_2$	SnO $\longrightarrow$ SnO$_2$	PbO $\longleftarrow$ PbO$_2$
GeS $\longrightarrow$ GeS$_2$	SnS $\longrightarrow$ SnS$_2$	PbS no PbS$_2$
GeCl$_2$ $\longrightarrow$ GeCl$_4$	SnCl$_2$ $\longrightarrow$ SnCl$_4$	PbCl$_2$ $\longleftarrow$ PbCl$_4$
$+2$ $\longrightarrow$ $+4$	$+2$ $\longrightarrow$ $+4$	$+2$ $\longleftarrow$ $+4$

Germanium, tin and lead form hydroxo- and halogeno-complexes particularly easily, e.g. $[GeF_6]^{2-}$, $[SnCl_6]^{2-}$ and $[Pb(OH)_6]^{4-}$.

4 MO oxides

GeO, SnO and PbO are all insoluble solids with similar crystal structures. They are all amphoteric, though predominantly basic, with lead(II) oxide the most basic. They form salts with acids, and solutions containing complex germanate(II), stannate(II) or plumbate(II) ions with alkalis. These ions are of uncertain composition but they are generally formulated as, for example, $[Sn(OH)_6]^{4-}$. Thus

$$SnO(s) + 4OH^-(aq) + H_2O(l) \longrightarrow [Sn(OH)_6]^{4-}(aq)$$

Germanium(II) and tin(II) oxides are strong reducing agents.

a Tin(II) oxide, SnO Anhydrous tin(II) oxide is a blue–black solid made by heating tin(II) ethanedioate out of contact with air. The hydrated oxide is precipitated on adding small amounts of alkali to a solution of a tin(II) salt.

Tin(II) oxide is stable in air at room temperature, but the conversion into tin(IV) oxide on warming is very exothermic.

b Lead(II) oxide, PbO (litharge and massicot) Lead(II) oxide is made by heating lead in air or oxygen, or by heating lead(IV) oxide or the carbonate, hydroxide or nitrate of lead. The oxide has two forms, which are known as litharge (red) and massicot (yellow); they have different

$2PbO_2 \longrightarrow 2PbO + O_2$

because electronegative

$PbO_2 + 2HCL \longrightarrow Cl_2 + PbCl_2$

densities and different crystal structures. The form obtained depends on the temperature used in the preparation.

On heating in air at about 400 °C lead(II) oxide is converted into red lead oxide (p. 226), but the lead(II) oxide is re-formed at 470 °C:

$$6PbO(s) + O_2(g) \underset{470\,°C}{\overset{400\,°C}{\rightleftharpoons}} 2Pb_3O_4(s)$$

Lead(II) oxide is used in making lead accumulators, glasses, glazes and enamels.

each metallic is surrounded by 6 oxygen atoms. and each oxygen is surrounded by 3 metallic atoms

5 MO$_2$ oxides

GeO$_2$

GeO$_2$, SnO$_2$ (cassiterite) and PbO$_2$ all exist as ionic crystals with the rutile structure (p. 16), but germanium(IV) oxide also has a covalent, silica-like structure (p. 214) above 1033 °C.

They are all amphoteric, though the acidic character predominates. They react with excess OH$^-$ ions to form solutions of germanates(IV), stannates(IV) or plumbates(IV), e.g.

$$PbO_2(s) + 2OH^-(aq) + 2H_2O(l) \longrightarrow [Pb(OH)_6]^{2-}(aq)$$

Lead(IV) oxide decomposes into the lead(II) oxide on heating, and it is a strong oxidising agent; SnO$_2$ and GeO$_2$ are both thermally stable, and they are not oxidising agents.

a Tin(IV) oxide, SnO$_2$ (cassiterite) This can be made by burning tin in air or oxygen. It is also formed in a hydrated form when tin reacts with concentrated nitric acid or when tin(IV) chloride is hydrolysed. The hydrated form varies in composition but gives the anhydrous form on heating.

Tin(IV) oxide is used as a mild abrasive, and for rendering enamels, glasses and glazes opaque.

b Lead(IV) oxide, PbO$_2$ This is made by heating a solution of lead(II) nitrate with sodium chlorate(I) solution or with bleaching powder; alternatively, red lead oxide can be treated with nitric acid:

$$Pb(NO_3)_2(aq) + ClO^-(aq) + H_2O(l)$$
$$\longrightarrow PbO_2(s) + 2HNO_3(aq) + Cl^-(aq)$$
$$Pb_3O_4(s) + 4HNO_3(aq) \longrightarrow 2Pb(NO_3)_2(aq) + PbO_2(s) + 2H_2O(l)$$

Lead(IV) oxide is a dark chocolate coloured, insoluble solid. It decomposes into lead(II) oxide and oxygen on heating above 300 °C, and oxidises warm concentrated hydrochloric acid to chlorine.

Lead(IV) oxide is used in lead–acid accumulators.

$PbO_2 + 2HCL \longrightarrow Cl_2 + Pb + O_2$

6 Dilead(II) lead(IV) oxide, Pb_3O_4 (red lead oxide)

This is made by heating lead(II) oxide in air at about 400 °C; the temperature must be limited for the reaction reverses at 470 °C:

$$6PbO(s) + O_2(g) \underset{470\,°C}{\overset{400\,°C}{\rightleftharpoons}} 2Pb_3O_4(s)$$

It is a brilliant red, insoluble powder, which functions chemically as the mixed oxide (p. 274), $PbO_2.2PbO$. The crystal contains, effectively, PbO_2 layers sandwiched between two PbO layers. This arises from $Pb^{IV}O_6$ octahedra linked through opposite edges, with three of the oxygen atoms, above and below, linked to Pb^{II} atoms.

Red lead oxide reacts with dilute nitric acid to form a solution of lead(II) nitrate and a precipitate of lead(IV) oxide, and with warm concentrated hydrochloric acid to liberate chlorine:

$$Pb_3O_4(s) + 2HNO_3(aq) \longrightarrow PbO_2(s) + 2Pb(NO_3)_2(aq) + H_2O(l)$$

$$Pb_3O_4(s) + 8HCl(aq) \longrightarrow 3PbCl_2(s) + 4H_2O(l) + Cl_2(g)$$

It is used in paint primers and in glass making.

7 Divalent halides

a Tin(II) chloride, $SnCl_2$ The anhydrous form of this salt is prepared by reaction between hydrogen chloride and hot tin. Concentrated hydrochloric acid and the metal, or tin(II) oxide and dilute hydrochloric acid, give a dihydrate, $SnCl_2.2H_2O$.

The anhydrous salt, which is predominantly covalent, is soluble in organic solvents and in water. It gives a cloudy solution with water, owing to the formation, by hydrolysis, of insoluble basic chlorides. Solutions acidified with hydrochloric acid are clear; they contain chloro-complexes, e.g. $[SnCl_3]^-$ and $[SnCl_4]^{2-}$. The chloro-complexes are also formed when tin(II) chloride dissolves in concentrated hydrochloric acid.

Tin(II) chloride is a strong reducing agent, reducing mercury(II) chloride to mercury(I) chloride and then to mercury, and iron(III) salts to iron(II) salts. In organic chemistry, it reduces nitrobenzene to phenylamine, and benzene diazonium chloride to phenylhydrazine hydrochloride.

$NO_2 - C_6H_6 \longrightarrow NH_2 - C_6H_6$.

b Lead(II) chloride, $PbCl_2$ This is a white, anhydrous, mainly ionic solid. It is insoluble in cold water but the solubility rises on heating. The chloride precipitates on adding Cl^- ions to cold solutions of lead(II) salts. Like tin(II) chloride, it forms chloro-complexes, e.g. $[PbCl_4]^{2-}$, with Cl^- ions and, because of this, it is soluble in concentrated hydrochloric acid. It is not a reducing agent.

$PbCl_2$ is not soluble however adding HCl excess giving $[PbCl_4]^{2-}$.

c Other divalent halides Germanium(II) chloride is made by reducing germanium(IV) chloride with germanium:

$$GeCl_4(l) + Ge(s) \longrightarrow 2GeCl_2(s)$$

It is a strong reducing agent, like $SnCl_2$.

Germanium, tin and lead all form MX_2 halides with all the halogens. The fluorides, bromides and iodides are very much like the chlorides, though there is an increase in ionic character in passing from the iodide to the fluoride.

Lead(II) iodide, PbI_2, is a golden-yellow crystalline solid which is easily precipitated in the cold and can be used as a test for Pb^{2+} or I^- ions.

8 Tetravalent halides

$GeCl_4$ and $SnCl_4$ are colourless liquids made by reaction between the hot element and chlorine. $PbCl_4$ is a yellow liquid; it is formed when lead(IV) oxide reacts with hydrochloric acid, or when chlorine is passed into a suspension of lead(II) chloride in concentrated hydrochloric acid, but it decomposes, even at room temperature, into lead(II) chloride and chlorine. All three tetrachlorides are covalent compounds that fume in moist air because of rapid hydrolysis. The solid formed is a hydrated dioxide, e.g.

$$SnCl_4(l) + (2+x)H_2O(l) \longrightarrow SnO_2.xH_2O(s) + 4HCl(g)$$

The halides form complexes with concentrated hydrochloric acid or ammonium chloride, e.g.

$$GeCl_4(l) + 2HCl(aq) \longrightarrow H_2[GeCl_6](aq)$$
$$SnCl_4(l) + 2NH_4Cl(aq) \longrightarrow (NH_4)_2[SnCl_6](aq)$$

The only tetrahalides that do not exist are $PbBr_4$ and PbI_4. Their non-existence is attributed to the ability of the Br^- and I^- ions to reduce Pb^{4+} to Pb^{2+}. Any attempt to make these two tetrahalides results, therefore, in the divalent halides.

a Tin(IV) iodide, SnI_4 This provides a good example of a compound which can conveniently be prepared in a non-aqueous solvent, such a solvent being chosen in order to prevent hydrolysis of the product. Granulated tin and iodine are refluxed in an organic solvent such as a mixture of ethanoic acid and ethanoic anhydride. After a time, red crystals of tin(IV) iodide are formed.

9 Hydrides

Nine germanium hydrides (germanes) are known; they have a formula, Ge_xH_{2x+2}, like that of the alkanes and silanes. The lower ones are made *the oxidation state of +4 decreases down the group.*

by reducing germanium(IV) chloride or germanium(IV) oxide with lithium tetrahydridoaluminate(III). To make the more reactive, higher germanes, it is necessary to pass an electrical discharge through germane, GeH_4. Germane is more reactive than methane, but less so than silane. It is oxidised by air but it is not spontaneously flammable, and it is surprisingly resistant to hydrolysis, not reacting, for example, with a 33 per cent solution of sodium hydroxide.

Stannane, SnH_4, is made by reaction between tin(IV) chloride and lithium tetrahydridoaluminate(III) in solution in an ether at a low temperature. It is much less thermally stable than methane, silane or germane, decomposing rapidly into its elements, even at $0\,°C$.

Plumbane, PbH_4, is even less stable, and its existence is open to doubt. It may be formed in very small quantities when magnesium–lead alloys are treated with dilute acids.

10 Salts of oxoacids

It is only lead that forms important salts of oxoacids, and the great majority of them have lead in the $+2$ oxidation state.

The only lead(IV) salt of any significance is lead(IV) ethanoate, $Pb(CH_3COO)_4$, which is made from red lead oxide and warm glacial ethanoic acid. It is covalent, having no measurable conductivity in ethanoic acid solution; it is rapidly hydrolysed; and it decomposes into lead(II) ethanoate on heating. It is used as an oxidising agent in organic chemistry.

The only two lead(II) salts that are soluble are the nitrate, $Pb(NO_3)_2$, and the ethanoate, $Pb(CH_3COO)_2$ (which is sometimes known as sugar of lead because it looks like sugar). These salts are generally used as sources of Pb^{2+} ions in solution, but hydrolysis means that the simple ions are converted to some extent into complex ions, e.g.

Similar to Al^{3+}

$$Pb^{2+}(aq) + H_2O(l) \longrightarrow [Pb(OH)]^+(aq) + H^+(aq)$$
$$4Pb^{2+}(aq) + 4H_2O(l) \longrightarrow [Pb_4(OH)_4]^{4+}(aq) + 4H^+(aq)$$

slightly acidic

Sn^{2+} and Ge^{2+} ions undergo similar, but more extensive, hydrolysis in solution. *because they have greater non metallic charac*

Insoluble lead(II) salts include the carbonate, the sulphate and the chromate; these can be made from solutions containing Pb^{2+} ions, by precipitation with HCO_3^-, SO_4^{2-} and CrO_4^{2-} ions respectively. Lead carbonate (cerrusite) and lead sulphate (anglesite) both occur naturally. Lead chromate is used in yellow paints, such as chrome yellow. *and PbI_2.*

Mg — impure metal → no hydrolysis
Pb — no hydrolysis due to pure metal.

Questions on Chapter 18

1 State *two* industrial uses of tin. Tin is said to be enantiotropic; explain what this means. Write down the electronic configuration of (a) Sn^{2+}, (b) Sn^{4+}. How would you convert a solution containing tin(II) into a solution containing tin(IV)?

$Tn^{2+} \longrightarrow Tn^{4+}$

$\left(Pb(OH)_3 \right)^{-2} \left(Pb(OH)_4 \right)^{-2}$

Pure tin was reacted with chlorine to give a colourless liquid that fumed in moist air. A solution of 0.1629 g of the liquid in water was equivalent to 24.00 cm³ of exactly 0.1 M sodium hydroxide solution using methyl orange as the indicator. Calculate the percentage purity of the liquid tin compound. (JMB)

2 Compare the properties of (a) the chlorides and (b) the oxides of lead and tin.

3 Outline one process for the extraction of lead from lead sulphide. Compare the chemistry of carbon, silicon and lead by considering the reactions of (a) their chlorides with water, (b) their dioxides with sodium hydroxide and (c) the elements with concentrated nitric acid. Explain any trends in reactivities which are shown. (JMB)

4 Write three equations showing Sn^{2+} reacting with oxidising agents and three showing lead(IV) oxide reacting with reducing agents.

5 What happens when solutions of (a) sodium hydroxide, (b) sodium carbonate, and (c) sodium chloride are added to solutions containing (i) Sn^{2+} ions and (ii) Pb^{2+} ions?

6 Compare (a) the physical properties, (b) the uses and (c) the methods of extraction of tin and lead.

7 How can germanium chlorides be made from germanium? How does $GeCl_2$ resemble $SnCl_2$, and how does $GeCl_4$ resemble $SiCl_4$?

8 Give a comparative account of the chemistry of the oxides and chlorides of carbon, silicon, germanium, tin and lead, paying particular attention to: (a) the formation, composition and stability of oxides; (b) the formation, hydrolytic behaviour and stability of chlorides. (L)

9 Choose examples from the chemistry of the Group 4B elements to illustrate the meaning of the word 'stability'.

10 Discuss the properties of the oxides and chlorides of the elements carbon, silicon, tin and lead. Suggest explanations for any variations in properties. Illustrate your answer by means of some carefully chosen examples. (OC)

11 How and under what conditions does silicon dioxide react with (a) carbon, (b) a mixture of calcium carbonate and sodium carbonate, (c) a mixture of calcium fluoride and sulphuric acid? Compare the chemistry of carbon, silicon and lead by considering the chemical properties of their chlorides and oxides. Explain any trends in reactivities which are shown. (JMB)

12 Compare and contrast the chemistry of silicon, germanium, tin and lead by referring to the properties and bond types of their oxides and chlorides. Give brief experimental details to indicate how you could prepare in the laboratory a sample of either tin(IV) chloride or tin(IV) iodide. How far does the chemistry of the oxides and chlorides of carbon support the statement that 'the head element of a group in the periodic table is not typical of that group'? (JMB)

13 Compare the properties of (a) CH_4 and SiH_4, (b) SiH_4 and GeH_4, (c) $PbCl_2$ and $SnCl_2$.

14 Draw up a list of five facts which point conclusively to the intermediate position of germanium between silicon and tin in Group 4B.

15 The elements carbon to lead in Group 4 of the periodic table illustrate differences and similarities between elements in a particular group. By reference to carbon and lead *only*, illustrate this statement by considering each of the following and explaining any differences: (a) methods of formation and structures of the dioxides, (b) the type of bonds that the elements form, (c) the relative stabilities of the lower and higher oxides, (d) the relative stabilities of the tetrachlorides, (e) the formation of a large number of hydrides of carbon. In each case, make a prediction for germanium, remembering that germanium is between carbon and lead in the periodic table. (L)

16 Illustrate the transition from non-metallic to metallic character of carbon, silicon and tin by describing the structure of, and giving one chemical property of, any *one* type of compound of these elements.

One of the products of the reaction between magnesium silicide (Mg_2Si) and sulphuric acid is a gas, X. A 0.620 g sample of X occupied 224 cm^3. When hydrolysed, this sample yielded 1568 cm^3 of hydrogen and a residue (SiO_2) which, after it had been strongly heated, weighed 1.200 g. (All gas volumes corrected to s.t.p.) What is the molecular formula of X? Write an equation for the hydrolysis of X. (OC)

17 Give a concise account of the chemistry of the elements in Group 4 of the periodic table (carbon, silicon, germanium, tin and lead), referring in your answer to *all* of the following aspects: (a) the variation in first ionisation energy, (b) the hydrolysis of the tetrachlorides, (c) the acid–base character of the oxides of the elements in oxidation states II and IV, (d) the relative stability of the II and IV oxidation states of the elements in the oxides and chlorides. Show how the electron pair repulsion theory can be used to predict the shape of the silicon tetrachloride molecule. (C)

18* Use standard enthalpies of formation to compare the decomposition of lead(IV) oxide into lead(II) oxide and oxygen with that of tin(IV) oxide into tin(II) oxide and oxygen.

19* Estimate the temperatures at which lead(IV) oxide and tin(IV) oxide might be expected to decompose into lead(II) oxide and tin(II) oxide and oxygen. Calculate the temperature at which ΔH becomes equal to $T\Delta S$. What assumptions do such a calculation make?

19
Nitrogen

1 Preparation and uses

There is about 78 per cent by volume of nitrogen in the air and it is obtained from that source by removal of water vapour and carbon dioxide and subsequent liquefaction and distillation. It is, therefore, readily available in cylinders.

Chemical reactions that yield nitrogen include the thermal decomposition of ammonium nitrite (as it is unstable, a mixture of ammonium sulphate and sodium nitrite in aqueous solution is generally used) or ammonium dichromate(VI), the oxidation of ammonia by hot copper(II) oxide, and the reduction of oxides of nitrogen by hot copper:

$$NH_4NO_2(aq) \longrightarrow N_2(g) + 2H_2O(l)$$
$$(NH_4)_2Cr_2O_7(s) \longrightarrow N_2(g) + Cr_2O_3(s) + 4H_2O(l)$$
$$2NH_3(g) + 3CuO(s) \longrightarrow 3Cu(s) + 3H_2O(l) + N_2(g)$$
$$N_2O(g) + Cu(s) \longrightarrow CuO(s) + N_2(g)$$

Nitrogen is used in making ammonia (p. 232), calcium cyanamide (p. 185) and cyanides. It may also be used to provide an inert atmosphere in certain processes, and liquid nitrogen serves as a coolant.

2 Compounds of nitrogen

Nitrogen is particularly unreactive at room temperature, largely because of the strength of the triple bond in the N_2 molecule ($945\,kJ\,mol^{-1}$). It will, however, react with many metals at high temperatures and pressures to form nitrides and, incompletely, with hydrogen and oxygen.

The gas becomes very reactive when subjected to an electric discharge at low pressure. The molecules are dissociated into atoms, and the unstable gas mixture (which is known as active nitrogen) glows yellow and will combine with many elements even in the cold.

The lack of reactivity in normal nitrogen does not mean, however, that there are few nitrogen containing compounds. There are a great many organic compounds of nitrogen, for example, amines, amides and proteins, and the wide range of inorganic compounds is shown in the following summary of possible oxidation states; -3, $+3$ and $+5$ are the commonest.

-3	-2	-1	0	$+1$	$+2$	$+3$	$+4$	$+5$
NH_3	N_2H_4	NH_2OH	N_2	N_2O	NO	N_2O_3	NO_2	N_2O_5
NH_4^+				$H_2N_2O_2$		HNO_2	N_2O_4	HNO_3
Mg_3N_2						NCl_3		
N^{3-}								

Most of the compounds of nitrogen are covalent, but nitrides of electropositive metals such as sodium and magnesium contain N^{3-} ions.

In NX_3 compounds, e.g. NH_3 and NCl_3, nitrogen forms three single covalent bonds. It can also form one single and one double bond, as in nitrous acid, $H—O—N{=}O$; or strong triple bonds as in N_2 and the $C{\equiv}N$ bond. In such cases, the nitrogen atom also has a lone pair which it can donate to form dative bonds as in NH_4^+ (p. 25) and ammine complexes (p. 234).   e's are completely given by an ahm

Nitrogen cannot form five single covalent bonds, i.e. there are no NX_5 compounds, for its maximum covalency is 4 (p. 36). It does, however, participate quite strongly in hydrogen bonding because of its high electronegativity.

A general comparison of the chemistry of nitrogen with that of phosphorus is given on p. 251, and with that of arsenic, antimony and bismuth on p. 260. The main trend is very much like that in Group 4B, with a change from the non-metallic nitrogen and phosphorus, through the semimetals arsenic and antimony, to bismuth, which is a metal.

3 Ammonia

Ammonia, NH_3, is the main hydride of nitrogen; others are hydrazine, N_2H_4 (p. 235), and hydrogen azide, HN_3 (p. 236).

a Manufacture and preparation   Ammonia is manufactured by the Haber process (p. 100). It is also formed when any ammonium salt is warmed with a base, or by the hydrolysis of ionic nitrides or calcium cyanamide (p. 186):

$$N_2(g) + 3H_2(g) \rightleftharpoons 2NH_3(g)$$
$$NH_4^+ + OH^- \rightleftharpoons NH_3(g) + H_2O(l)$$
$$N^{3-} + 3H_2O(l) \longrightarrow NH_3(g) + 3OH^-(aq)$$
$$CaCN_2(s) + 3H_2O(l) \longrightarrow 2NH_3(g) + CaCO_3(s)$$

b Uses   The gas is used in making fertilisers and nitric acid. In solution, it is a common household cleanser.

c Physical properties   Ammonia is a colourless, poisonous gas with a characteristic smell. It has a lower density than air. It is one of the most water soluble gases; a saturated solution (which is known as 880 ammonia because it has a density of $0.880\,\mathrm{g\,cm^{-3}}$) contains about 35 per cent by mass of the gas.

The high relative permittivity (dielectric constant) of ammonia also means that liquid ammonia is a good ionising solvent. Many reactions commonly carried out in aqueous solution can take place, in a similar way, in liquid ammonia solution.

d Structure   The NH_3 molecule is pyramidal, with H–N–H angles of $107°$; the molecule turns inside-out (inverts) like an umbrella.

C_8H_{10} : 106   octane can pass through
a tiny hole whereas
H_2O : 18   water can't due to the
hydrogen bonding the H_2O
molecules are all linked.

The bonds in the molecule are highly polar, and this accounts for the high dipole moment (1.48 D or 4.94×10^{-30} C m). The N—H bond can also participate in hydrogen bonding because of its polarity. This accounts for the high solubility of ammonia in water, and the fact that it is associated, like water and hydrogen fluoride (p. 53).

e Thermal decomposition   Although ammonia is made in the Haber process by combination between hydrogen and nitrogen, the reaction is reversible and the gas decomposes into its elements at high temperatures. The decomposition can be brought about by passing an electric spark through ammonia:

$$2NH_3(g) \longrightarrow N_2(g) + 3H_2(g)$$

f Oxidation   The oxidation state of nitrogen in ammonia is -3, and the gas can be oxidised in a number of different ways, giving nitrogen in more positive oxidation states.

Oxidation by sodium chlorate(I), for example, produces hydrazine (p. 235):

$$\underset{-3}{2NH_3(g)} + NaClO(aq) \longrightarrow \underset{-2}{N_2H_4(l)} + NaCl(aq) + H_2O(l)$$

Burning ammonia in oxygen (it will not burn in air) produces nitrogen:

$$\underset{-3}{4NH_3(g)} + 3O_2(g) \longrightarrow \underset{0}{2N_2(g)} + 6H_2O(l)$$

In the presence of a hot platinum catalyst at 900 °C, ammonia is oxidised to NO, as in the manufacture of nitric acid from ammonia (p. 242):

$$\underset{-3}{4NH_3(g)} + 5O_2(g) \longrightarrow \underset{+2}{4NO(g)} + 6H_2O(g)$$

g As a reducing agent   As ammonia can be oxidised, it has some reducing powers. It is not a strong reducing agent, but it will reduce hot metallic oxides to the metals, and it forms nitrogen trichloride, NCl_3 (a dangerously explosive, oily liquid) with excess chlorine:

$$2NH_3(g) + 3CuO(s) \longrightarrow 3Cu(s) + N_2(g) + 3H_2O(l)$$
$$NH_3(g) + 3Cl_2(g) \longrightarrow NCl_3(l) + 3HCl(g)$$

h Reaction with water   It is sometimes said that ammonia reacts with water to form ammonium hydroxide which is a weak base and, therefore, slightly ionised,

$$NH_3(g) + H_2O(l) \rightleftharpoons NH_4OH(aq) \rightleftharpoons NH_4^+(aq) + OH^-(aq)$$

but ammonium hydroxide does not really exist as a hydroxide. The ammonia actually combines with the water by hydrogen bonding,

$$NH_3(g) + H_2O(l) \rightleftharpoons H_3N \cdots H{-}O{-}H \rightleftharpoons NH_4{}^+(aq) + OH^-(aq)$$

so it is really aqueous ammonia that is formed and which is partially ionised. Hydrates of ammonia, $NH_3.H_2O$ and $2NH_3.H_2O$, can, in fact, be isolated by crystallisation from a concentrated aqueous solution.

It is this reaction of ammonia with water that accounts for its high solubility.

i Basic and acidic nature   Ammonia is a weak base in aqueous solution because it can accept protons to form ammonium salts:

$$NH_3 + H_3O^+ \longrightarrow NH_4{}^+(aq) + H_2O(l)$$

Its K_b value is about equal to the K_a value of ethanoic acid, so that the one is as weak as a base as the other is as an acid.

In the liquid state, ammonia is very slightly ionised, like water:

$$H_2O(l) + H_2O(l) \rightleftharpoons H_3O^+ + OH^-$$
$$NH_3(l) + NH_3(l) \rightleftharpoons NH_4{}^+ + NH_2{}^-$$
$$\text{(acid)} \qquad \text{(base)}$$

The amide ion, $NH_2{}^-$, is formed by the loss of a proton from an ammonia molecule, so that ammonia is acting as an acid. Liquid ammonia is, however, a very, very weak acid (even weaker than water), and the only chemical manifestation of the acidity of gaseous ammonia is found in the reactions with hot sodium or potassium to form amides, e.g.

$$2Na(s) + 2NH_3(g) \longrightarrow 2NaNH_2(s) + H_2(g)$$

Liquid ammonia is less reactive towards electropositive metals than water. It dissolves in, rather than reacts with, them to give blue solutions that conduct electricity because they contain solvated metal ions and solvated electrons, e.g.

$$Na(s) + (x+y)NH_3(l) \longrightarrow Na(NH_3)_x{}^+ + e^-(NH_3)_y$$

j Formation of complexes   The ability of the lone pair on the nitrogen atom of the ammonia molecule to form dative bonds by donation, and/or the high polarity of the N—H bond, enables the formation of a multitude of stable complex ions in which NH_3 is the ligand; they are known as ammines, e.g. $[Cu(NH_3)_4(H_2O)_2]^{2+}$ and $[Co(NH_3)_6]^{2+}$.

k Summary   A summary of the properties of ammonia is given on p. 254.

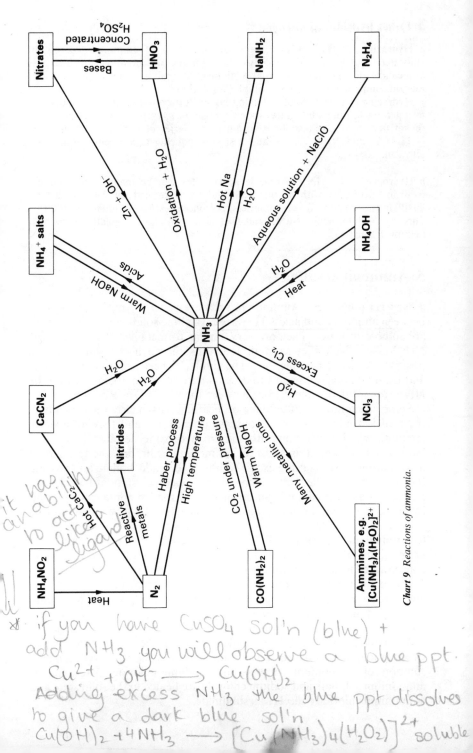

Chart 9 Reactions of ammonia.

it has an ability to act like a ligand

* if you have CuSO₄ sol'n (blue) +
add NH₃ you will observe a blue ppt.
$Cu^{2+} + OH^- \longrightarrow Cu(OH)_2$
Adding excess NH₃ the blue ppt dissolves
to give a dark blue sol'n
$Cu(OH)_2 + 4NH_3 \longrightarrow [Cu(NH_3)_4(H_2O_2)]^{2+}$ soluble

4 Other hydrides of nitrogen

a **Hydrazine, N_2H_4** This is made by oxidation of ammonia by sodium chlorate(I), but good yields are only obtained in the presence of gelatin. This acts as a catalyst for the reaction and also inhibits the catalytic action of any impurity ions on competing reactions, by complex formation.

Hydrazine is a water soluble, fuming liquid. It decomposes into ammonia and nitrogen on heating, and ignites spontaneously in oxygen; it has been used as a rocket fuel. It is a di-acid base, forming two series of hydrazinium salts, e.g. $N_2H_5^+Cl^-$ and $N_2H_6^{2+}Cl^-_2$, and a strong reducing agent, being oxidised to nitrogen.

b **Hydrogen azide, HN_3 (hydrazoic acid)** This can be made in solution by oxidising hydrazine with nitrous acid, and the pure acid can be obtained by distilling the solution. It is a colourless, poisonous, explosive liquid which forms salts, e.g. sodium azide, NaN_3, and lead· azide, $Pb(N_3)_2$. The latter is used in making detonators.

5 Ammonium salts

a **General properties** Ammonium salts are generally made by reaction between ammonia and acids. They are crystalline solids, and most of them are soluble in water. They are very similar to metallic salts because the NH_4^+ ion which they contain functions in much the same way as positively charged metallic ions. The similarity between ammonium salts and those of potassium and rubidium is particularly marked, because the NH_4^+ ion is intermediate in size between the K^+ and the Rb^+ ion.

Some ammonium salts sublime on heating, particularly those in which the acid is volatile. Others decompose on heating, and they all liberate ammonia gas on warming with sodium hydroxide or other bases. The latter reaction is used as a test for ammonium salts, and the test can be made quantitative by using excess sodium hydroxide, boiling and passing the ammonia expelled into a measured excess of standard acid. The amount of acid remaining after absorption of the ammonia can be found by titration, so that the amount of ammonia evolved can be measured.

b **Ammonium sulphate, $(NH_4)_2SO_4$** This is the most important ammonium salt, owing to its extensive use as a fertiliser and, in more concentrated form, as a weedkiller.

It is made from ammonia and sulphuric acid, or by passing ammonia and carbon dioxide into a suspension of calcium sulphate in water and filtering off the calcium carbonate.

c **Ammonium chloride, NH_4Cl (sal-ammoniac)** Made by reaction between ammonia and hydrochloric acid or by crystallisation from a mixed solution of ammonium sulphate and sodium chloride. Alternatively, ammonium sulphate and sodium chloride are heated in the solid state and ammonium chloride collects as a sublimate.

It is used in dry cells and as a flux in soldering, tin plating and galvanising.

d Ammonium carbonate, $(NH_4)_2CO_3$ Commercial ammonium carbonate is made by heating a mixture of solid calcium carbonate and ammonium sulphate. The white solid which collects as a sublimate is a mixture of ammonium hydrogencarbonate, ammonium carbamate and ammonium carbonate. Pure ammonium carbonate is obtained by treating a solution of the commercial product with ammonia.

It is used in smelling salts because it smells of ammonia, since decomposition takes place slightly, even at room temperature.

e Ammonium nitrate, NH_4NO_3 Made by direct combination between ammonia and nitric acid. It decomposes on heating, explosively if the temperature gets too high:

$$NH_4NO_3(s) \longrightarrow N_2O(g) + 2H_2O(g)$$

It is used as a fertiliser and in making explosives; it is mixed with TNT in *amatol*, with TNT and aluminium in *ammonal*, and with fuel oils in *ANFO*.

f Ammonium sulphides Saturation of a solution of ammonia with hydrogen sulphide produces a solution of ammonium hydrogensulphide, from which colourless crystals can be obtained:

$$NH_3(aq) + H_2S(g) \longrightarrow NH_4HS(aq)$$

Treatment of the acid salt solution with an equal volume of ammonia solution produces the normal salt:

$$NH_4HS(aq) + NH_3(g) \longrightarrow (NH_4)_2S(aq)$$

Both the acid and the normal salt give colourless solutions in water but they rapidly oxidise in air to yellow ammonium sulphide, which contains polysulphides with a formula of $(NH_4)_2S_x$, with x varying from 2 to 7. Yellow ammonium sulphide can also be made by warming the colourless solution with powdered sulphur; free sulphur may then be present in the solution.

g Ammonium molybdate(VI), $(NH_4)_2MoO_4$ This is made by dissolving molybdenum(VI) oxide in ammonia solution and treating a solution of the crystals obtained with ammonia. It is used as a test for phosphates(V) (p. 398).

h Ammonium thiocyanate, NH_4SCN Made from hydrogen cyanide and yellow ammonium sulphide solution, or by heating carbon disulphide and alcoholic ammonia solution:

$$HCN + (NH_4)_2S_2 \longrightarrow NH_4SCN + NH_4HS$$
$$CS_2 + 2NH_3 \longrightarrow NH_4SCN + H_2S$$

It is used as a test for Fe^{3+} ions, in making explosives, in photography, and in volumetric analysis.

6 Oxides of nitrogen

The various oxides formed by nitrogen are summarised below, together with the relevant oxidation states and the standard molar enthalpies of formation in $kJ\,mol^{-1}$.

+1	N_2O	dinitrogen oxide	+81.6
+2	NO	nitrogen monoxide	+90.4
+3	N_2O_3	dinitrogen trioxide	+92.9
+4	N_2O_4	dinitrogen tetraoxide	+9.7
+4	NO_2	nitrogen dioxide	+33.9
+5	N_2O_5	dinitrogen pentoxide	+15.0

The standard molar enthalpies of formation are all positive, and the oxides are both easily decomposed into their elements and are inter-convertible.

a Dinitrogen oxide, N_2O This is made by carefully heating a mixture of sodium nitrate and ammonium sulphate in aqueous solution, which functions as a solution of ammonium nitrate:

$$NH_4NO_3(aq) \xrightarrow{\text{heated}} N_2O(g) + 2H_2O(l)$$

The gas may be collected over warm water.

It decomposes into nitrogen and oxygen at about 600 °C,

$$2N_2O(g) \longrightarrow 2N_2(g) + O_2(g)$$

so that many hot elements will burn in it, or react, to form oxides, e.g.

$$2N_2O(g) + C(s) \longrightarrow 2N_2(g) + CO_2(g)$$
$$N_2O(g) + Cu(s) \longrightarrow N_2(g) + CuO(s)$$

A glowing splint is hot enough to bring about the decomposition, so that the splint relights in the gas as it does in oxygen. Dinitrogen oxide can be distinguished from oxygen by the fact that it produces nitrogen on passing over hot copper.

Dinitrogen oxide is used as a mild anaesthetic. The anaesthetic pro-perties were discovered by Davy over 100 years ago, and as it caused him, in some doses, to be mildly hysterical he named it 'laughing gas'.

The N_2O molecule is collinear and involves delocalised π bonding: $N\equiv N\equiv O$.

b Nitrogen monoxide, NO Copper reduces half-concentrated nitric acid to nitrogen monoxide as the main product (p. 243):

$$3Cu(s) + 8HNO_3(aq) \longrightarrow 3Cu(NO_3)_2(aq) + 4H_2O(l) + 2NO(g)$$

because it is insoluble ↗

If the gas is collected over water, any other oxides of nitrogen that are present as by-products dissolve, so that a reasonably pure product is obtained. Alternatively, the nitrogen monoxide can be absorbed in iron(II) sulphate solution, with which it forms a brown compound (see below). On heating in the absence of air, this compound decomposes to give pure nitrogen monoxide. The gas is also obtained by the catalytic oxidation of ammonia (p. 233) and by direct combination of nitrogen and oxygen on passing air through an electric arc at around 3000 °C. The yield in the latter process is very low, but it has been tried as an industrial process for making nitric acid, the formation of NO being followed by oxidation to NO_2 and then by reaction with water.

The gas is colourless and insoluble, but it reacts with oxygen or air at room temperature to form coloured nitrogen dioxide:

$$2NO(g) + O_2(g) \longrightarrow 2NO_2(g)$$

Like dinitrogen oxide, it decomposes into the elements,

$$2NO(g) \longrightarrow N_2(g) + O_2(g)$$

but a higher temperature is required (about 1000 °C), so that only strongly burning substances, such as magnesium and phosphorus, will burn in the gas. It can, however, be reduced to nitrogen by hot metals.

Nitrogen monoxide reacts with iron(II) salt solutions to form a brown complex ion. The Fe^{2+} ion is hydrated, $[Fe(H_2O)_6]^{2+}$; with nitrogen monoxide it is converted into the $[Fe(NO)(H_2O)_5]^{2+}$ ion. It is the formation of this ion that causes the brown ring in the well-known test for a nitrate (p. 246). Nitrogen monoxide also acts as a ligand in other complexes, e.g. $[Fe(CN)_5(NO)]^{2-}$ (p. 380).

It is not easy to represent the bonding in the NO molecule accurately, for an odd number of electrons is involved, and it is often referred to as an 'odd' molecule. The over-simplified structures given in Fig. 77(a) and (b) show the existence of an odd unpaired electron on either the nitrogen or the oxygen atom. The existence of such an electron is shown by the fact that nitrogen monoxide is paramagnetic (p. 352), and by the ease with which it can lose one electron to form the nitrosyl cation, NO^+, which has a triple bond similar to that in N_2 (Fig. 77(d)). But the best simple representation of the bonding in NO shows a triple bond consisting of one σ bond and two π bonds, together with an unpaired electron that must be thought of as belonging to the molecule as a whole rather than to either one of the atoms. The unpaired electron causes repulsion between the two atoms, so that the bond strength is weaker than the triple bond in N_2.

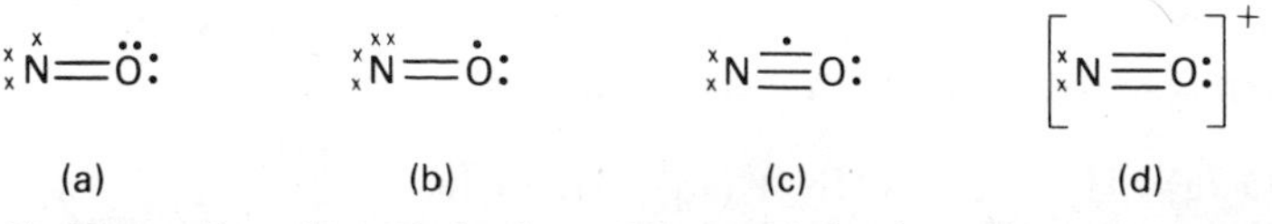

| (a) | (b) | (c) | (d) |

Fig. 77 (a), (b) and (c) The bonding in NO. (d) The bonding in the nitrosyl cation, NO^+.

c Dinitrogen trioxide, N_2O_3   This is made by cooling an equimolecular mixture of nitrogen monoxide and nitrogen dioxide. It only exists as a solid or liquid below $-25\,°C$; above that temperature it decomposes into the gases from which it is made. Dinitrogen trioxide is the acid anhydride of nitrous acid and it forms this acid with water, and nitrites with alkalis.

d Dinitrogen tetraoxide, N_2O_4, and nitrogen dioxide, NO_2   Nitrogen dioxide is obtained as a reddish-brown gas on heating lead nitrate (p. 246), reducing concentrated nitric acid with copper, or mixing nitrogen oxide and oxygen:

$$2Pb(NO_3)_2(s) \longrightarrow 2PbO(s) + 4NO_2(g) + O_2(g)$$
$$Cu(s) + 4HNO_3(aq) \longrightarrow Cu(NO_3)_2(aq) + 2NO_2(g) + 2H_2O(l)$$
$$2NO(g) + O_2(g) \longrightarrow 2NO_2(g)$$

The gas is best collected by passing through a U-tube immersed in a freezing mixture which liquefies the gas.

The reddish-brown gas has a nasty smell and is poisonous. It liquefies into a yellow liquid at $21\,°C$ and becomes a pale yellow solid at $-11.2\,°C$. The solid contains N_2O_4 molecules and is really dinitrogen tetraoxide. It melts at $-11.2\,°C$ into a liquid which also contains N_2O_4 molecules, but on raising the temperature, some of the molecules dissociate into NO_2 molecules which give an increasing reddish-brown colour to the liquid. There are, however, very few NO_2 molecules in the liquid state. In the vapour phase (above $21\,°C$), the dissociation increases as the temperature rises, so that the gas consists of a mixture of NO_2 and N_2O_4 molecules. The proportion of each at any one temperature can be calculated from density measurements. The dissociation is complete at $140\,°C$, when the gas is a very dark colour. Above that temperature, the NO_2 molecules begin to decompose into NO and O_2 molecules, so that the colour lightens. This decomposition is complete at about $600\,°C$, when the mixture is colourless. The reverse changes take place on cooling:

$$N_2O_4(l) \rightleftharpoons 2NO_2(g) \rightleftharpoons 2NO(g) + O_2(g)$$

Like other oxides of nitrogen, nitrogen dioxide can be reduced to nitrogen by hot metals, and it will support the combustion of elements such as strongly burning sulphur, phosphorus and magnesium, which are hot enough to decompose it:

$$2NO_2(g) \longrightarrow N_2(g) + 2O_2(g)$$

It is a mixed acid oxide, functioning as a mixture of N_2O_3 and N_2O_5 molecules. Thus it gives a mixture of nitrous and nitric acids with water and a mixture of nitrites and nitrates with alkalis:

$$2NO_2(g) + H_2O(l) \longrightarrow HNO_2(aq) + HNO_3(aq)$$
$$2NO_2(g) + 2NaOH(aq) \longrightarrow NaNO_3(aq) + NaNO_2(aq) + H_2O(l)$$

The bonding in the N_2O_4 molecule is best represented with delocalised bonds (Fig. 78(a)). In the NO_2 molecule, there is an unpaired electron (Fig. 78(b)), and that is why nitrogen dioxide, unlike dinitrogen tetraoxide, is paramagnetic. The loss of the unpaired electron in NO_2 results in the formation of the nitryl cation, NO_2^+ (Fig. 78(c)).

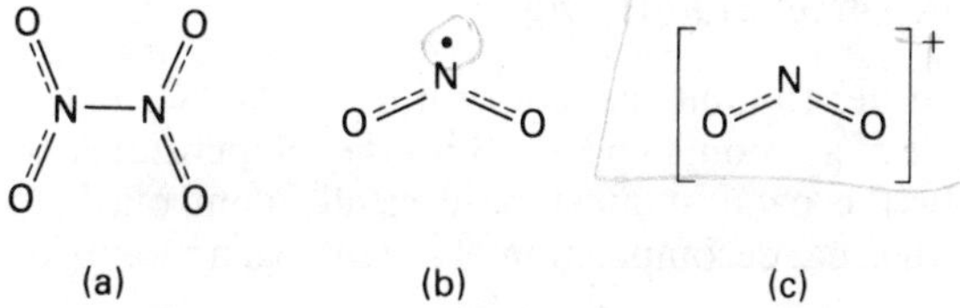

(a)   (b)   (c)

Fig. 78 *Bonding in (a) N_2O_4, (b) NO_2 and (c) the nitryl cation, NO_2^+.*

e Dinitrogen pentoxide, N_2O_5 This is made by dehydrating nitric acid with phosphorus(V) oxide. It is the acid anhydride of nitric acid, and reverts to the acid on adding water. The pentoxide is thermally unstable, decomposing at about $0\,°C$ into nitrogen dioxide and oxygen:

$$2N_2O_5(g) \longrightarrow 4NO_2(g) + O_2(g)$$

The N_2O_5 molecule is present in the vapour state, but the structure is ionic in the solid state, $NO_2^+NO_3^-$, i.e. nitryl nitrate.

7 Nitric acid, HNO_3

Though nitrogen forms five oxides it forms only two important oxoacids; nitric acid, HNO_3, and nitrous acid, HNO_2. A less important oxoacid, hyponitrous acid, $H_2N_2O_2$, is also known.

The two important acids can be regarded as being formed from N_2O_3 and N_2O_5, or from $N(OH)_3$ and $N(OH)_5$:

$$N_2O_3 \xrightarrow{H_2O} 2HNO_2 \qquad N_2O_5 \xrightarrow{H_2O} 2HNO_3$$

$$N(OH)_3 \xrightarrow{-H_2O} HNO_2 \qquad N(OH)_5 \xrightarrow{-2H_2O} HNO_3$$

a Manufacture and preparation Metallic nitrates produce nitric acid on heating with concentrated sulphuric acid, e.g.

$$NaNO_3(s) + H_2SO_4(aq) \longrightarrow NaHSO_4(aq) + HNO_3(g)$$

The nitric acid formed must be condensed, and as the vapour is very corrosive, this must be done in an all-glass apparatus.

The acid used to be manufactured by such a reaction, but it is now made by the catalytic oxidation of ammonia gas by excess air in the presence of a platinum–rhodium catalyst. The catalyst is used in the form of a fine gauze through which the gas mixture is passed under slight pressure and at about 900 °C.

The first reaction is a rapid exothermic oxidation of ammonia by the air:

$$4NH_3(g) + 5O_2(g) \longrightarrow 4NO(g) + 6H_2O(g)$$

Excess air is required in order to limit the temperature to 900 °C, and to avoid an explosive mixture of ammonia and air. Both the temperature and the time of contact with the catalyst must be carefully controlled, to prevent the reduction or thermal decomposition of the nitrogen monoxide to nitrogen.

The gas mixture, which contains about 10 per cent of nitrogen monoxide, is cooled, and the resulting nitrogen dioxide is absorbed in dilute nitric acid until the concentration of acid reaches about 65 per cent:

$$2NO(g) + O_2(g) \longrightarrow 2NO_2(g)$$
$$3NO_2(g) + H_2O(l) \longrightarrow 2HNO_3(aq) + NO(g)$$

The nitrogen monoxide released in this process is recycled, so that only a small amount of it is lost.

b Acidic and basic properties   Pure nitric acid is a very poor conductor of electricity but it is slightly ionised:

$$HNO_3 + HNO_3 \longrightarrow H_2NO_3^+ + NO_3^-$$
$$\text{(acid)} \quad \text{(base)} \qquad\qquad \downarrow$$
$$H_2O + NO_2^+$$
$$\text{nitryl cation}$$

In the presence of a protophilic solvent such as water, which is a proton acceptor, the nitric acid functions as a strong acid,

$$HNO_3(l) + H_2O(l) \longrightarrow H_3O^+(aq) + NO_3^-(aq)$$

being about 93 per cent ionised in molar solution. But in protogenic solvents, such as concentrated sulphuric or chloric(VII) acid, which are strong proton donors, the basic nature of the nitric acid predominates:

$$2H_2SO_4 \longrightarrow 2H^+ + 2HSO_4^-$$
$$HNO_3 + H^+ \longrightarrow H_2NO_3^+ \longrightarrow H_2O + NO_2^+$$
$$H_2O + H^+ \longrightarrow H_3O^+$$
$$\overline{HNO_3 + 2H_2SO_4 \longrightarrow H_3O^+ + NO_2^+ + 2HSO_4^-}$$

The existence of the nitryl cation, NO_2^+, is supported by the isolation of salts, e.g. $NO_2^+ClO_4^-$, and by its existence in solid dinitrogen pentoxide, $NO_2^+NO_3^-$ (p. 241).

Aqueous nitric acid reacts as an acid in the usual way with bases, and in the formation of esters with alcohols, but the fact that it is also a very strong oxidising agent complicates many of its reactions.

c As an oxidising agent   Nitric acid, particularly when it is hot and concentrated, is a very strong oxidising agent. When it does oxidise, it is reduced to many different compounds (all with nitrogen in an oxidation state lower than $+5$), such as N_2O_4, HNO_2, NO, N_2O, N_2, NH_2OH, N_2H_4 and NH_4^+. Mixtures of these products are commonly formed, and the major product is probably influenced more by unknown kinetic factors than by energy considerations.

The commonest processes that occur are summarised below:

i	$NO_3^-(aq) + 2H^+(aq) + e^- \longrightarrow H_2O(l) + NO_2(g)$
ii	$NO_3^-(aq) + 4H^+(aq) + 3e^- \longrightarrow 2H_2O(l) + NO(g)$
iii	$2NO_3^-(aq) + 10H^+(aq) + 8e^- \longrightarrow 5H_2O(l) + N_2O(g)$
iv	$NO_3^-(aq) + 7H^+(aq) + 6e^- \longrightarrow 2H_2O(l) + NH_2OH(l)$
v	$NO_3^-(aq) + 10H^+(aq) + 8e^- \longrightarrow 3H_2O(l) + NH_4^+(aq)$

and, as a broad generalisation, reaction i is most likely with highly concentrated acid and with mild reducing agents. But, as the nature of the reaction is affected by concentration of the acid, the strength of the reducing agent and the temperature, it is not possible to make detailed theoretical predictions.

d Reactions with metals   Calcium and magnesium react with very dilute nitric acid to form hydrogen gas, but other metals give a variety of products depending, as above, on the acid concentration, the strength of the metal as a reducing agent and the temperature. Reaction i above is favoured by concentrated acids, and both reactions i and ii by mild reducing agents, e.g. lead, copper, mercury and silver. The other reactions tend to take place with more dilute acid and with stronger reducing agents, e.g. magnesium, zinc, tin and iron.

Some metals, including tin, antimony and arsenic, form hydrated oxides, e.g. $SnO_2.xH_2O$, rather than nitrates; others, including iron, aluminium and chromium, are rendered passive by concentrated nitric acid, probably due to the formation of an impervious layer of oxide. Gold and platinum are not affected by the hot, concentrated acid (*aqua fortis*), but they are attacked by *aqua regia* (a mixture of three parts of concentrated hydrochloric acid and one part of concentrated nitric acid), mainly because the Cl^- ion forms complexes with the metals.

e Reaction with non-metals   Hot concentrated nitric acid will also oxidise many non-metals to their highest oxide, which may then react with water to form an acid. Thus

$$S \longrightarrow SO_3 \longrightarrow H_2SO_4$$
$$C \longrightarrow CO_2 \longrightarrow H_2CO_3$$
$$P_4 \longrightarrow P_4O_{10} \longrightarrow H_3PO_4$$
$$I_2 \longrightarrow I_2O_5 \longrightarrow HIO_3$$

f Nitration A mixture of concentrated nitric and sulphuric acids is known as nitrating mixture, and is widely used for replacing —H groups by —NO$_2$ groups. It is the nitryl cation, NO$_2$$^+$, that is the effective nitrating agent:

$$R\text{—}H + NO_2{}^+ \longrightarrow R\text{—}NO_2 + H^+$$

g Structure of nitric acid The HNO$_3$ molecule is planar with dimensions, in nm, as shown in Fig. 79(a). There is some hydrogen bonding between the molecules.

The NO$_3$$^-$ ion has bond lengths of 0.124 nm and bond angles of 120°; delocalised bonds are involved (Fig. 79(b)).

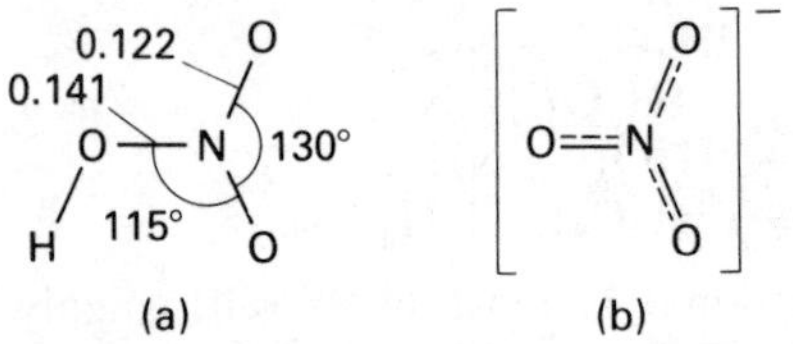

Fig. 79 (a) Structure of the HNO$_3$ molecule. (b) Structure of the NO$_3$$^-$ ion.

h Uses of nitric acid The acid is used as an oxidising and nitrating agent and as an acid. It is required for making both inorganic and organic nitrates, e.g. ammonium nitrate, ethyl nitrate and cellulose nitrate, and organic nitro-derivatives, e.g. trinitrotoluene (methyl-2, 4, 6-trinitrobenzene) and propane-1, 2, 3-triyl trinitrate (nitroglycerine). These products are used as fertilisers or explosives, and nitro-derivatives are also needed in making dyes.

8 Nitrous acid, HNO$_2$

A pale blue solution of nitrous acid is formed when a solution of a metallic nitrite is acidified or when dinitrogen trioxide (the acid anhydride) reacts with water:

$$NO_2{}^-(aq) + H^+(aq) \longrightarrow HNO_2(aq)$$
$$N_2O_3(l) + H_2O(l) \longrightarrow 2HNO_2(aq)$$

As the acid decomposes even at room temperature,

$$3HNO_2(aq) \longrightarrow HNO_3(aq) + 2NO(g) + H_2O(l)$$
$$2NO(g) + O_2(g) \longrightarrow 2NO_2(g)$$

the colder the solutions used, the better.

Nitrous acid is known only in solution and it is an unstable, weak, monobasic acid. It can act both as a reducing agent and as an oxidising agent. As a reducing agent it is oxidised to nitric acid,

$$NO_2{}^-(aq) + H_2O(l) \longrightarrow NO_3{}^-(aq) + 2H^+(aq) + 2e^-$$

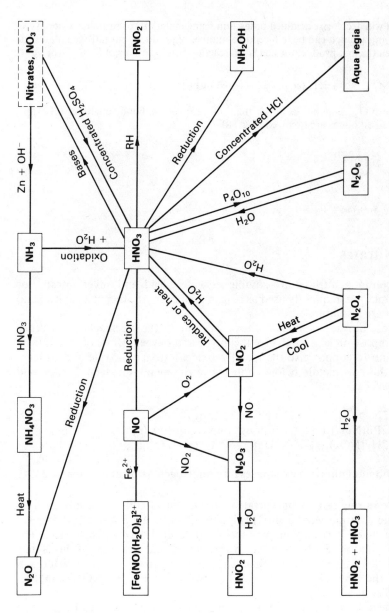

Chart 10 Reactions of nitric acid.

and it will decolorise acidified potassium manganate(VII) or bromine water. As an oxidising agent, it can be reduced to a number of products, but nitrogen monoxide is often the first product formed, as in reaction with solutions of I^-, Sn^{2+} or Fe^{2+}, i.e.

$$NO_2^-(aq) + 2H^+(aq) + 2e^- \longrightarrow NO(g) + H_2O(l)$$

The NO_2^- ion has a bond angle of $115°$ and a bond length of $0.124 nm$; delocalised bonds are involved (Fig. 80).

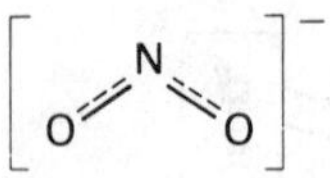

Fig. 80 *Structure of the NO_2^- ion.*

9 Nitrates

All common nitrates are soluble in water; lead and silver nitrate, for example, are frequently used as reagents, since most other salts of lead and silver are insoluble.

Most nitrates of metals above calcium in the electrochemical series decompose on heating to give the nitrite and oxygen; those of metals from calcium to copper give the metallic oxide, nitrogen dioxide and oxygen; and those of metals below copper give the metal, nitrogen dioxide and oxygen, e.g.

$$2NaNO_3(s) \longrightarrow 2NaNO_2(s) + O_2(g)$$
$$2Pb(NO_3)_2(s) \longrightarrow 2PbO(s) + 4NO_2(g) + O_2(g)$$
$$2Hg(NO_3)_2(s) \longrightarrow 2Hg(l) + 4NO_2(g) + 2O_2(g)$$

Ammonium nitrate decomposes into dinitrogen oxide and water (p. 238).

a Brown ring test Concentrated sulphuric acid added to a mixture of a nitrate solution with iron(II) sulphate sinks to the bottom and forms a lower layer. It is only at the junction that all three chemicals meet. The nitrate and the acid react to form some nitric acid and some of this is then reduced by the iron(II) sulphate to form nitrogen monoxide, which reacts with the iron(II) ions to form a brown complex, $[Fe(NO)(H_2O)_5]^{2+}$ (p. 377), as a brown ring.

The test is unreliable in the presence of a nitrite, which must be removed by acidification and addition of excess carbamide (urea). Metallic radicals that form insoluble sulphates also interfere with the test, and bromides and iodides give red and purple rings respectively.

b Reduction of nitrates Solutions of nitrates are reduced by zinc or, better, by Devarda's alloy (Al, Cu, Zn) in alkaline solution, e.g.

$$NO_3^-(aq) + 4Zn(s) + 7OH^-(aq) + 6H_2O(l)$$
$$\longrightarrow NH_3(g) + 4[Zn(OH)_4]^{2-}(aq)$$

The reaction can be used quantitatively for estimating nitrates by passing the ammonia formed into a measured excess of standard acid.

10 Halides of nitrogen

Nitrogen forms trihalides, NX_3, with fluorine, chlorine and bromine; the compounds have pyramidal molecules. Because nitrogen cannot expand its octet, there are no NX_5 compounds.

Nitrogen trifluoride, NF_3, is made by electrolysis of a solution of ammonium fluoride in anhydrous hydrogen fluoride. It is a very stable gas, and is not hydrolysed by water.

Nitrogen trichloride, NCl_3, is made by reaction between excess chlorine and ammonia, or between an acidic solution of ammonium chloride and chlorine:

$$NH_3(g) + 3Cl_2(g) \longrightarrow NCl_3(l) + 3HCl(g)$$
$$NH_4Cl(aq) + 3Cl_2(g) \longrightarrow NCl_3(l) + 4HCl(aq)$$

It is an oily liquid, which is liable to explode and is hydrolysed in an unusual way (p. 317) to form chloric(I) acid:

$$NCl_3(l) + 3H_2O(l) \longrightarrow NH_3(g) + 3HClO(aq)$$

Nitrogen tribromide is similar to the trichloride.

What is commonly called nitrogen triiodide is formed, as a black solid, when concentrated ammonia solution reacts with solid iodine. It is very unstable and is liable to explode on touching when dry. The compound has the formula $(NI_3 \cdot NH_3)_x$; it consists of NI_4 tetrahedra linked through the corners into zig-zag chains, with NH_3 molecules linking the chains together.

Questions on Chapter 19

(See also p. 261.)

1 Compare the methods of preparation, and the properties, of the hydrides of nitrogen and calcium.

2 'The fixation of nitrogen is vital to the progress of civilised humanity.' What does this mean and how far do you consider it to be true?

3 Write down the electronic configurations of the nitrogen and oxygen atoms and describe the bonding in a molecule of (a) ammonia and (b) water. Compare the chemistry of ammonia with that of water by describing and interpreting their reactions with (i) sodium, (ii) chlorine, (iii) calcium chloride, (iv) anhydrous copper(II) sulphate. What types of bonding are present in the reaction products from (iv)? (JMB)

4 (a) Describe the Haber process for the manufacture of ammonia

23¹
17

(without industrial details), explaining carefully the reasons for the physical and chemical conditions under which the synthesis is carried out. Give two important uses of ammonia. (b) How does ammonia react with (i) chlorine, (ii) water, (iii) aqueous copper(II) sulphate? (O)

5 Outline the laboratory preparation of a sample of dinitrogen tetraoxide. Describe and explain what happens when it is heated from $-10°$ to $600°C$. Suggest electronic structures for dinitrogen tetraoxide and the other nitrogen containing molecules formed from it on heating to $600°C$. Point out any unusual structural features. (C)

6 How does nitrogen dioxide (a) resemble and (b) differ from sulphur dioxide?

7 Compare and contrast the reactions of nitric acid and sulphuric acid with typical metals and non-metals.

8 Describe and explain the effect of heat on the following: ammonium nitrite, ammonium nitrate, copper(II) nitrate, mercury(II) nitrate, nitric acid, potassium nitrate, ammonium chloride, ammonium dichromate(VI).

9 Outline the similarities and differences between carbon monoxide and nitrogen monoxide.

10 What do the ozone molecule and the nitrite ion have in common?

11 Illustrate from the chemistry of the element nitrogen, its oxides and chlorides, the anomalous character of the first element in a periodic group. (OC)

12 'Nitric acid can behave in various ways according to the conditions under which it is used'. Illustrate this statement by classifying the reactions of nitric acid and taking examples from all branches of chemistry. Give the electronic structure of nitric acid and comment on it. (L)

13 Compare and contrast the structures of (a) CO and N_2, (b) NO and NO^+, (c) NO_2 and NO_2^+.

14* Use bond enthalpy values (p. 38) to explain why ammonia is an exothermic compound whereas hydrazine is endothermic.

15* Concentrated solutions of hydrazine and hydrogen peroxide have been tried as rocket fuels. Use standard enthalpies of formation to calculate the enthalpy change in the reaction

$$N_2H_4(l) + 2H_2O_2(l) \longrightarrow N_2(g) + 4H_2O(g)$$

20
Phosphorus, arsenic, antimony and bismuth

1 Manufacture and uses of phosphorus

Phosphorus occurs naturally as rock phosphate, which contains fluoro-apatite, $CaF_2.3Ca_3(PO_4)_2$; it is mined mainly in Florida. There are two main allotropic forms of phosphorus (p. 250), which are known as red and white phosphorus, and a much less common black allotrope.

a Manufacture of white phosphorus Rock phosphate is heated with granite chippings and anthracite in an electric furnace at about 1400 °C. The calcium phosphate(V) in the ore functions as a mixture of calcium oxide and phosphorus(V) oxide. The basic calcium oxide reacts with the acidic silicon(IV) oxide in the granite to form calcium silicate(IV), and the phosphorus(V) oxide is reduced by the carbon:

$$Ca_3(PO_4)_2(s) = 3CaO(s) + P_2O_5(s)$$
$$3CaO(s) + 3SiO_2(s) \longrightarrow 3CaSiO_3(s)$$
$$2P_2O_5(s) + 10C(s) \longrightarrow P_4(g) + 10CO(g)$$

The calcium silicate(IV) and calcium fluoride drop to the bottom of the furnace as a molten slag, which can be tapped off periodically. Hot phosphorus vapour and carbon monoxide pass out of the top of the furnace. This mixture is passed through warm water sprays to liquefy the phosphorus. The liquid phosphorus is then pumped away through steam-heated pipes into cold water, which solidifies it. The carbon monoxide passes through the sprays and is used as a fuel. The whole process is continuous.

White phosphorus catches fire spontaneously in a warm room or on touching. It must be stored under water, and treated with care.

b Manufacture of red phosphorus In making red phosphorus, the natural, very slow, change from white to red phosphorus (p. 251) is accelerated by heating white phosphorus up to 250 °C in pots with only a small pipe outlet to the atmosphere to limit the oxidation of the phosphorus. After four days or so, the white phosphorus is almost fully converted into the red form. Any unchanged white phosphorus is removed by distillation and by treatment with sodium carbonate or sodium hydroxide solution. This reacts (p. 252) with white but not with red phosphorus; it also serves to remove any phosphorus(V) oxide which may have formed.

c Uses of phosphorus Most phosphorus is converted into phosphoric(V) acid and its salts, the products being required for use in fertilisers, detergents and soft drinks. Smaller quantities of phosphorus are used in matches (red phosphorus) and in alloys, e.g. phosphor-bronze.

Organo-phosphorus compounds find use in weedkillers, insecticides, plasticisers and oil additives. Phosphate esters are involved in many essential biological processes (p. 257).

2 Allotropes of phosphorus

Nitrogen is an insoluble gas, made up of N_2 molecules, whereas phosphorus has at least three allotropes, with very different structures.

White, or yellow, phosphorus is a yellowish-white, waxy, poisonous solid which melts at 44.2 °C and boils, in the absence of air, at 280.5 °C. It is insoluble in water (under which it is stored) but very soluble in carbon disulphide and soluble in benzene and other organic solvents. White phosphorus consist of tetrahedral P_4 molecules (Fig. 81) in a molecular crystal. The same molecule also occurs in phosphorus vapour, though a P_2 molecule, similar to the N_2 molecule, exists above about 750 °C.

Fig. 81 The P_4 molecule.

Red phosphorus has a higher density than white; it is a violet–red powder which is not poisonous and which is insoluble in water, carbon disulphide and most common solvents. Red phosphorus sublimes at temperatures over about 400 °C; white phosphorus forms on cooling the vapour from it. The structure of red phosphorus consists of cross-linked chains of P_4 tetrahedra, as shown in Fig. 82.

Fig. 82 Chains of P_4 groups in red phosphorus crystals.

Red phosphorus is stabler than white at room temperature,

$$P_4(\text{white}) \longrightarrow P_4(\text{red}) \qquad \Delta H_m^{\ominus}(298\,\text{K}) = -73.64\,\text{kJ}\,\text{mol}^{-1}$$

and white phosphorus slowly changes into red. At room temperature, however, the change is so slow that the white allotrope can be stored for long periods without any significant change to the red form.

A black allotrope of phosphorus, which is still more stable than the red form, can be made from white phosphorus, either at high temperature and very high pressure, or at about 250 °C in the presence of a mercury catalyst and a little black phosphorus to act as a 'seed'. Black phosphorus is a dark grey, flaky solid not unlike graphite; it is a conductor of heat and electricity, and has a layer structure. Similar structures occur in the main allotropes of arsenic and antimony.

3 Compounds of phosphorus

Nitrogen is singularly unreactive at room temperature, but phosphorus, particularly the white allotrope, is reactive. The P—P bond in the P_4 molecules of white phosphorus has a bond enthalpy of $209\,kJ\,mol^{-1}$, which is much lower than the bond enthalpy of $N\equiv N$ ($945\,kJ\,mol^{-1}$), though higher than that of the N—N bond ($163\,kJ\,mol^{-1}$). There is also considerable strain in the P_4 molecule, for the bond angle is only 60°, which is much lower than would be expected if phosphorus atoms formed three bonds on the general basis of electron pair repulsion (p. 27).

a Reactions of phosphorus Dry white phosphorus will react spontaneously with oxygen to form P_4O_6 or P_4O_{10}; with chlorine to form PCl_3 or PCl_5; with sulphur to form sulphides; with metals to form phosphides; with sodium hydroxide solution to form phosphine, PH_3; and with concentrated nitric acid to form phosphoric(V) acid, H_3PO_4.

Red phosphorus reacts in similar ways, at higher temperatures, but it will not react with sodium hydroxide solution.

b Comparison with nitrogen Like nitrogen (p. 231), phosphorus forms compounds with many different oxidation states, as shown in the following summary.

−3	−2	−1	0	+1	+2	+3	+4	+5
PH_3	P_2H_4		P_4	H_3PO_2		P_4O_6	$H_4P_2O_6$	P_4O_{10}
PH_4^+						HPO_2		$(HPO_3)_n$
Mg_3P_2						H_2PHO_3		H_3PO_4
P^{3-}						PCl_3		$H_4P_2O_7$
								PCl_5

Phosphides of electropositive metals, e.g. Na_3P, contain P^{3-} ions, just as nitrides contain N^{3-} ions, but most compounds of phosphorus, as of nitrogen, are covalent.

In PX_3 compounds, e.g. PH_3 and PCl_3, phosphorus forms three single bonds, as nitrogen does in its NX_3 compounds. There are, however, no stable phosphorus analogues of $N\equiv N$ or $C\equiv N$ bonds. Moreover, the lone pair of a PX_3 molecule is donated to another atom or group much less readily than that of an NX_3 molecule.

Another important difference between nitrogen and phosphorus arises because phosphorus has a maximum covalency (p. 36) of 6, whereas that of nitrogen is only 4. As a result, compounds such as PCl_5 and $POCl_3$ have no nitrogen counterparts.

The similarities and differences between nitrogen and phosphorus and the other three Group 5B elements, arsenic, antimony and bismuth, are summarised on p. 260.

4 Phosphine, PH_3

Phosphorus forms two main hydrides, phosphine, PH_3, and diphosphane, P_2H_4, which correspond to ammonia, NH_3, and hydrazine, N_2H_4.

a Preparation of phosphine Phosphine is generally made by heating white phosphorus with a concentrated solution of sodium hydroxide in a flask in which the air has been displaced by natural gas:

$$P_4(s) + 3NaOH(aq) + 3H_2O(l) \longrightarrow 3NaH_2PO_2(aq) + PH_3(g)$$

A side reaction produces some diphosphane, P_2H_4, and as this is spontaneously flammable in air, the gas coming from the flask produces smoke rings of phosphorus(V) oxide when bubbled through water. The diphosphane can be removed by passing the gas mixture through a freezing mixture, for it is a liquid with a boiling point of 52 °C.

Phosphine can also be prepared by the reduction of phosphorus trichloride with lithium tetrahydridoaluminate(III) or by the action of dilute acids on most metallic phosphides.

b Physical properties and structure Phosphine is a colourless, very poisonous gas. Compared with ammonia, it is only slightly soluble in water, is not associated, has lower melting and boiling temperatures, has a lower dipole moment (0.55 D or 1.83×10^{-30} C m), and is not ionised in the liquid state, so it is not a good ionising solvent.

The PH_3 molecule is pyramidal and inverts, like the NH_3 molecule, but the H–P–H angle is 93°, against 107° for the H–N–H angle. The smaller angle in phosphine is due to the lower electronegativity of phosphorus as compared with nitrogen. This causes lower polarity in the P—H bond than in the N—H bond, so that repulsion between hydrogen atoms in PH_3 is less than in NH_3 molecules. It is the lower polarity in PH_3 that mainly accounts for the other physical differences between it and ammonia.

c Comparison with ammonia The chemical differences between phosphine and ammonia arise because *i* phosphorus has a lower electronegativity than nitrogen,

ii phosphorus has a maximum covalency of 6,

iii the P—H bond ($322 \, kJ \, mol^{-1}$) is weaker than the N—H bond ($391 \, kJ \, mol^{-1}$),

iv the P—H bond is less polar than the N—H bond, so it participates in hydrogen bonding less readily, and

v phosphine does not donate its lone pair of electrons so readily as ammonia.

Some of the chemical consequences of these differences are outlined below.

i Phosphine is an endothermic compound, whereas ammonia is exothermic:

$$4PH_3(g) \longrightarrow P_4(g) + 6H_2(g) \qquad \Delta H_m^{\ominus}(298 \, K) = -36.8 \, kJ \, mol^{-1}$$
$$4NH_3(g) \longrightarrow 2N_2(g) + 6H_2(g) \qquad \Delta H_m^{\ominus}(298 \, K) = +184.8 \, kJ \, mol^{-1}$$

As a result, phosphine decomposes into its elements, on heating, at a much lower temperature than ammonia.

ii Phosphine is more readily oxidised than ammonia, so it is a stronger reducing agent. Unlike ammonia, it will burn in air, igniting, and possibly exploding, at about 150 °C:

$$4PH_3(g) + 8O_2(g) \longrightarrow P_4O_{10}(s) + 6H_2O(l) \longrightarrow 4H_3PO_4(aq)$$

It also catches fire in chlorine, producing phosphorus trichloride, and it is a strong enough reducing agent to convert copper(II) or silver salts into the metals or phosphides.

iii Phosphine donates its lone pair of electrons to a hydrogen ion, H^+, much less readily than ammonia does, so that it is much less basic. An aqueous solution of phosphine is, in fact, neutral to litmus, but the slight basic character is shown by the formation of some phosphonium, PH_4^+, salts, which correspond to the ammonium salts.

Phosphonium iodide, which is made from dry phosphine and hydrogen iodide, is the commonest and most stable of these, but it decomposes at a low temperature and is readily hydrolysed by water:

$$PH_4I(s) + H_2O(l) \longrightarrow PH_3(g) + H_3O^+(aq) + I^-(aq)$$

The chlorides and bromides are even less stable.

iv Just as phosphine donates its lone pair to the H^+ ion less readily than ammonia does, so it forms many fewer complexes. The enormous number of ammines formed with NH_3 as a ligand is not matched by the number of complexes with PH_3 as ligand; but substituted phosphines, e.g. $(C_6H_5)_3P$ are excellent as ligands.

d Summary A summary of the differences between phosphine and ammonia is given below.

ammonia	phosphine
colourless, smelly, poisonous gases	
decomposes on strong heating	decomposes at lower temperature
burns in oxygen	burns in air above $150\,°C$
N—H bond enthalpy, $391\,kJ\,mol^{-1}$	P—H bond enthalpy, $322\,kJ\,mol^{-1}$
strong base	weaker base
forms many NH_4^+ salts	forms a few PH_4^+ salts
weak reducing agent	strong reducing agent
pyramidal molecule: bond angle = 107°	pyramidal molecule: bond angle = 93°
hydrogen bonded	not hydrogen bonded
very soluble in water	almost insoluble
high dipole moment (1.48 D)	lower dipole moment (0.55 D)
high relative permittivity	low relative permittivity
good ionising solvent	poor ionising solvent
forms many ammines	forms fewer complexes

5 Oxides of phosphorus

The two main oxides of phosphorus, together with their standard enthalpies of formation and oxidation states are

+5	P_4O_{10}	phosphorus(V) oxide	$-3012\,kJ\,mol^{-1}$
+3	P_4O_6	phosphorus(III) oxide	$-1640\,kJ\,mol^{-1}$

When it is necessary to emphasise the molecular nature of the oxides, P_4O_6 is named tetraphosphorus hexaoxide, and P_4O_{10}, tetraphosphorus decaoxide.

The oxides of phosphorus have high negative enthalpies of formation, unlike the oxides of nitrogen, so that they are neither decomposed nor reduced easily. There are, in fact, few likenesses between the oxides of phosphorus and nitrogen, but those in oxidation states +3 and +5 are all acidic.

a Phosphorus(III) oxide, P_4O_6 This is made by burning excess white phosphorus in a slow stream of air to lower the oxygen supply. A mixture of phosphorus(III) and phosphorus(V) oxides is formed. The latter is filtered out by a glass wool plug in a U-tube maintained at 50 °C; the phosphorus(III) oxide which passes through is collected in an ice-cooled container.

Phosphorus(III) oxide is a white, waxy, poisonous solid which catches fire on heating, being converted into the higher oxide:

$$P_4(s) + 3O_2(g) \longrightarrow P_4O_6(s) \qquad \Delta H_m^{\ominus}(298\,K) = -1640\,kJ\,mol^{-1}$$
$$P_4O_6(s) + 2O_2(g) \longrightarrow P_4O_{10}(s) \qquad \Delta H_m^{\ominus}(298\,K) = -1372\,kJ\,mol^{-1}$$

It is the acid anhydride of phosphonic acid, H_2PHO_3, which it forms with cold water, but with hot water this acid decomposes to give phosphoric(V) acid, H_3PO_4, and phosphine, sometimes explosively:

$$P_4O_6(s) + 6H_2O(l) \xrightarrow{\text{cold}} 4H_2PHO_3(aq) \xrightarrow{\text{hot}} PH_3(g) + 3H_3PO_4(aq)$$

Density measurements in the vapour state and relative molecular mass measurements in organic solvents show that the molecular formula of phosphorus(III) oxide is P_4O_6. The structure is based on the tetrahedral arrangement of four phosphorus atoms found in the element (Fig. 83(a)).

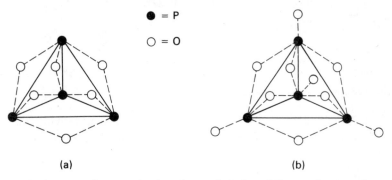

● = P

○ = O

(a)   (b)

Fig. 83 *Structures of (**a**) tetraphosphorus hexaoxide, P_4O_6, and (**b**) tetraphosphorus decaoxide, P_4O_{10}.*

b Phosphorus(V) oxide, P_4O_{10} This is a white powder made by burning white phosphorus or phosphorus(III) oxide, in excess oxygen.

It reacts vigorously with water to produce a mixture of acids (p. 256):

$$P_4O_{10}(s) + 2H_2O(l) \longrightarrow 4HPO_3(aq)$$
$$2HPO_3(s) + H_2O(l) \longrightarrow H_4P_2O_7(aq)$$
$$HPO_3(s) + H_2O(l) \longrightarrow H_3PO_4(aq)$$

The affinity of phosphorus(V) oxide for water is such that it is a very strong drying and dehydrating agent. It will dehydrate nitric and sulphuric acids to dinitrogen pentoxide and sulphur(VI) oxide respectively, and acid amides to nitriles.

Relative molecular mass measurements show that the molecule is P_4O_{10}, and the structure is based on that of P_4O_6 (Fig. 83(b)).

6 Common oxoacids of phosphorus

a Nomenclature There are many oxoacids of phosphorus (Fig. 84) and the nomenclature is difficult because many older names still persist.

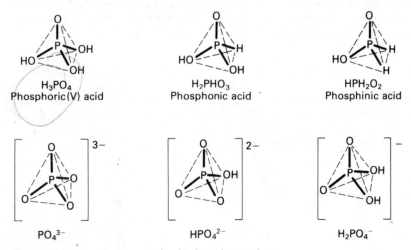

Fig. 84 *The shapes of some oxoacid molecules and oxoacid ions.*

The acids with oxidation state $+5$ can all be regarded as deriving from the dehydration of the non-existent $P(OH)_5$ in the following ways.

$$P(OH)_5 \xrightarrow{-H_2O} H_3PO_4$$

$$2P(OH)_5 \xrightarrow{-3H_2O} H_4P_2O_7$$

$$nP(OH)_5 \xrightarrow{-2nH_2O} (HPO_3)_n$$

H_3PO_4, *phosphoric(V) acid*, is the commonest. Loss of water from between two molecules of this acid gives $H_4P_2O_7$, which is a condensed acid called *diphosphoric(V) acid*. The polymer, $(HPO_3)_n$, contains $(PO_3)_n^{n-}$ ions, and these are called *polytrioxophosphate(V) ions*.

The commonest oxoacid with oxidation state $+3$ is H_2PHO_3, *phosphonic acid*. The formula is written as H_2PHO_3 to denote the presence of a P—H bond. *Phosphinic acid*, HPH_2O_2, contains two P—H bonds; the oxidation state is $+1$.

b Phosphoric(V) acid, H_3PO_4 This is made by reacting phosphorus(V) oxide with hot water, by oxidising phosphorus with concentrated nitric acid or by treating calcium phosphate(V) with sulphuric acid:

$$P_4O_{10}(s) + 6H_2O(l) \longrightarrow 4H_3PO_4(aq)$$
$$P_4(s) + 10HNO_3(aq) + H_2O(l) \longrightarrow 4H_3PO_4(aq) + 5NO(g) + 5NO_2(g)$$
$$Ca_3(PO_4)_2(s) + 3H_2SO_4(aq) \longrightarrow 3CaSO_4(s) + 2H_3PO_4(aq)$$

Pure phosphoric(V) acid is a colourless solid which is very soluble in water, giving a syrupy solution. The tetrahedral H_3PO_4 molecules (Fig. 84) are hydrogen bonded to some extent in the solid, and it is probably the hydrogen bonding even in solution that causes the syrupy nature. Phosphoric(V) acid is a tribasic acid, forming PO_4^{3-}, HPO_4^{2-} and $H_2PO_4^{-}$ ions (Fig. 84).

The pure acid melts at 42 °C, and further heating causes loss of water to give diphosphoric(V) acid. Still stronger heating yields $(HPO_3)_n$:

$$2H_3PO_4 - H_2O \longrightarrow H_4P_2O_7$$
$$nH_4P_2O_7 - nH_2O \longrightarrow 2(HPO_3)_n$$

Phosphoric(V) acid is mainly used in making salts, in rustproofing processes and in making soft drinks.

c Phosphates(V) Phosphoric(V) acid forms one normal and two acid salts, i.e.

$NaH_2PO_4 . H_2O$	$Na_2HPO_4 . 12H_2O$	$Na_3PO_4 . 12H_2O$
sodium dihydrogen phosphate(V)	disodium hydrogen phosphate(V)	trisodium phosphate(V)

They all contain tetrahedral anions (Fig. 84).

Calcium phosphate(V), which occurs naturally, is used in making phosphorus (p. 249) and the important fertilisers, superphosphate and triplephosphate. The acid salts, monoammonium phosphate (MAP) and diammonium phosphate (DAP) are also widely used as fertilisers; they are both nitrogenous and phosphatic.

H_3PO_4 contains three hydroxyl groups, so that it can form mono-, di- or tri-esters with alcohols. Many of these play essential roles in biological processes. Phospholipids or phosphatides occur in cell membranes; adenosine phosphates are involved in many biosynthetic processes with important energy changes; and DNA and RNA both contain phosphate ester linkages.

7 Halides of phosphorus

a Trihalides, PX_3 Phosphorus forms PX_3 trihalides with all the halogens. The compounds are thermally stable, and have pyramidal molecules. They react with water, the fluoride only slowly, e.g.

$$PCl_3(l) + 3H_2O(l) \longrightarrow H_2PHO_3(aq) + 3HCl(g)$$

in a different way from the nitrogen halides, which produce HClO and not HCl (p. 317).

b Phosphorus trichloride, PCl_3 This is made by reaction of red phosphorus with a limited supply of chlorine; with excess chlorine, PCl_5 is the main product.

A slow stream of dry chlorine is led over hot red phosphorus in a combustion tube, and the phosphorus trichloride vapour is condensed in a receiver immersed in a freezing mixture. Access to the atmosphere is through a soda lime bulb.

Phosphorus trichloride is a fuming liquid which boils at 76 °C. It reacts with water and with other compounds containing —OH groups, the —OH groups being replaced by —Cl, e.g.

$$PCl_3(l) + 3H_2O(l) \longrightarrow H_2PHO_3(aq) + 3HCl(g)$$
$$PCl_3(l) + 3ROH(l) \longrightarrow H_2PHO_3 + 3RCl$$

It also reacts with chlorine to form the pentachloride and with oxygen to form phosphorus trichloride oxide, $POCl_3$. This is another fuming liquid, which boils at 108 °C and is hydrolysed to phosphoric(V) acid:

$$POCl_3(l) + 3H_2O(l) \longrightarrow H_3PO_4(aq) + 3HCl(g)$$

It is used in making phosphate esters (p. 257):

$$POCl_3(l) + 3ROH(l) \longrightarrow PO(OR)_3 + 3HCl(g)$$

c Pentahalides, PX_5 Unlike nitrogen, phosphorus forms pentahalides with all the halogens except iodine, the non-existence of the pentaiodide probably being due to the large size of the iodine atom.

The pentahalides can be made by treating phosphorus, or the trihalide, with excess halogen. They decompose into the trihalide and halogen on heating, and they are hydrolysed as described for phosphorus pentachloride below.

d Phosphorus pentachloride, PCl_5 This is made by dropping phosphorus trichloride into a flask through which dry chlorine is passing. This ensures that there is an excess of chlorine, and phosphorus pentachloride collects in the flask.

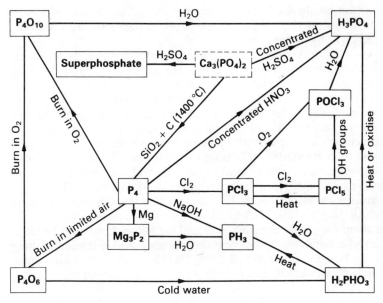

Chart 11 *Phosphorus and its simple compounds. $Ca_3(PO_4)_2$ occurs naturally.*

It is a white solid which sublimes and dissociates on heating,

$$PCl_5(g) \rightleftharpoons PCl_3(g) + Cl_2(g)$$

and which is readily hydrolysed in two stages:

$$PCl_5(s) + H_2O(l) \longrightarrow POCl_3(l) + 2HCl(g)$$
$$POCl_3(l) + 3H_2O(l) \longrightarrow H_3PO_4(aq) + 3HCl(g)$$

It also reacts with other compounds containing —OH groups, e.g.

$$PCl_5(s) + ROH \longrightarrow POCl_3(l) + RCl + HCl(g)$$

Phosphorus pentachloride contains trigonal bipyramidal PCl_5 molecules in the vapour state (p. 28) but it is made up of $[PCl_4]^+$ and $[PCl_6]^-$ ions in the solid, the $[PCl_4]^+$ ion being tetrahedral and the $[PCl_6]^-$ ion octahedral.

ARSENIC, ANTIMONY AND BISMUTH

8 The elements

The elements occur in a number of ores, with the sulphides orpiment (As_2S_3), stibnite (Sb_2S_3) and bismuth glance (Bi_2S_3) being typical. The elements can be obtained either by heating these sulphide ores with iron, or by first roasting them to form the oxides and then reducing the oxides with carbon or hydrogen.

They are all dark solids with a shiny, metallic appearance and some electrical conductivity, which decreases from arsenic to bismuth. They all have layer structures, similar to that of black phosphorus (p. 251), in their main allotropic forms, but arsenic and antimony (but not bismuth) have other unstable allotropes, which contain As_4 and Sb_4 molecules. Arsenic and antimony are best described as semi-metals; bismuth is metallic.

The elements form trivalent oxides on heating with oxygen and either trihalides or pentahalides on heating with halogens. Concentrated nitric acid converts arsenic into arsenic(V) acid, H_3AsO_4; antimony into antimony(V) oxide, Sb_2O_5; and bismuth into bismuth(III) nitrate, $Bi(NO_3)_3$. The nature of the product in the last reaction illustrates the metallic nature of bismuth.

The main use of the elements is in making alloys. Lead shot contains arsenic; type metal and pewter contain antimony; and many low melting alloys, e.g. Wood's metal (50% Bi : 25% Pb : 12.5% Sn : 12.5% Cd), which melts at 71 °C, contain bismuth.

9 The compounds of arsenic, antimony and bismuth

a Comparison with nitrogen and phosphorus The main oxidation states of arsenic and antimony are -3, $+3$ and $+5$; bismuth is similar, but the $+5$ state is less common. The following summary shows some typical examples.

-3		$+3$		$+5$		
Mg_3As_2 AsH_3		$AsCl_3$	As_4O_6	AsF_5	As_2O_5	H_3AsO_4
As^{3-} SbH_3	Sb^{3+}	$SbCl_3$	Sb_4O_6	SbF_5	Sb_2O_5	
BiH_3	Bi^{3+}	$BiCl_3$	Bi_2O_3	BiF_5	Bi_2O_5	

Arsenic forms arsenides with some metals, e.g. Mg_3As_2, and these contain As^{3-} ions, which are similar to the nitride and phosphide ions. But such ions are not formed by antimony or bismuth.

The covalent compounds of arsenic, antimony and bismuth resemble those of nitrogen and phosphorus, but there is a general decrease in the bond energy of the covalent bonds formed with other atoms in passing from nitrogen to bismuth.

Bismuth and, to a lesser extent, antimony can also form ionic compounds containing trivalent ions, but there are no Bi^{5+} or Sb^{5+} ions.

b The hydrides Arsine, AsH_3, and stibine, SbH_3, are made by reducing the necessary trihalide with tetrahydridoaluminate(III). Bismuthine, BiH_3, is formed when dilute acids react with an alloy of bismuth and magnesium. The hydrides are unpleasant-smelling, poisonous gases.

The molecules of the hydrides are all pyramidal in shape, like those of NH_3 and PH_3, but with decreasing bond angles in passing from NH_3 to

BiH_3. The covalent bond energies in the molecules fall in passing from nitrogen to bismuth, and this means that there is a fall in thermal stability from NH_3 to BiH_3. BiH_3 is particularly unstable, and arsine decomposes into its elements on gentle heating.

Neither AsH_3, SbH_3 nor BiH_3 exhibits any basic properties, i.e. they do not make use of their lone pairs to form ions such as NH_4^+ and PH_4^+.

c The halides All three elements form trihalides with fluorine, chlorine, bromine and iodine; they are made by direct combination of the elements, using an excess of arsenic, antimony or bismuth to limit the formation of any pentahalide. With excess halogen it is possible to make $SbCl_5$ and the pentafluorides of arsenic, antimony and bismuth, but the other pentahalides do not exist.

Bismuth(III) fluoride is essentially ionic and is not affected by water. The other halides are mainly covalent and they are hydrolysed by water in much the same way as PCl_3 and PCl_5, but the reactions are reversible; with bismuth(III) chloride the product is a white, insoluble solid, bismuth(III) chloride oxide, $BiClO$:

$$BiCl_3(aq) + H_2O(l) \rightleftharpoons BiClO(s) + 2HCl(aq)$$

d The oxides Arsenic, antimony and bismuth all combine with oxygen on heating to form trioxides. Arsenic(III) and antimony(III) oxides are like phosphorus(III) oxide and have similar structures; they are formulated as As_4O_6 and Sb_4O_6 respectively. Bismuth(III) oxide does not contain Bi_4O_6 molecules within its crystals, so it is written as Bi_2O_3.

As_4O_6 is predominantly acidic, like P_4O_6, and it reacts with alkalis to form the arsenate(III) ion, AsO_3^{3-}. Sb_4O_6 is amphoteric; it gives antimonate(III) ions, SbO_3^{3-}, with alkalis, and Sb^{3+} ions with acids. Bi_2O_3 is basic; it does not react with alkalis, but reacts with acids to form salts containing Bi^{3+} ions.

Oxidation of arsenic or antimony by concentrated nitric acid and dehydration of the products gives As_2O_5 and Sb_2O_5 respectively. But Bi_2O_5 is only known in an impure form made by oxidising Bi_2O_3 with peroxodisulphate(VI) ions. All the pentaoxides decompose into the trioxide and oxygen on heating.

Arsenic(V) oxide is an acidic oxide, like phosphorus(V) oxide. It reacts with water to form arsenic(V) acid, H_3AsO_4, which gives a series of arsenate(V) salts containing AsO_4^{3-} ions.

Questions on Chapter 20

1 What is meant by saying that phosphorus pentachloride dissociates? Give six other examples of dissociation that are similar.

2 How are the various allotropes of phosphorus made, and how can they be interconverted?

3 Compare the chemistry of the chlorides of nitrogen and phosphorus.

4 In what important ways do nitrogen and its compounds differ from phosphorus and its compounds?

5 Give a brief comparative account of the chemistry of nitrogen and phosphorus. Liquid ammonia ionises slightly according to the equation

$$2NH_3 \longrightarrow NH_4^+ + NH_2^-$$

What would you expect to happen if sodium is added to liquid ammonia and, when reaction is complete, a small quantity of zinc powder is then added? (CS)

6 In what general ways does bismuth differ from nitrogen?

7 Draw up a table, similar to the one on p. 254, to compare the properties of AsH_3 and SbH_3.

8 How do the general trends in passing down Group 4B resemble those in passing down Group 5B?

9 Use phosphorus and bismuth as examples to discuss the general differences between non-metals and metals.

10 For nitrogen, phosphorus and arsenic, give (a) the electronic configurations of the elements, (b) a different type of reaction in each case for the formation of the trihydrides of the elements, and (c) the structures, relating them to the physical states and stabilities of the elements.

11 Give concise explanations for *each* of the following: (a) NH_3 readily accepts a proton but AsH_3 does not; (b) nitrogen is much less reactive than phosphorus and arsenic; (c) nitrogen forms N^{3-} but not N^{3+}, whilst bismuth forms Bi^{3+} but not Bi^{3-}; (d) the bond angle in ammonia is 107° but in phosphine it is 93°; (e) ammonia has a higher dissociation constant than phosphine.

12 (a) Give *one* example of a difference, apart from colour, between the physical properties and *one* example of a difference between the chemical properties of the red and white allotropes of phosphorus. What is the reason for these differences? (b) How may red phosphorus be converted into white phosphorus? (c) Why does nitrogen not form allotropes similar to those of phosphorus? (d) A compound PBr_x contains 88.6 per cent by mass of bromine. Deduce the molecular formula for PBr_x and sketch the shape of the molecule, giving your reasons. (e) Under certain conditions, PBr_x reacts with bromide ions to form the ion $[PBr_{x+1}]^-$. Predict the shape of this ion, giving your reasons. (OC)

13 Nitrogen, like boron, aluminium and phosphorus, forms trivalent compounds. Like phosphorus but unlike boron and aluminium it can form compounds with oxidation states of $+5$. Explain these observations and give *one* suitable illustrative example for each element.

Explain, giving *one* example for each element, why all the elements can form compounds in which the element has a coordination number of four but in some of their compounds aluminium and phosphorus have higher coordination numbers.

Explain the differences in the acidic behaviour of (i) boric(III) acid and hydrated Al^{3+} and (ii) nitric acid and phosphoric(V) acid. (W)

14* Compare the combustion of ammonia and phosphine, and the hydrolysis of nitrogen and phosphorus trichlorides.

15* Use standard enthalpies of formation to explain why phosphine will burn so much more readily than ammonia.

16* Use the enthalpies of atomisation of nitrogen, phosphorus and chlorine and the bond enthalpies of the N—Cl and P—Cl bonds to calculate the enthalpies of formation of $NCl_3(g)$ and $PCl_3(g)$. Comment on the two values.

17* Comment on the enthalpy changes in the reactions between ammonia, hydrazine and phosphine with oxygen.

18* Calculate the enthalpy change for the formation of 1 mol of H_3PO_4 from P_4O_{10} and water.

21
Oxygen

1 Sources of oxygen

Oxygen is the most abundant element in the earth's crust (approximately 50 per cent). It occurs free in the atmosphere (20.8 per cent by volume or 23 per cent by weight) and in combination with hydrogen as water, and with many other elements as oxides. It was discovered in 1774 by Scheele, who heated potassium nitrate, and by Priestley, who heated mercury(II) oxide.

Oxygen can be obtained by heating oxides of metals below copper in the reactivity series, or by heating many higher oxides, e.g. Pb_3O_4 and PbO_2, many peroxides, e.g. BaO_2, and many salts containing ions rich in oxygen, e.g. nitrates, chlorates(VII) and manganates(VII).

Oxygen is manufactured, however, by the distillation of liquid air and is readily available in cylinders.   ▷ on industrial basis.

2 Properties and uses of oxygen

Oxygen is a colourless, odourless gas, which is only very slightly soluble in water but soluble in alkaline benzene-1, 2, 3-triol (pyrogallol) solutions. It condenses at $-183\,°C$ to a pale blue liquid, and solidifies into a pale blue solid at $-218\,°C$.

The oxygen molecule is diatomic. The simplest formulation is $O{=}O$ but both gaseous and liquid oxygen are paramagnetic (p. 352) to an extent that shows the presence of two unpaired electrons.

Oxygen will react with most other elements and, once initiated, the reactions generally form oxides (p. 273) with much heat; they may be explosive if gases or finely powdered solids are involved. The conditions under which the reactions take place are, however, very varied. Hydrogen and oxygen will not react unless a catalyst is present or the mixture is sparked; phosphorus and amalgamated aluminium react in the cold; magnesium reacts on heating and produces a vivid white flame; iron in bulk reacts very slowly but dry, finely powdered iron can be pyrophoric; the reaction between nitrogen and oxygen (p. 239) is endothermic and produces a yield of only 2 per cent, even at $2000\,°C$.

Many compounds, as well as elements, will also react with oxygen with the evolution of heat and, very often, the emission of light. Such reactions are examples of combustion reactions, and oxygen is referred to as a good supporter of combustion. It is widely used in this role to support the combustion of fuels. Slow combustion can also take place as, for example, in the reaction between food and oxygen in the body, in the 'drying' of paint, and in the rotting of vegetation.

Oxygen is used in treating lung diseases and other breathing difficulties, and in mountaineering, high altitude flying, underwater diving, mine rescue and fire fighting when there is a shortage or complete absence of air. It is also used in the oxy-acetylene (ethyne) or oxy-hydrogen blow-pipe flames used for cutting and welding metals, in rocket engines, in making steel (p. 370) and in the manufacture of nitric acid (p. 242).

3 Preparation of ozone

Ozone, O_3, is a relatively unstable allotrope of oxygen, O_2, and the names trioxygen and dioxygen are used in this context. Ozone is made from oxygen by subjecting it to a silent electric discharge in an ozoniser (Fig. 85).

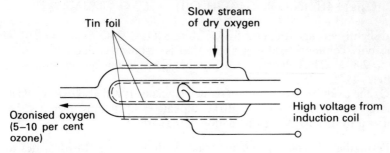

Fig. 85 *A typical ozoniser.*

The electric discharge passes through the glass of the apparatus as well as through the oxygen. This prevents any large rise in temperature such as would result from an ordinary electric spark, and which would tend to decompose any ozone formed. The electric discharge is, nevertheless, sufficient to provide the necessary energy to convert dioxygen into trioxygen:

$$3O_2(g) \rightleftharpoons 2O_3(g) \qquad \Delta H_m^{\ominus}(298\,\text{K}) = +284.6\,\text{kJ mol}^{-1}$$

The ozonised oxygen coming out of the ozoniser contains 5–10 per cent of ozone. To obtain pure ozone the mixture must be liquefied and fractionally distilled.

4 Properties and uses of ozone

a Physical properties Ozone is a pale blue gas with a characteristic smell; it is only slightly soluble in water. The ozone molecule is triangular with a bond angle of 116.8° and two equal bond lengths of 0.128 nm; it is best represented with delocalised bonds:

b As an oxidising agent Ozone decomposes on heating, possibly explosively, and particularly in the presence of catalysts such as finely powdered metals or oxides:

$$2O_3(g) \longrightarrow 3O_2(g) \qquad \Delta H_m^{\ominus}(298\,K) = -284.6\,kJ\,mol^{-1}$$

It is the readiness with which ozone can provide oxygen that makes it one of the strongest known oxidising agents:

$$O_3(g) + 2H^+(aq) + 2e^- \longrightarrow O_2(g) + H_2O(l) \qquad E^{\ominus} = +2.07\,V$$

It is a much stronger oxidising agent than oxygen,

$$O_2(g) + 4H^+(aq) + 4e^- \longrightarrow 2H_2O(l) \qquad E^{\ominus} = +1.23\,V$$

and slightly stronger than the $S_2O_8^{2-}$ ion; $S_2O_8^{2-}/2SO_4^{2-}$, $E^{\ominus} = 2.01\,V$. The only common oxidising agent that is stronger is fluorine; F_2/F^-, $E^{\ominus} = 2.87\,V$. Thus fluorine will produce some ozone in reaction with water.

Ozone oxidises most common reducing agents. It will also oxidise black lead(II) sulphide to white lead(II) sulphate, and it 'tails' mercury, i.e. causes it to stick to glass by oxidation of the surface.

c Formation of ozonides Ozone reacts with alkali metal hydroxides to form ozonides, MO_3, which contain the O_3^- ion; these are yellow or red solids.

A different type of compound, also called an ozonide, is formed when ozone reacts with many organic compounds containing C=C bonds. These ozonides contain a peroxide bond, —O—O—; they are oily, explosive liquids, which are not generally isolated. They can be decomposed into mixtures of ketones or aldehydes by hydrogen, or into mixtures of ketones, aldehydes and acids by water, with the overall effect of splitting the double bond. The process is known as *ozonolysis* and the nature of the products allows the position of a multiple bond in a compound to be determined. Ozone also slowly rots rubber because the rubber molecules contain C=C bonds.

d Uses of ozone Ozone is used as a bleaching agent and as a disinfectant, on account of its oxidising powers. It is used in water and sewage purification and in air conditioning plants, though there is no evidence to support the widely held view that there is more ozone by the seaside than inland. Ozone is also used in ozonolysis and in testing rubbers.

5 Compounds of oxygen

Oxygen occurs in very many compounds, generally in an oxidation state of -2.

It can form the O^{2-} ion, as in ionic oxides, but the ion does not exist in aqueous solutions, due to hydrolysis:

$$O^{2-} + H_2O(l) \longrightarrow 2OH^-(aq)$$

In its covalent compounds, it either forms two single covalent bonds, as in water, or a double covalent bond, as in the $C{=}O$ bond in ketones and aldehydes. Having formed two covalent bonds, an oxygen atom still has two lone pairs of electrons. It uses one of these to form dative bonds as, for example, in H_3O^+ (p. 25) and in the many aqua-complexes (p. 268). But the use of *both* lone pairs is rare, and the maximum covalency of 4 is only attained in compounds such as basic beryllium ethanoate, $Be_4O(CH_3COO)_6$, in which a central oxygen atom is surrounded by four beryllium atoms arranged tetrahedrally. It is one of the few compounds in which oxygen acts as a central atom with ligands around it.

In peroxides, which contain either the $(O{-}O)^{2-}$ ion (p. 274) or the $-O-O-$ bond (p. 272), the oxidation state is -1. In oxygen fluorides, the oxidation state may be $+2$, as in OF_2, or $+1$ as in O_2F_2, because fluorine is more electronegative than oxygen.

6 Water   associated liquid due to H bonds.

a Physical properties and structure   Ice, water and steam all consist of triangular H_2O molecules with bond angles of 104.5° and bond lengths of 0.096 nm. The high electronegativity of oxygen makes the $O-H$ bond very polar, which leads to strong hydrogen bonding, and this accounts for many of the 'odd' properties of ice and water.

Ice is particularly unusual for it has several different modifications depending on the temperature and pressure. Ordinary ice, at around 0 °C and 1 atmosphere pressure, is known as ice I. Its structure contains H_2O molecules surrounded by four other H_2O molecules arranged tetrahedrally and linked by hydrogen bonds (Fig. 36, p. 54). The structure is very 'open', giving ice a lower density (0.917 g cm^{-3} at 0 °C) than water. As the ice melts at 0 °C, the structure begins to break down and the H_2O molecules pack more tightly together so that the density increases. The packing is at its tightest at 4 °C, when the density is 1 g cm^{-3}; above that temperature, normal expansion causes a density decrease (to 0.997 g cm^{-3} at 25 °C).

There is still considerable hydrogen bonding in liquid water, so it is associated, and this accounts for its relatively high melting and boiling temperatures (p. 53), for its high relative permittivity (dielectric constant) (81), for its high dipole moment (1.85 D or 6.14×10^{-30} C m), and for its conductivity (4×10^{-6} S m^{-1}), which is caused by the slight ionisation:

$$2H_2O \rightleftharpoons H_3O^+ + OH^-$$

These properties make water an extremely good ionising solvent, and it is the medium in which many chemical and biological processes take place. It is also good as a solvent because it is thermally stable, only decomposing slightly into its elements at temperatures around 2000 °C, and not readily oxidised or reduced.

b Hydrates and aqua-complexes   Many solids exist in hydrated forms, i.e. they contain water within their structures. The water molecules may be present as ligands in aqua-cations or aqua-anions, or bound more loosely within a crystal structure.

Water is a common ligand, and it forms hosts of aqua-complexes. With a metallic cation, it is the negative charge and the lone pair, on the oxygen atom in a water molecule that are responsible for the bonding as, for example, in $[Cu(H_2O)_6]^{2+}$ and $[Fe(H_2O)_6]^{3+}$ (p. 17). With anions, the bonding involves hydrogen bonding through the hydrogen atoms in the water molecule as, for example, in $[SO_4\cdots H_2O]^{2-}$ (p. 55).

It is generally possible to remove water from hydrates containing aqua-complexes by gentle heating. The crystal structure changes completely and lower hydrates or anhydrous compounds are formed.

In noble gas hydrates, the water molecules form 'cages' which trap noble gas atoms; the products are examples of clathrates (p. 143). In zeolites (p. 219), water molecules are trapped within complex silicate structures.

7 Hardness of water   (Revision material)
                                        study

Because calcium and magnesium octadecanoates (stearates) are insoluble, any calcium or magnesium compounds dissolved in water will cause hardness and prevent lathering. The hydrogencarbonates and the sulphates are the salts most commonly present. Any water passing through gypsum deposits will contain some calcium sulphate, and calcium hydrogen-carbonate gets into water supplies because rain water (which contains carbonic acid) dissolves the calcium carbonate in chalk or limestone deposits. Magnesium sulphate and hydrogencarbonate originate in similar ways.

The hardness caused by hydrogencarbonates is said to be *temporary*, for it can be removed by boiling, the soluble hydrogencarbonates being removed from solution by conversion into insoluble carbonates. Sulphates cannot be removed by boiling; they are said to cause *permanent hardness*.

a Removal of temporary hardness   Temporary hardness can be removed by boiling or by treatment with calcium hydroxide.

i Boiling   This forms insoluble calcium carbonate which can be a great nuisance when it forms in kettles or boilers or pipes as a hard deposit of fur or boiler scale. This may result in a blockage of a pipe, and always causes loss of heat since it is such a bad thermal conductor. A similar conversion of hydrogencarbonates into carbonates takes place slowly on evaporation of temporarily hard water, and this can form stalactites and stalagmites.

ii Clarke's method In Clarke's method, slaked lime reacts with calcium and magnesium hydrogencarbonates, but not with sulphates, to form the corresponding insoluble carbonate, e.g.

$$Ca(HCO_3)_2(aq) + Ca(OH)_2(aq) \longrightarrow 2CaCO_3(s) + 2H_2O(l)$$

The correct amount of slaked lime must be used, for any excess would cause hardness just as any other soluble calcium compound does.

b Removal of total hardness Distillation removes all dissolved solids from water so that distilled water is very soft. Both temporary and permanent hardness can, however, be removed more economically, by the following methods.

i Use of sodium carbonate (washing soda) Both hydrogencarbonates and sulphates of calcium and magnesium react with sodium carbonate to produce insoluble calcium or magnesium carbonates. In this way the calcium and magnesium are removed from solution.

ii Use of ion exchange materials Certain insoluble solids are known which can effectively absorb one ion and release another; the materials are known as ion exchangers. Naturally occurring greensands and certain zeolites (aluminosilicates) were first used for this purpose. They can absorb Ca^{2+} or Mg^{2+} ions and release Na^+ ions, and can, therefore, be used as water softeners. They are known as *permutites*, and the natural substances have now been replaced by more effective synthetic ones made by fusing clay, sand and sodium carbonate together. After a time, the ion exchanger has to be regenerated by treatment with concentrated sodium chloride solution. This replaces the absorbed Ca^{2+} and Mg^{2+} ions by the original Na^+ ions so that the material can be used again and again.

Other synthetic polymers have also been developed which will exchange any cations in a solution by H^+ ions (base or cation exchangers) or any anions by OH^- ions (acid or anion exchangers):

$HR(s) + X^+(aq) \rightleftharpoons H^+(aq) + XR(s)$
(cation
exchanger)

$HOR(s) + Y^-(aq) \rightleftharpoons OH^-(aq) + YR(s)$
(anion
exchanger)

These ion exchangers can be used to replace all the X^+ or Y^- ions in a solution by H^+ or OH^- ions; the product is known as *demineralised* or *deionised water* and it is purer than distilled water.

Cation exchangers can be regenerated using acids, and anion exchangers using alkalis.

8 Deuterium oxide (heavy water)

a Preparation Ordinary water contains D_2O and DOH molecules as well as H_2O molecules; the ratio of D_2O to H_2O is about $1:4500$. During electrolysis, hydrogen is evolved more readily than deuterium, so electrolytes become richer in deuterium oxide as electrolysis proceeds. Deuterium oxide is therefore made by repeated electrolysis of a sodium hydroxide solution. Starting with about $35\,dm^3$ of solution from old electrolytic cells it is possible to obtain about $1\,cm^3$ of 99 per cent pure deuterium oxide, and the compound is available commercially.

b Properties and uses Some of the physical properties of deuterium oxide as compared with those of water are summarised below.

	D_2O	H_2O
boiling temperature/°C	101.42	100
melting temperature/°C	3.72	0
temperature at maximum density/°C	11.2	4
density at 25 °C/g cm^{-3}	1.105	0.997

The proportion of deuterium oxide in a sample of water is usually determined by measuring the density.

Deuterium oxide undergoes the same reactions as water (though there may be some difference in the rates of reaction), so that many deutero-compounds can easily be made, e.g.

$$Mg_3N_2(s) + 6D_2O(l) \longrightarrow 3Mg(OD)_2(aq) + 2ND_3(g) \text{ (deuteroammonia)}$$
$$CaC_2(s) + 2D_2O(l) \longrightarrow Ca(OD)_2(aq) + C_2D_2(g) \text{ (deuteroethyne)}$$
$$SO_3(s) + D_2O(l) \longrightarrow D_2SO_4(l) \text{ (deuterosulphuric acid)}$$

Deuterium oxide is used as a moderator, for reducing the speed of neutrons, in nuclear reactors.

9 Preparation of hydrogen peroxide

a From metallic peroxides Metallic peroxides react with acids to form hydrogen peroxide, and both hydrated barium peroxide and sodium peroxide give reasonably concentrated aqueous solutions of hydrogen peroxide on treatment with ice-cold sulphuric acid:

$$BaO_2.8H_2O(s) + H_2SO_4(aq) \longrightarrow BaSO_4(s) + H_2O_2(aq) + 8H_2O(l)$$
$$Na_2O_2(s) + H_2SO_4(aq) \longrightarrow Na_2SO_4(aq) + H_2O_2(aq)$$

The temperature must be kept low to minimise the decomposition of the hydrogen peroxide. In the first reaction, hydrated barium peroxide is used because it gets coated with a layer of insoluble barium sulphate less readily than anhydrous barium peroxide. In the second reaction, the crystallisation of sodium sulphate helps to concentrate the hydrogen peroxide solution.

b By electrolysis Electrolysis of cold 50 per cent sulphuric acid using a high current density produces peroxodisulphuric(VI) acid (p. 298) at the anode,

$$HSO_4^-(aq) \longrightarrow HSO_4 + e^-$$
$$2HSO_4 \longrightarrow H_2S_2O_8(aq)$$

and similar electrolysis of a mixture of ammonium sulphate and sulphuric acid produces ammonium peroxodisulphate(VI).

Both the acid and the salt can be hydrolysed to hydrogen peroxide:

$$H_2S_2O_8(aq) + 2H_2O(l) \longrightarrow H_2O_2(aq) + 2H_2SO_4(aq)$$

c Manufacture of hydrogen peroxide Hydrogen peroxide used to be made electrolytically but nowadays it is mainly made by an autoxidation process using substituted anthraquinones. 2-butyl anthraquinone, for example, is reduced by hydrogen in the presence of a palladium catalyst to the corresponding anthraquinol. Blowing air through the solution of the anthraquinol produces hydrogen peroxide and liberates the anthraquinone for re-use. The peroxide is made, effectively, from hydrogen and oxygen:

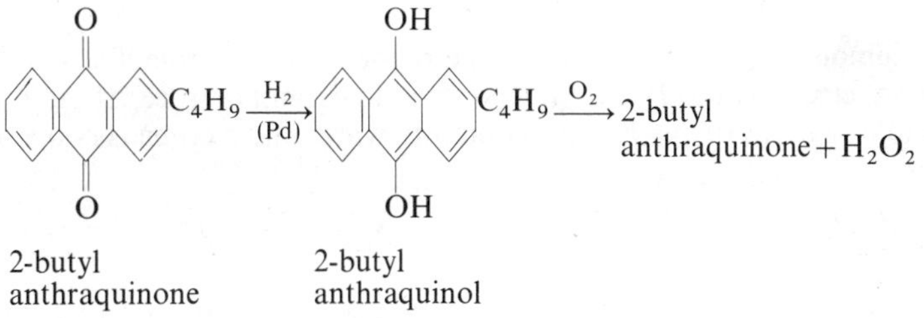

2-butyl
anthraquinone

2-butyl
anthraquinol

In a similar process, propanone can be reduced to propan-2-ol, and this can be made to react with oxygen to produce hydrogen peroxide and re-liberate propanone:

$$(CH_3)_2CHOH(l) + O_2 \longrightarrow (CH_3)_2CO(l) + H_2O_2(l)$$

The solutions of hydrogen peroxide obtained can be concentrated by distillation under reduced pressure.

10 Properties and uses of hydrogen peroxide

a Physical properties and structure Pure hydrogen peroxide is a pale blue, syrupy liquid which solidifies at $-0.4\,°C$. Like water, it has a high relative permittivity (dielectric constant) (89), is a good ionising solvent, and is associated, due to hydrogen bonding. It is not, however, very useful as a solvent because it decomposes so readily and is easily oxidised and reduced.

It dissolves in water and is generally used in aqueous solution, the concentration being expressed as a percentage of hydrogen peroxide by weight or by volume, or as a volume-strength. Thus, 10-volume hydrogen peroxide solution (3 per cent by volume) is such that $1\,cm^3$ of the solution will give $10\,cm^3$ (at s.t.p.) of oxygen on complete decomposition.

The hydrogen peroxide molecule contains single O—O and O—H bonds of lengths 0.149 and 0.097 nm respectively and has the shape shown in Fig. 86.

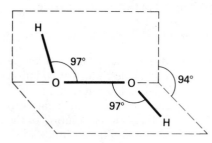

Fig. 86 Molecule of H_2O_2.

b Decomposition   Pure hydrogen peroxide is fairly stable if stored carefully in clean, smooth containers at a low temperature, though it is unstable so far as disproportionation into water and oxygen is concerned:

$$2H_2O_2(l)\longrightarrow 2H_2O(l)+O_2(g) \qquad \Delta H_m^{\ominus}(298\,K) = -196\,kJ\,mol^{-1}$$

The exothermic decomposition is greatly accelerated by rise of temperature, by alkalis, and by minute amounts of catalysts such as dust, finely powdered metals or metallic oxides, and some metallic ions. The catalytic decomposition can be explosive, and if any organic material is present it may be ignited.

c Oxidation and reduction   The oxidation state of oxygen in hydrogen peroxide is -1 (p. 83), and it can be reduced or oxidised. When reduced, the peroxide acts as an oxidising agent and forms water; when oxidised it acts as a reducing agent and forms oxygen.

$$\underset{-2}{H_2O} \xleftarrow{\text{reduction}} \underset{-1}{H_2O_2} \xrightarrow{\text{oxidation}} \underset{0}{O_2}$$

It is particularly strong as an oxidising agent:

$$H_2O_2(l) + 2H^+(aq) + 2e^- \longrightarrow 2H_2O(l) \qquad E^\ominus = +1.77\,V$$

Of special interest are the oxidation of black lead(II) sulphide to white lead(II) sulphate (used in restoring the colour of old lead paint pictures) and the oxidation of acidified potassium dichromate(VI) to chromium peroxide, $CrO(O_2)_2$:

$$4H_2O_2(aq) + Cr_2O_7{}^{2-}(aq) + 2H^+(aq) \longrightarrow 2CrO(O_2)_2 + 5H_2O(l)$$

The chromium peroxide is unstable in aqueous solution but forms a blue compound with an ether. Its formation serves as a test for both hydrogen peroxide and dichromates(VI).

Hydrogen peroxide is not a strong reducing agent,

$$2H^+(aq) + O_2(g) + 2e^- \longrightarrow H_2O_2(l) \qquad E^\ominus = +0.68\,V$$

and it will only react with strong oxidising agents such as potassium manganate(VII).

d As an acid Hydrogen peroxide can function as an acid. It is stronger than water, but still very weak, so that aqueous solutions will not affect indicators:

$$H_2O_2(l) + H_2O(l) \longrightarrow H_3O^+(aq) + HO_2{}^-(aq)$$
$$H_2O(l) + H_2O(l) \longrightarrow H_3O^+(aq) + OH^-(aq)$$

Concentrated hydrogen peroxide solutions will form salts (peroxides) with some bases and carbonates, e.g.

$$Ba(OH)_2(s) + H_2O_2(l) + 6H_2O(l) \longrightarrow BaO_2 . 8H_2O(s)$$
$$Na_2CO_3(s) + H_2O_2(l) \longrightarrow Na_2O_2(s) + H_2O(l) + CO_2(g)$$

e Uses of hydrogen peroxide It is used, because of its oxidising powers, as a mild bleaching agent. As water is the only product formed, it can be used to bleach delicate materials such as silk, straw and human hair. It is also used in restoring old pictures, as a mild antiseptic in mouth washes and for dressing cuts and scratches, and in making sodium peroxoborate(III) (used in washing powders as a source of hydrogen peroxide) and sodium chlorate(III) (used as a bleaching agent).

11 Types of oxide

All the elements, excepting some of the noble gases, combine with oxygen; even XeO_3 is known (p. 143). As many elements form more than one oxide, the total number of oxides is large. Classification is based on the nature of the bonding, the crystal structure and the acid–base character.

a Normal oxides   In normal oxides, oxygen atoms or ions are linked directly, and solely, to atoms or ions of the other element concerned, either by ionic bonds involving O^{2-} ions or by covalent bonds, $O=$ or $-O-$. There are no bonds between oxygen and oxygen or between element and element.

b Peroxides   Peroxides contain the $(O-O)^{2-}$ ion. They liberate hydrogen peroxide when treated with water or dilute acids and they may be regarded as 'salts' of hydrogen peroxide.

The stablest peroxides are formed by the s-block elements, excepting beryllium; sodium peroxide, Na_2O_2 (p. 168), is typical. Other metals form less stable peroxides, e.g. ZnO_2.

The peroxides are made by heating the metal or its normal basic oxide in air or oxygen,

$$2Na(s) + O_2(g) \longrightarrow Na_2O_2(s)$$
$$2BaO(s) + O_2(g) \longrightarrow 2BaO_2(s)$$

or by precipitation methods:

$$H_2O_2(aq) + Ca(OH)_2(aq) \longrightarrow CaO_2(s) + 2H_2O(l)$$

Peroxides decompose on heating to give normal basic oxides and oxygen; they are powerful oxidising agents; they react with water and dilute acids to liberate hydrogen peroxide; and they often crystallise as hydrates, e.g. $CaO_2.8H_2O$.

c Hyperoxides and suboxides   Hyperoxides, containing the $(O-O)^-$ ion, are formed on heating the alkali metals (excepting lithium) or their basic oxides or peroxides. The heavier metals form the hyperoxides most easily, and pressure is required to make sodium hyperoxide. The hyperoxides are paramagnetic (p. 352) because of the unpaired electron in the $\left[\, \ddot{\overset{..}{O}} - \overset{..}{\underset{..}{O}} \, \right]^-$ ion. They react with water to give hydrogen peroxide and oxygen.

In suboxides, e.g. tricarbon dioxide, $O=C=C=C=O$, elements are linked to themselves as well as to oxygen and the element concerned is exhibiting a valency less than its normal group valency.

d Mixed oxides   These are individual substances but they act, chemically, as though they were mixtures. Iron(II) diiron(III) oxide, for example, reacts with acids to give a mixture of iron(II) and iron(III) salts:

$$Fe_3O_4(s) + 8HCl(aq) \longrightarrow FeCl_2(aq) + 2FeCl_3(aq) + 4H_2O(l)$$

It acts like a mixture of FeO and Fe_2O_3. The crystal, in fact, contains Fe^{2+}, Fe^{3+} and O^{2-} ions. Dilead(II) lead(IV) oxide (red lead oxide), Pb_3O_4, $(2PbO + PbO_2)$ is similar, but its crystal structure is more covalent and contains PbO_6 units (p. 226).

Mixed oxides of non-metallic oxides also occur; they are sometimes known as *mixed anhydrides*. Nitrogen dioxide is a typical example. On reaction with water it gives a mixture of acids; with alkalis it gives a mixture of salts:

$$2NO_2(g) + H_2O(l) \longrightarrow HNO_2(aq) + HNO_3(aq)$$
$$2NO_2(g) + 2NaOH(aq) \longrightarrow NaNO_2(aq) + NaNO_3(aq) + H_2O(l)$$

It functions as a mixture of N_2O_3 and N_2O_5, though it exists as NO_2 or N_2O_4 molecules.

The oxides of two different metals occur as a third type of mixed oxide (a *mixed metal oxide*). The spinels, AB_2O_4, are an example of this type, spinel itself (a naturally occurring mineral) being $MgO.Al_2O_3$; it contains Mg^{2+}, Al^{3+} and O^{2-} ions in the crystal.

12 Acid–base character of normal oxides

Of all the common oxides, only carbon monoxide, dinitrogen oxide and nitrogen monoxide are neutral. Others range from strongly acidic (e.g. P_4O_{10}, SO_3), to weakly acidic (e.g. CO_2, B_2O_3), through amphoteric (e.g. Al_2O_3, ZnO, SnO, PbO), to weakly basic (e.g. CuO, Ag_2O) and strongly basic (e.g. Na_2O, BaO).

The nature of any oxide, MO_x, depends mainly on the electronegativity of M and on the value of x. Metals with low electronegativity, exerting a low valency, form basic oxides; metals with higher electronegativity may form amphoteric oxides; acidic oxides are formed by non-metals or by metals of high electronegativity when exerting a very high valency.

a Basic oxides   Oxides are basic if they will accept H^+ ions or, in aqueous solution, produce OH^- ions. They can function in these ways either because they are ionic and contain O^{2-} ions, or because the M—O bond, if covalent, has considerable ionic character, $M^{\delta+}—O^{\delta-}$, because of the higher electronegativity of the oxygen atom as compared with that of M.

$$O^{2-} + H^+(aq) \longrightarrow OH^-(aq)$$
$$O^{2-} + H_2O(l) \longrightarrow 2OH^-(aq)$$
$$CuO(s) + 2H^+(aq) \longrightarrow Cu^{2+}(aq) + H_2O(l)$$

The lower the electronegativity of M in the oxide, the more ionic will be the bonding and the stronger the base formed. The strongest basic oxides are, therefore, formed by the highly electropositive s-block metals (p. 125).

Basic character of oxides increases, in keeping with electronegativity values, in passing down a vertical group of the periodic table, or in passing from right to left along a horizontal period (p. 135).

b Acidic oxides Acidic oxides are covalent and their acidic nature is due to the positive charge built up on the M atom in the $M^{\delta+}—O^{\delta-}$ bond because of the high electronegativity of the oxygen atom. The positively charged M atom is open to attack by the negatively charged oxygen atom in a water molecule, e.g.

The greater the number of oxygen atoms in an oxide, the higher the positive charge on M. This explains why the highest oxides of some metals, e.g. CrO_3 and Mn_2O_7, are acidic.

The effect of increasing the number of oxygen atoms is also shown in the change from basic to acidic oxides of the same element as the oxidation number increases. In the following summary, n stands for neutral, b for basic, am for amphoteric and ac for acidic.

+2	+3	+4	+5	+6	+7
CO(n)		CO_2(ac)			
		SO_2(ac)		SO_3(ac)	
NO(n)	N_2O_3(ac)		N_2O_5(ac)		
CrO(b)	Cr_2O_3(am)			CrO_3(ac)	
MnO(b)	Mn_2O_3(am)	MnO_2(am)			Mn_2O_7(ac)

———→ ——————— increase in acidic nature —————→

c Amphoteric oxides An amphoteric oxide can act as a base by accepting protons or as an acid by giving protons. Zinc oxide, for example, is basic in its reaction with acids,

$$ZnO(s) + 2H^+(aq) \longrightarrow Zn^{2+}(aq) + H_2O(l)$$

and acidic in its reaction with alkalis,

$$ZnO(s) + H_2O(l) + 2OH^-(aq) \longrightarrow [Zn(OH)_4]^{2-}(aq)$$

The $[Zn(OH)_4]^{2-}$ ion is sometimes written, in an over-simplified dehydrated form, as ZnO_2^{2-}:

$$ZnO(s) + H_2O(l) \longrightarrow ZnO_2^{2-}(aq) + 2H^+(aq)$$

[handwritten: M—O low electronegativity. M+ and O²⁻ → forming OH⁻ easier bond to break]

13 Metallic hydroxides

Metallic hydroxides, like the corresponding oxides (p. 275), are either basic or amphoteric. They are basic when the metal, M, in MOH has a low electronegativity and more likely to be amphoteric when M has higher electronegativity. Thus

typical basic hydroxides	NaOH	KOH	$Mg(OH)_2$	$Ba(OH)_2$
electronegativity of metal	0.9	0.8	1.2	0.9

typical amphoteric hydroxides	$Al(OH)_3$	$Zn(OH)_2$	$Sn(OH)_2$	$Cr(OH)_3$
electronegativity of metal	1.5	1.6	1.8	1.6

Basic hydroxides react with acids,

$$M—O—H + H^+ \longrightarrow M^+ + H_2O$$

and this is encouraged if M has a low electronegativity, i.e. if M^+ can form easily.

Amphoteric hydroxides can react with acids as above, but they will also react with bases:

$$M—O—H + OH^- \longrightarrow M—O^- + H_2O$$

This is encouraged when M has a high electronegativity so that the formation of M^+ is not easy. Indeed, when M is a non-metal, with high electronegativity, the formation of MO^- is the only process that will take place, so that non-metallic hydroxides are acidic.

[handwritten: difference in electro. between M and O is not so great so M holds on to O it does not want to release e's thus it gives H+]

14 Non-metallic hydroxides and oxoacids

The hydroxides of non-metals are acidic, typical examples being

$Cl(OH)$	$B(OH)_3$ or H_3BO_3	$Si(OH)_4$ or H_4SiO_4
chloric(I) acid	boric acid	silicic(IV) acid

These acids are, however, somewhat unusual. With the great majority of non-metals, the non-metallic hydroxide is only hypothetical for it loses water to form an oxoacid. Typical oxoacids that can be considered as being formed in this way are as follows:

$B(OH)_3$	$C(OH)_4$	$N(OH)_5$	$N(OH)_3$	$S(OH)_4$	$Cl(OH)_7$
$\downarrow$	$\downarrow -H_2O$	$\downarrow -2H_2O$	$\downarrow -H_2O$	$\downarrow -H_2O$	$\downarrow -3H_2O$
H_3BO_3	H_2CO_3	HNO_3	HNO_2	H_2SO_3	$HClO_4$

$Al(OH)_3$	$Si(OH)_4$	$P(OH)_5$	$P(OH)_3$	$S(OH)_6$
$\downarrow -H_2O$	$\downarrow -H_2O$	$\downarrow -H_2O$	$\downarrow$	$\downarrow -2H_2O$
$[HAlO_2]$	H_2SiO_3	H_3PO_4	H_2PHO_3	H_2SO_4

Further dehydration can give rise to other acids, or to the non-metallic oxide, i.e. the acid anhydride, e.g.

$$H_3PO_4 \xrightarrow{-H_2O} HPO_3 \qquad H_2PHO_3 \xrightarrow{-H_2O} HPO_2$$

$$H_2SO_4 \xrightarrow{-H_2O} SO_3 \qquad 2HNO_3 \xrightarrow{-H_2O} N_2O_5$$

a Simple oxoacids   An oxoacid containing only one central non-metallic atom is sometimes referred to as a simple oxoacid. It can be represented by the general formula $XO_m(OH)_n$. When an element forms a number of different oxoacids, their names all end in -ic but they are distinguished by using oxidation numbers, e.g.

$HClO_4$	$HClO_3$	$HClO_2$	$HClO$
chloric(VII) acid	chloric(V) acid	chloric(III) acid	chloric(I) acid

But an older usage has been retained for the simple oxoacids of nitrogen and sulphur. Hence, HNO_3 is nitric acid and HNO_2, with less oxygen, is nitrous; H_2SO_4 is sulphuric acid and H_2SO_3 is sulphurous.

b Condensed or poly-acids   The loss of a molecule of water between two molecules of an acid is a simple example of condensation polymerisation, and the resulting acid is called a di-ic acid, e.g.

$$2H_2SO_4 \xrightarrow{-H_2O} H_2S_2O_7$$
sulphuric acid   disulphuric(VI) acid

$$2H_2CrO_4 \xrightarrow{-H_2O} H_2Cr_2O_7$$
chromic(VI) acid   dichromic(VI) acid

The condensation polymerisation can take place between many more than two molecules of acid, so that large polymers can be formed. When the acids concerned in forming the polymer are all alike, an *isopoly-acid* results; if different acids are concerned the product is a *heteropoly-acid*.

c Peroxoacids   These are acids containing —O—O— bonds and they may be regarded as derivatives of hydrogen peroxide, e.g.

HO—O—SO$_2$.OH   HO.O$_2$S—O—O—SO$_2$.OH
peroxosulphuric(VI) acid   peroxodisulphuric(VI) acid

15 The strength of oxoacids

Any simple oxoacid can be represented by the generalised formula, $XO_m(OH)_n$, and the species may be a molecule or an anion, e.g.

$SO_2(OH)_2$ $SO_3(OH)^-$
sulphuric acid   hydrogensulphate ion

The strength of the acid depends on the electronegativity of X, on the charge on the species and on the value of m.

a The electronegativity of X For acids with the same values of m and n, the strength of the acid depends mainly on the electronegativity of X. The lower the electronegativity, the weaker the acid, as illustrated by the pK values of the acids below.

$B(OH)_3$	$pK = 9.2$	$SO(OH)_2$	$pK = 1.9$
$Al(OH)_3$	$pK = 11.2$	$SeO(OH)_2$	$pK = 2.6$
		$TeO(OH)_2$	$pK = 2.7$
$PO(OH)_3$	$pK = 2.1$	$Cl(OH)$	$pK = 7.2$
$AsO(OH)_3$	$pK = 2.3$	$Br(OH)$	$pK = 8.7$
		$I(OH)$	$pK = 10.6$

This is to be expected, for a more electronegative X will withdraw electrons from the O—H bond more strongly, and hence facilitate the release of a proton.

b The charge on the species Similarly, loss of a proton from a doubly charged ion is more difficult than from a singly charged one. The greater the negative charge on an acidic ion, the weaker the acid strength, e.g.

H_2SO_3	$pK = 1.9$	H_3PO_4	$pK = 2.1$
HSO_3^-	$pK = 7.2$	$H_2PO_4^-$	$pK = 7.2$
		HPO_4^{2-}	$pK = 12.4$

Pauling proposed a general rule that the successive dissociation constants, $K_1, K_2, K_3, \ldots$, of an acid are in the approximate ratio $1 : 10^{-5} : 10^{-10}, \ldots$, and this is broadly true.

c The value of m Pauling also suggested that the dissociation constant of an acid depends on m as follows:

value of m	0	1	2	3
approx. value of K	$<10^{-7}$	10^{-2}	10^3	10^8
approx. value of pK	>7	2	-3	-8
strength of acid	very weak	weak	strong	very strong
examples (with pK values)	$Cl(OH)$ 7.2	$ClO(OH)$ 2	$ClO_2(OH)$ -1	$ClO_3(OH)$ $c. -10$
	$Br(OH)$ 8.7	$NO(OH)$ 3.3	$NO_2(OH)$ -1.4	
	$I(OH)$ 10.6	$SO(OH)_2$ 1.9	$SO_2(OH)_2$ -3	
	$B(OH)_3$ 9.2	$PO(OH)_3$ 2.1		

The central atom, X, withdraws electrons from the O—H bond in the molecule but it will do this all the more effectively if it has attached oxygen atoms, because of the high electronegativity of the oxygen atom.

Questions on Chapter 21

1 How is oxygen obtained from air on a large scale? What products can be obtained, and under what conditions, by the action of gaseous oxygen on the following: phosphorus, sodium, methanol, barium oxide? (OC)

2 How, and under what conditions, does water react with the following: carbon, magnesium, chromium(VI) dichloride dioxide, calcium dicarbide, chlorine, sodium peroxide, tin(IV) chloride?

3 'If the substance water had not been known previously and had only recently been discovered and investigated, it would be thought to be a most unusual and interesting compound.' Discuss this statement in the context of the physico-chemical properties of water. (JMB)

4 How does water (or steam) react with each of the following? (Give conditions, name the products formed and give equations): (a) copper sulphate, (b) sodium peroxide, (c) ethanoyl chloride, (d) bismuth chloride, (e) calcium dicarbide, (f) iron, (g) ammonia. As far as possible classify the types of reactions that are taking place. (L)

5 Give as many varied examples as you can of reactions in which water is a reagent. For any three of these examples, comment on any significant points.

6 How can a reasonably pure sample of hydrogen peroxide be prepared? How would you determine the composition of an aqueous solution of hydrogen peroxide? How does hydrogen peroxide react with (a) potassium iodide, (b) manganese dioxide, and (c) iron(II) sulphate?

7 Compare and contrast the chemistry of hydrogen peroxide and ozone.

8 You are provided with a mixture of oxides believed to be those of copper(II), lead(IV) and iron(III). How would you confirm the presence of copper, lead and iron in the mixture? Describe by means of equations and brief notes how, starting from this mixture, you would obtain pure samples of each of the three oxides. (JMB)

9 Give examples to illustrate the fact that wide differences in bonding can occur in oxides of formula MO_2. Contrast the chemical properties of MnO_2 and BaO_2 and compare the acidic nature of MnO, MnO_2 and Mn_2O_7. (W)

10 Water plays a vital role in chemistry. Consider the structure of this compound in the three physical states, and give reasons for any physical behaviour you believe to be anomalous. Survey the action of water as a solvating agent, complexing agent, and agent of hydrolysis. (L)

11 Give a general account of the hydrides of the elements, relating the nature of the bonding in the compounds to the position of the element in the periodic table, and their properties to the nature of the bonding. Given a supply of 'heavy water', D_2O, how would you prepare specimens of *four* of the following: ND_3, D_2SO_4, D_2O_2, NaD, DI? (L)

12 (a) How may XeF_4 be prepared? What is the spatial arrangement of the atoms in XeF_4 and how may this be accounted for in terms of the Sidgwick–Powell theory of electron pair repulsion? (b) How may the following be prepared, starting in each case from D_2O as the source of deuterium: (i) C_2D_2, (ii) D_2S, (iii) ND_3? Your answers should be confined to giving the reactions involved (with equations) and the necessary conditions. (O)

13 Give an account of the oxides and hydrides of the elements in the second short period, i.e. from sodium to chlorine, in the periodic table. You should concern yourself particularly with (i) their structures and bonding, (ii) the acid–base behaviour of the oxides, and (iii) the hydrolytic behaviour of the hydrides, but your answer should not be confined to these points. (L)

14 (a) Name and give the formulae of the oxides of sodium, magnesium, aluminium, silicon and phosphorus. (b) For each oxide state whether it is acidic, amphoteric or basic and in each case give an equation for one reaction to support your statement. (c) Discuss briefly the bonding and structure in each oxide. (AEB)

15 The $O{=}O$ bond length in oxygen is 0.121 nm and has a bond enthalpy of 497 kJ mol^{-1}. The $O{-}O$ bond length in hydrogen peroxide is 0.149 nm and has an enthalpy of 146 kJ mol^{-1}. From the reaction

$$3O_2(g) \longrightarrow 2O_3(g) \qquad \Delta H = 284.6\,\text{kJ mol}^{-1}$$

calculate the average oxygen–oxygen bond enthalpy in ozone. What light does your result throw on the nature of the bonding in hydrogen peroxide?

16 The enthalpies of atomisation of hydrogen and oxygen are 436 and 498 kJ (mol of H_2 or O_2)$^{-1}$ respectively. The enthalpy of formation of $H_2O(g)$ is -241.9 kJ mol^{-1}. Calculate the mean bond enthalpy for the $O{-}H$ bond. If the enthalpy of formation of $H_2O_2(g)$ is -133 kJ mol^{-1} calculate the bond enthalpy of the $O{-}O$ bond. What assumption is necessary in making the calculation?

17* Use the relevant molar enthalpies of formation to show why hydrogen peroxide decomposes into water and oxygen rather than into hydrogen and oxygen.

22
Sulphur

1 Sources and uses of sulphur

Sulphur occurs widely, both in the combined state as sulphides and sulphates, and as the free element.

The main supplies of natural sulphur occur underground in Texas, Louisiana, Mexico and Poland. It is extracted by the Frasch process (Fig. 87). A hole (diameter, 20 cm) is bored down to the sulphur deposit and three concentric pipes (diameters, 2.5, 7.5 and 15 cm) are then lowered into the hole. Water, heated under pressure to 165 °C, is passed down the outermost pipe and melts the sulphur (melting temperature 113 °C) at the bottom. The molten sulphur is forced up the 7.5 cm pipe by the pressure of the water and by hot compressed air passed down the central pipe. The molten sulphur solidifies on cooling and is stored in large blocks out in the open.

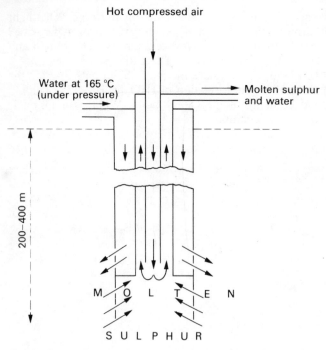

Fig. 87 *The Frasch process for extracting sulphur.*

Considerable amounts of sulphur are also obtained from the hydrogen sulphide that occurs along with natural gas, particularly in France and

$$2H_2S + O_2 \rightarrow SO_2 + H_2O$$
$$SO_2 + H_2S \rightarrow S + H_2O$$

Canada, and in the gases produced on refining some types of crude oil. The hydrogen sulphide is first burnt to form sulphur dioxide, and this is then reacted with more hydrogen sulphide:

$$2H_2S(g) + 3O_2(g) \longrightarrow 2H_2O(g) + 2SO_2(g)$$
$$2H_2S(g) + SO_2(g) \longrightarrow 2H_2O(l) + 3S(s)$$

Sulphur is used mainly in making sulphuric acid (p. 104), but it is also needed in the manufacture of carbon disulphide (p. 207), sodium thiosulphate (p. 175), black powder (p. 172), calcium hydrogensulphite (p. 292) and matches (p. 250). The free element is used in sulphur ointment and as a fungicidal spray. It is also incorporated into rubber in the process of vulcanisation; this toughens the rubber and makes it less sticky in hot weather and less brittle in cold.

2 Forms of sulphur

Sulphur exists in a number of allotropic varieties depending on the method of preparation and the temperature. In the *cyclo-sulphurs*, sulphur atoms are arranged in rings containing from 6 to 20 atoms; in the *catena-sulphurs*, the atoms are arranged in chains.

a α- and β-sulphur   There are two main solid allotropes: α- or rhombic sulphur and β- or monoclinic sulphur. The former is stable below a transition temperature of 95.5 °C; the latter is stable above that temperature up to its melting temperature of 119 °C:

$$\alpha\text{-S} \rightleftharpoons \beta\text{-S} \qquad \Delta H^{\ominus}(298\,\text{K}) = +0.3\,\text{kJ}\,\text{mol}^{-1}$$

The change from one form to the other is slow.

Crystals of α-sulphur can be made from a solution of sulphur in carbon disulphide or disulphur dichloride at a temperature *below* 95.5 °C. Crystals of β-sulphur result on cooling molten sulphur or a hot solution of sulphur in turpentine *above* 95.5 °C.

Both α- and β-sulphur crystals are made up of S_8 molecules (Fig. 88), but the α-sulphur crystal is basically octahedral in shape whilst that of β-sulphur is long and thin. The two allotropes have different melting temperatures and densities.

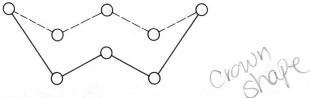

crown shape

Fig. 88 *The S_8 molecule.*

b Effect of heat   The melting temperature of α-sulphur is 113 °C and that of β-sulphur is 119 °C. Solid sulphur therefore melts between these two temperatures, depending on how rapidly it is heated. A honey-coloured liquid is formed; it consists of S_8 molecules. On raising the temperature, the liquid darkens in colour and it begins to get very viscous at about 160 °C, reaching a maximum viscosity at about 200 °C; above that temperature, the liquid becomes more fluid again. These changes take place because some of the S_8 molecules, which can 'slip' over each other, split up into forms of catena-sulphur that contain spiral chains of intertwining sulphur atoms. Other cyclo-sulphurs, containing from 6 to 20 sulphur atoms, are also formed.

The liquid boils at 444.5 °C. The vapour contains S_8 molecules just above the boiling point but these dissociate into S_6, S_4 and S_2 molecules on further heating. At about 1000 °C the vapour consists of S_2 molecules and these are paramagnetic, like those of O_2 (p. 264). At still higher temperatures, dissociation into atoms occurs.

c Other forms   Plastic sulphur is formed by pouring molten sulphur at temperatures above 200 °C into cold water. The plastic sulphur consists mainly of intertwining chains of sulphur atoms. It is unstable, mainly reverting to α-sulphur on standing, and insoluble in carbon disulphide.

On cooling sulphur vapour, it sublimes, i.e. solidifies directly, to a finely divided form of the element, known as flowers of sulphur. Colloidal sulphur results from many reactions in which sulphur is a product, e.g. the reaction between acids and sodium thiosulphate solution.

3 Compounds of sulphur

In so far as it forms S^{2-} and HS^- ions, and $S=$ or $-S-$ covalent bonds, sulphur is similar to oxygen. But it differs from it in being less electronegative so that it participates much less in hydrogen bonding, in having a higher degree of catenation, and in having a maximum covalency of 6, which enables it to form compounds, e.g. SF_6, that have no oxygen analogues. Sulphur also forms many oxides, oxoacids and oxoanions.

The wide range of oxidation states in sulphur compounds is shown in the following summary.*

−2	−1	0	+2	+3	+4	+5	+6
H_2S	H_2S_2	S	$S_2O_3{}^{2-}$	$S_2O_4{}^{2-}$	SO_2	$S_2O_6{}^{2-}$	SO_3
HS^-			SCl_2		$SO_3{}^{2-}$		H_2SO_4
S^{2-}					$SOCl_2$		$SO_4{}^{2-}$
							SF_6
							SO_2Cl_2

*'Average' oxidation states (p. 83) are used.

[handwritten: as size increases, electronegativity decreases, the valence electrons are further apart, attractive forces decreases]

a Formation of S^{2-} Because sulphur is a larger atom than oxygen, the S^{2-} ion is formed less readily than O^{2-}, and ionic sulphides are less common than ionic oxides. Also, the lattice enthalpies of sulphides are smaller than those of comparable oxides. As the sulphides are less ionic, the agreement between their measured and calculated lattice enthalpies is less good than for the oxides.

It is only the metals of Groups 1A and 2A that form sulphides that are mainly ionic, containing S^{2-} ions. In solution this ion is hydrolysed:

$$S^{2-} + H_2O(l) \longrightarrow HS^-(aq) + OH^-(aq)$$

[handwritten: calculated theoretically is due to ionic atoms. Measured is due to the born Haber cycle.]

b Formation of HS^- Alkali and alkaline earth metals can form ionic hydrogensulphides, containing HS^- ions, but only in liquid ammonia solution; the compounds are like the corresponding hydroxides, which contain OH^- ions.

Many other metallic hydroxides are also known, but they have no hydrogensulphide counterparts. This is because the hydroxides have structures that involve hydrogen bonding, but sulphur cannot form similar bonds.

c Formation of covalent bonds Both sulphur and oxygen can form two single covalent bonds, as in H_2S and H_2O, or one double bond as in CO_2 and CS_2. The sulphur bonds are much less polar than the oxygen bonds, and this accounts for the many differences between pairs of compounds such as hydrogen sulphide and water, and carbon dioxide and carbon disulphide.

The maximum covalency of oxygen is 4 and this is achieved very rarely (p. 267), but sulphur can form up to six bonds as, for example, in H_2SO_4, SO_4^{2-} and SF_6.

d Catenation Oxygen will only form three-atom chains, as in ozone, O_3, but sulphur catenates much more freely. There is a ring of eight sulphur atoms in the solid element; rings of up to twenty atoms in liquid sulphur; a chain of up to six atoms in H_2S_x hydrides (p. 288); chains of up to six atoms in the polysulphides of the alkali metals and alkaline earths, e.g. Na_2S_x and CaS_x; chains of up to twenty atoms in the thionic acids, $H_2S_xO_6$; and chains of up to five atoms in some halides, e.g. S_xCl_2.

This greater catenation of sulphur is due to the strength of the S—S bond ($266\,kJ\,mol^{-1}$) as compared with that of the O—O bond ($146\,kJ\,mol^{-1}$). Compounds with O—O bonds, such as ozone, peroxides and peroxodisulphuric(VI) acid, decompose readily. *[handwritten: due to their weakness]*

e Thio compounds A rather formal similarity is shown between oxygen and sulphur when the former is replaced in compounds by the latter without any very extensive change in property. The compounds are

[handwritten: when O is replaced by a S.]

known as thio compounds, e.g.

| Na_2SO_4 | NaOCN |
| sodium sulphate | sodium cyanate |

| $Na_2S_2O_3$ | NaSCN |
| sodium thiosulphate | sodium thiocyanate |

There are, too, some analogous reactions. Carbon dioxide, for example, reacts with sodium hydroxide to form sodium carbonate; carbon disulphide reacts with sodium hydroxide to form sodium thiocarbonate, Na_2CS_3. Tin(IV) oxide reacts with sodium monoxide to form sodium stannate(IV), Na_2SnO_3; tin(IV) sulphide reacts with yellow ammonium polysulphide, $(NH_4)_2S_x$, to form ammonium trithiostannate(IV), $(NH_4)_2[SnS_3]$.

4 Hydrogen sulphide

a Preparation   Both water and hydrogen sulphide can be made by direct combination. The reaction between oxygen and hydrogen can be explosive, but only small yields of hydrogen sulphide are obtainable either by passing hydrogen through molten sulphur or by burning sulphur vapour in hydrogen gas.

$$2H_2(g) + O_2(g) \longrightarrow 2H_2O(g) \qquad \Delta H_m^{\ominus}(298\,K) = -484\,kJ\,mol^{-1}$$
$$2H_2(g) + 2S(s) \longrightarrow 2H_2S(g) \qquad \Delta H_m^{\ominus}(298\,K) = -40.4\,kJ\,mol^{-1}$$

Hydrogen sulphide is generally made by reaction between a metallic sulphide and an acid, iron(II) sulphide and hydrochloric acid being the commonest choices:

$$FeS(s) + 2HCl(aq) \longrightarrow FeCl_2(aq) + H_2S(g)$$

The gas is collected over warm water or by upward delivery; it can be dried by phosphorus(V) oxide. Hydrogen sulphide is also obtainable from natural gas and from oil refining (p. 282).

b Physical properties and structure   The H_2S molecule is triangular, like the H_2O molecule, but the S—H bond is longer (0.154 nm as against 0.096 nm), weaker (344 kJ mol^{-1} as against 464 kJ mol^{-1}) and very much less polar than the O—H bond.

The lower polarity in the S—H bond, which is due to the lower electronegativity of sulphur as compared with oxygen, causes the bond angle in H_2S to be lower (92°) than that in H_2O (104.5°).

The low polarity of the S—H bond also means that, unlike water, hydrogen sulphide is not associated through hydrogen bonding. Therefore it has lower melting and boiling temperatures than water, which is why it is a gas at room temperature. For similar reasons, it has low relative permittivity and dipole moment; it is only very slightly ionised in the liquid state, so that liquid hydrogen sulphide is not a good ionising solvent; it does not feature as a ligand in the same way that water does, so it forms very few complexes; it has a close-packed structure in the solid form; and it does not have similar density changes with temperature to the odd changes that are shown by water.

Hydrogen sulphide is also a smelly, poisonous gas, whereas water has no smell and is essential to life.

c Thermal stability   Water is very thermally stable, though it is slightly decomposed into its elements at about 2000 °C, but hydrogen sulphide decomposes on passing through a red-hot tube. That is why it is not easy to make hydrogen sulphide from its elements.

The trend in stability is related to the bond enthalpies (O—H, 464 kJ mol^{-1} and S—H, 344 kJ mol^{-1}) and to the standard enthalpies of formation (H_2O, -242 kJ mol^{-1} and H_2S, $-20.2 \text{ kJ mol}^{-1}$).

d Acidity   Hydrogen sulphide is a much stronger acid than water:

$$H_2O(l) + H_2O(l) \longrightarrow H_3O^+(aq) + OH^-(aq) \qquad pK_1 = 15.75$$
$$OH^-(aq) + H_2O(l) \longrightarrow H_3O^+(aq) + O^{2-}(aq) \qquad pK_2 = \text{very high}$$
$$H_2S(aq) + H_2O(l) \longrightarrow H_3O^+(aq) + HS^-(aq) \qquad pK_1 = 7.04$$
$$HS^-(aq) + H_2O(l) \longrightarrow H_3O^+(aq) + S^{2-}(aq) \qquad pK_2 = 13.92$$

As the sulphur atom is larger than that of oxygen it would be expected that the formation of S^{2-} ions would be more difficult than the formation of O^{2-} ions. It is, however, the hydrogen bonding that limits the ionisation of water and this makes hydrogen sulphide the stronger acid. The greater strength of the O—H bond as compared with the S—H bond also contributes.

e Oxidation of hydrogen sulphide   Hydrogen sulphide burns readily in air, being oxidised to sulphur or sulphur dioxide, depending on the amount of oxygen:

$$2H_2S(g) + 3O_2(g) \longrightarrow 2H_2O(g) + 2SO_2(g)$$
$$2H_2S(g) + O_2(g) \longrightarrow 2H_2O(g) + 2S(s)$$

It can be oxidised to sulphur by many other oxidising agents because it is quite a strong reducing agent; much stronger than water, which only reacts with very strong oxidising agents:

$$S(s) + 2H^+(aq) + 2e^- \longrightarrow H_2S(g) \qquad E^\ominus = +0.141 \text{ V}$$
$$O_2(g) + 4H^+(aq) + 4e^- \longrightarrow 2H_2O(l) \qquad E^\ominus = +1.23 \text{ V}$$

f Comparison with water   A summary of the differences between hydrogen sulphide and water is given below.

hydrogen sulphide	water
colourless, smelly, poisonous gas (no H-bonds)	colourless, non-poisonous liquid (H-bonds)
burns in air	does not burn in air
H—S bond enthalpy = 344 kJ mol^{-1}	H—O bond enthalpy = 464 kJ mol^{-1}
bond angle = 92°	bond angle = 104.5°
not hydrogen bonded	hydrogen bonded
close-packed structure as a solid	'open' structure in ice
low dipole moment	high dipole moment
low relative permittivity	high relative permittivity
poor solvent	good ionising solvent
does not form complexes	forms many aqua-complexes
decomposes at red heat	thermally stable
strong reducing agent	very weak reducing agent

5 Other hydrides of sulphur

Oxygen forms only two hydrides: H_2O and H_2O_2. Hydrogen sulphide corresponds to water, and hydrogen persulphide, H_2S_2, to hydrogen peroxide. But, because sulphur catenates more than oxygen, polysulphides (sulphanes), e.g. H_2S_4, H_2S_5 and H_2S_6, can also be made. They contain chains of sulphur atoms.

The hydrogen polysulphides are made by acidifying solutions of metallic polysulphides. A yellow oil is produced and this can be separated into the different hydrides by careful fractional distillation under reduced pressure. The hydrogen polysulphides are unstable, decomposing into hydrogen sulphide and sulphur.

6 Halides of sulphur

The main halides of sulphur are S_2F_2, SF_4, SF_6, S_2Cl_2, SCl_2, SCl_4 and S_2Br_2.

a Chlorides of sulphur The various chlorides of sulphur are made as follows:

$$S(l) \xrightarrow[Cl_2]{dry} S_2Cl_2 \xrightarrow[Cl_2]{excess} SCl_2 \xrightarrow[-80°C]{Cl_2(l)} SCl_4$$

Disulphur dichloride, S_2Cl_2, is a yellow, smelly liquid that fumes in moist air because it is hydrolysed to form hydrogen chloride. It is a very good solvent for sulphur and is used as such in vulcanising rubber.

Sulphur dichloride, SCl_2, is a dark red liquid, which quickly dissociates into

S_2Cl_2 and sulphur. Sulphur tetrachloride, SCl_4, is a yellow crystalline solid at low temperatures, but it dissociates above about $-30\,°C$.

b Fluorides of sulphur   Sulphur tetrafluoride, SF_4, is made by reaction between sulphur dichloride and sodium fluoride in ethanenitrile at $75\,°C$. It is a very reactive gas, which is used as a fluorinating agent. It reacts very readily with water:

$$SF_4(g) + 2H_2O(l) \longrightarrow SO_2(g) + 4HF(aq)$$

The SF_4 molecule contains a trigonal bipyramidal arrangement of electron pairs, with the lone pair occupying one of the equatorial positions (p. 29).

Sulphur hexafluoride, SF_6, is made by burning sulphur in fluorine. It is a colourless gas that is only slightly soluble in water and very unreactive. It is not affected by molten potassium hydroxide or hot steam. The 'theoretical' hydrolysis is highly exothermic,

$$SF_6(g) + 4H_2O(l) \longrightarrow H_2SO_4(aq) + 6HF(aq)$$
$$\Delta H^{\ominus}(298\ K) = -638\ kJ\ mol^{-1}$$

and the failure of this reaction to take place must be due to kinetic factors, which are probably associated with the fact that all the electrons around the sulphur atom are fully shared (compare tetrachloromethane, p. 216). The SF_6 molecule is octahedral.

7 Sulphides

Nearly all metals, and most non-metals, will combine with sulphur to form sulphides.

Reactive metals, like sodium and potassium, and unreactive ones like copper, silver and mercury, will combine to some extent without heating. Other metals require heating to initiate the reaction but, once started, the combination with sulphur may produce much heat, and if the metal and the sulphur are powdered, the reaction may be explosive. Non-metals require heating before they will combine with sulphur.

a Ionic sulphides   Metallic sulphides are less ionic than the oxides and have lower lattice enthalpies, and only the sulphides of the s-block metals are essentially ionic. The sulphides of lithium, sodium, potassium and the alkaline earths are, in fact, isomorphous with the oxides. The sulphides are also like the oxides in that they are readily soluble giving alkaline solutions:

$$S^{2-} + H_2O(l) \longrightarrow HS^-(aq) + OH^-(aq)$$
$$O^{2-} + H_2O(l) \longrightarrow OH^-(aq) + OH^-(aq)$$

Hydrolysis of the S^{2-} ion is incomplete in cold water, but goes to completion on boiling, for hydrogen sulphide gas is evolved:

$$HS^-(aq) + H_2O(l) \longrightarrow H_2S(g) + OH^-(aq)$$

By le Chatlier Principle

they don't combine in small molecules (handwritten margin note)

Solutions of these ionic sulphides react on heating with sulphur to form polysulphides containing S_x^{2-} ions with chains of sulphur atoms in the ions:

$$S^{2-}(aq) + (x-1)S(s) \longrightarrow S_x^{2-}(aq)$$

b Covalent sulphides   Most metallic sulphides have considerable covalent character and there are wide discrepancies between calculated and measured lattice enthalpies (p. 13). They are insoluble in water, and many of them occur naturally and are highly coloured with a metallic lustre. They may be non-stoicheiometric; the composition of iron(II) sulphide, for instance, varies between $Fe_{0.86}S$ and $Fe_{0.89}S$.

The precipitation of coloured sulphides on adding hydrogen sulphide to a solution of a metallic salt is used in qualitative analysis. If the precipitation is brought about in acidic solution, only the most insoluble sulphides will be precipitated, for the acid present retards the ionisation of the hydrogen sulphide, i.e. it pushes the equilibrium

$$H_2S(g) \rightleftharpoons 2H^+(aq) + S^{2-}(aq)$$

By le Chatelier (handwritten margin note)

to the left. The S^{2-} ion concentration is, therefore, low. In alkaline solutions, the equilibrium is shifted to the right, giving a higher S^{2-} ion concentration which will precipitate less insoluble sulphides. Thus

sulphides precipitated in acidic solution		sulphides precipitated in alkaline solution
HgS (black)	SnS$_2$ (yellow)	NiS (black)
PbS (black)	SnS (brown)	CoS (black)
CuS (black)	As$_2$S$_3$ (yellow)	MnS (pink)
CdS (yellow)	Sb$_2$S$_3$ (orange)	ZnS (white)

Sulphides of non-metals, e.g. CS_2, H_2S, SO_2 and SF_6, are covalent.

8 Sulphur dioxide, SO_2

a Preparation   Sulphur dioxide can be made by treating sulphites or hydrogensulphites with dilute acids, or by heating copper with concentrated sulphuric acid:

$$SO_3^{2-} + 2H^+(aq) \longrightarrow SO_2(g) + H_2O(l)$$
$$HSO_3^- + H^+(aq) \longrightarrow SO_2(g) + H_2O(l)$$
$$Cu(s) + 2H_2SO_4(aq) \longrightarrow CuSO_4(aq) + SO_2(g) + 2H_2O(l)$$

Industrially, it is obtained by burning sulphur, or by roasting metallic sulphides in air (p. 103). It is liquefied by slight compression, and stored in cylinders.

Coal and oil may contain sulphur as an impurity, and burning them introduces a lot of sulphur dioxide into the atmosphere. It is a major pollutant, as, for example, in 'acid rain'.

b Properties The gas is colourless and poisonous, and has a characteristic, choking smell. It is very soluble in water, giving an acidic solution of sulphurous acid. It is easily liquefied at room temperature and about 3 atm pressure. Liquid sulphur dioxide is a useful solvent.

The sulphur dioxide molecule is triangular with a bond angle of 119°; the two covalent bonds between sulphur and oxygen atoms are 0.143 nm long. The molecule is best represented with delocalised bonds:

S
O   O

(reasonating structures)

Dry gaseous sulphur dioxide can be oxidised by oxygen to sulphur(VI) oxide, by chlorine to sulphur dichloride dioxide, SO_2Cl_2, and by lead(IV) oxide to lead(II) sulphate, $PbSO_4$. It can be reduced to sulphur by strong reducing agents such as burning magnesium.

c Uses of sulphur dioxide Most sulphur dioxide is used in the manufacture of sulphuric acid (p. 103). It is used, too, for bleaching silk, straw and sponges (it reduces many coloured substances to colourless products); to kill bacteria and fungi in preserving fruits, fruit juices and grain; and in making calcium hydrogensulphite and refining petrol.

9 Sulphurous acid and sulphites

a Sulphurous acid When sulphur dioxide is passed into water, an acidic solution known as sulphurous acid is formed. This is sometimes written, in an oversimplified way, as H_2SO_3, but there is no evidence for the existence of such a molecule, and the solution contains hydrated sulphur dioxide, $SO_2 . xH_2O$, together with H_3O^+, HSO_3^- and SO_3^{2-} ions.

The solution decomposes into sulphur dioxide and water on heating. It is slowly oxidised to sulphuric acid on standing in air, and more rapidly by many oxidising agents. It can also be reduced to sulphur by some reducing agents, e.g. hydrogen sulphide.

b Preparation of sulphites Saturation of a solution of sodium hydroxide with sulphur dioxide produces a solution in which HSO_3^- ions predominate, but solid sodium hydrogensulphite cannot be obtained because crystallisation of the solution produces mainly disulphate(IV), $S_2O_5^{2-}$, ions:

$$2HSO_3^-(aq) \longrightarrow S_2O_5^{2-}(aq) + H_2O(l)$$

Addition of further alkali builds up the concentration of $SO_3^{2-}(aq)$ ions,

$$HSO_3^-(aq) + OH^-(aq) \longrightarrow SO_3^{2-}(aq) + H_2O(l)$$

so that sodium sulphite, $Na_2SO_3.7H_2O$, can be obtained by crystallisation.

Insoluble sulphites can be made by precipitation methods. Calcium sulphite, for example, is precipitated on passing excess sulphur dioxide into a solution containing Ca^{2+} ions. This explains why sulphur dioxide can give a white precipitate with limewater.

c Properties and uses of sulphites Apart from those of the alkali metals, sulphites are generally insoluble in water, although they are soluble, through reaction, in acids. They are ionic solids that contain the pyramidal SO_3^{2-} ion (Fig. 89).

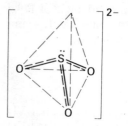

Fig. 89 *The sulphite ion, SO_3^{2-}.*

The oxidation state of sulphur in sulphites is $+4$, and it is possible to oxidise or reduce them in a variety of ways. For example

0	+1	+2	+3	+4	+5	+6
			$S_2O_4^{2-} \xleftarrow{\text{Zn}} SO_3^{2-} \xrightarrow{\text{MnO}_2} S_2O_6^{2-}$			
		$S_2O_3^{2-} \xleftarrow{\quad S \quad} SO_3^{2-} \xrightarrow{\text{many oxidising agents}} SO_4^{2-}$				
$S_8 \xleftarrow{\qquad\qquad H_2S \qquad\qquad} SO_3^{2-}$						

The commonest change is that from sulphite to sulphate,

$$SO_3^{2-}(aq) + H_2O(l) \longrightarrow SO_4^{2-}(aq) + 2H^+(aq) + 2e^-$$

and this is brought about slowly by air and rapidly by many oxidising agents.

Calcium hydrogensulphite is used in making paper; it removes the lignins from wood pulp and releases the cellulose. Sodium hydrogensulphite reacts with aldehydes and ketones to form addition compounds; sodium disulphate(IV) is used in photography; and various sulphite solutions are used as reducing agents, particularly in making dyes.

10 Sulphur(VI) oxide (sulphur trioxide)

a Preparation Sulphur(VI) oxide is made by oxidising sulphur dioxide by oxygen:

$$2SO_2(g) + O_2(g) \rightleftharpoons 2SO_3(g) \qquad \Delta H^{\ominus}(298\,K) = -197.6\,kJ$$

The reaction is exothermic and reversible. A good yield of sulphur(VI) oxide requires a catalyst (platinum or vanadium(V) oxide), a slight pressure and a temperature of 400–450°C. The reaction is carried out industrially as the contact process (p. 104).

Sulphur(VI) oxide can also be obtained by dehydrating sulphuric acid with phosphorus(V) oxide, or by heating some sulphates and hydrogen-sulphates.

b Properties Sulphur(VI) oxide can exist in three forms (see section **c** below) but the stable form at room temperature is a white solid with needle-shaped crystals. It has a low boiling temperature (45 °C), so it is readily vaporised on heating.

It is an acidic oxide, forming sulphuric acid with water and sulphates with bases. It dissolves in concentrated sulphuric acid to form *oleum* or *fuming sulphuric acid*. In equimolecular proportions, disulphuric(VI) acid is formed,

$$SO_3(s) + H_2SO_4(l) \longrightarrow H_2S_2O_7(l)$$

but larger amounts of sulphur(VI) oxide react further to give other condensed acids, $H_2SO_4 . xSO_3$.

c Structure Sulphur(VI) oxide exists in three forms, all of which are polymers, $(SO_3)_x$. The white crystals that exist at room temperature have an asbestos-like appearance and a structure that consists of SO_4 tetrahedra linked into chains and cross-linked in layers.

In the vapour state, the sulphur(VI) oxide molecule is planar, with bond angles of 120° and three equal bond lengths (0.143 nm). The molecule contains delocalised bonds:

11 Sulphuric acid

Sulphuric acid is manufactured by the contact process described on p. 104. It owes its old name, *oil of vitriol*, to the fact that it can be made by passing the gases obtained by heating green vitriol or other vitriols (p. 377), into water.

Pure sulphuric acid is an oily liquid with a high density; it forms colourless crystals below 10.4 °C. The concentrated acid contains about 98 per cent acid and 2 per cent water; it is about 18 M.

a Acidity Sulphuric acid is a strong dibasic acid:

$$H_2SO_4(l) + H_2O(l) \rightleftharpoons H_3O^+(aq) + HSO_4^-(aq)$$
$$HSO_4^-(aq) + H_2O(l) \rightleftharpoons H_3O^+(aq) + SO_4^{2-}(aq)$$

It forms hydrogensulphates and sulphates with bases, and it dissolves sulphur(VI) oxide to form *oleum* or *fuming sulphuric acid*, which contains disulphuric(VI) acid, $H_2S_2O_7$ (p. 298).

b As a dehydrating agent Concentrated sulphuric acid has a very great affinity for water, and when the two liquids are mixed much heat is evolved. This might be sufficient to boil the mixture, so care must be taken when mixing them.

The concentrated acid will absorb water vapour from moist air or other moist substances (i.e. it is hygroscopic), and it is used as a drying agent for gases and in desiccators. It will, furthermore, decompose (dehydrate) many compounds by removing, in the form of water, the hydrogen and oxygen that they contain. Glucose, ethanedioic (oxalic) acid, methanoic acid, ethanol and hydrated copper(II) sulphate are all dehydrated in this way:

$$C_6H_{12}O_6(s) \xrightarrow{-6H_2O} 6C(s)$$

$$H_2C_2O_4.2H_2O(s) \xrightarrow{-3H_2O} CO(g) + CO_2(g)$$

$$HCOOH(l) \xrightarrow{-H_2O} CO(g)$$

$$C_2H_5OH(l) \xrightarrow{-H_2O} C_2H_4(g)$$

$$CuSO_4.5H_2O(s) \xrightarrow{-5H_2O} CuSO_4(s)$$

c Oxidation and reduction The oxidation state of sulphur in sulphuric acid is $+6$. The acid can be oxidised to peroxodisulphuric(VI) acid, $H_2S_2O_8$,

$$2SO_4^{2-}(aq) \longrightarrow S_2O_8^{2-}(aq) + 2e^-$$

but the process can only be brought about electrolytically.

Sulphuric acid can be reduced much more easily and, as with nitric acid, a wide range of products is possible, depending on the concentration of the acid, the temperature and the strength of the reducing agent used. Hot concentrated sulphuric acid is a particularly strong oxidising agent. Sulphurous acid solution, or sulphur dioxide gas, are the commonest reduction products:

$$SO_4^{2-}(aq) + 4H^+(aq) + 2e^- \longrightarrow SO_2(aq) + 2H_2O(l)$$

$$E^\ominus = +0.17\,V$$

Most metals, for example, will produce some sulphur dioxide gas on heating with concentrated sulphuric acid, and copper is commonly used in this method of making the gas (p. 290). Reactive metals, e.g. zinc, may cause further reduction, either to sulphur or to hydrogen sulphide:

$$SO_4^{2-}(aq) + 8H^+(aq) + 6e^- \longrightarrow S(s) + 4H_2O(l)$$
$$SO_4^{2-}(aq) + 10H^+(aq) + 8e^- \longrightarrow H_2S(g) + 4H_2O(l)$$

The acid will oxidise carbon to carbon dioxide, sulphur to sulphur dioxide and phosphorus to phosphorus(V) oxide, and will also react with many common reducing agents.

d As a sulphonating agent In organic chemistry, concentrated sulphuric acid will replace hydrogen atoms by $-SO_3H$ groups; the process is known as *sulphonation*, and *sulphonic acids*, e.g. $C_6H_5.SO_3H$, are formed.

e Reaction with salts Because sulphuric acid is a strong acid and not easily vaporised, it will displace other, more volatile, acids from their salts, e.g.

$$NaNO_3(s) + H_2SO_4(aq) \longrightarrow NaHSO_4(aq) + HNO_3(g)$$
$$NaCl(s) + H_2SO_4(aq) \longrightarrow NaHSO_4(aq) + HCl(g)$$

These reactions are used in qualitative analysis to detect many acid radicals or to obtain free acids from their salts. The acid formed may be dehydrated or oxidised. Thus, methanoates give carbon monoxide, ethanedioates (oxalates) give carbon monoxide and carbon dioxide, and bromides and iodides give bromine and iodine respectively.

f Structure The H_2SO_4 molecule is tetrahedral (Fig. 90(a)). There is extensive hydrogen bonding between these molecules, both in the solid and in the liquid, which is why the liquid is so viscous. The sulphate ion is also tetrahedral; the bond distances are all equal (0.151 nm) and the ion is best represented with delocalised bonding, as in Fig. 90(b).

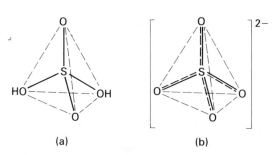

(a)   (b)

Fig. 90 (a) The tetrahedral shape of the H_2SO_4 molecule (the delocalised bonds are not shown). *(b)* The delocalised bonds in the tetrahedral SO_4^{2-} ion.

g Uses of sulphuric acid   The best-known direct use of sulphuric acid is in the lead–acid batteries in motor cars, but that uses very little of the total production. It is a major industrial chemical, essential in the manufacture of many basic requirements. The level of usage of the acid in a country gives a good general indication of the industrial activity and prosperity of that country. The main uses in the United Kingdom are fertilisers (30 %), chemicals (16 %), paints and pigments (15 %), soap and detergents (12 %), textiles and fibres (9 %), metallurgy (2 %) and dyestuffs (2 %).

12 Acid chlorides of sulphurous and sulphuric acids

The acids (even if they are hypothetical, as in the case of sulphurous acid) can form acid chlorides as follows:

HO   Cl
  \   \
   S=O   S=O
  /   /
HO   Cl

'sulphurous acid'   sulphur
                      dichloride oxide

HO   O   Cl   O   Cl   O
  \   ‖   \   ‖   \   ‖
   S   S   S
  /   ‖   /   ‖   /   ‖
HO   O   HO   O   Cl   O

sulphuric acid   chlorosulphonic acid   sulphur
                                        dichloride dioxide

a Sulphur dichloride oxide, $SOCl_2$ (thionyl chloride)   This is made by re-action between phosphorus pentachloride and dry sulphur dioxide:

$$PCl_5(s) + SO_2(g) \longrightarrow POCl_3(l) + SOCl_2(l)$$

The sulphur dichloride oxide is separated by fractional distillation.

It is a colourless liquid which fumes in moist air, due to the formation of hydrogen chloride by hydrolysis:

$$SOCl_2(l) + H_2O(l) \longrightarrow 2HCl(g) + SO_2(g)$$

It is used in organic chemistry to replace —OH groups by —Cl, and as a dehydrating agent (p. 315). The molecule has a triangular pyramidal shape.

b Sulphur dichloride dioxide, SO_2Cl_2 (sulphuryl chloride)   Replacement of both —OH groups in sulphuric acid can be brought about by phosphorus pentachloride:

$$H_2SO_4(aq) + 2PCl_5(s) \longrightarrow SO_2Cl_2(l) + 2POCl_3(l) + 2HCl(g)$$

It can also be made by direct combination of sulphur dioxide and chlorine in sunlight and in the presence of a catalyst such as iron(III) chloride:

$$SO_2(g) + Cl_2(g) \longrightarrow SO_2Cl_2(l)$$

Sulphur dichloride dioxide is a colourless liquid, and has a distorted tetrahedral molecule. It is hydrolysed by water, but only slowly:

$$SO_2Cl_2(l) + 2H_2O(l) \longrightarrow H_2SO_4(aq) + 2HCl(g)$$

It is used as a sulphonating and as a chlorinating agent.

13 Other oxoacids of sulphur

Sulphur forms a range of oxoacids, other than 'sulphurous' and sulphuric acids, although some of them are only known as their salts.

In *thioacids*, one oxygen atom is replaced by a sulphur atom, e.g. thiosulphuric acid, $H_2S_2O_3$; in *thionic acids*, $H_2S_xO_6$, there are —S—S— chains; in *peroxoacids*, there are —O—O— bonds; and in *condensed* acids, water is eliminated from two molecules of simple acid.

a Thiosulphuric acid, $H_2S_2O_3$ The pure acid has not been isolated but its salts, particularly sodium thiosulphate (*hypo*), are well known. This salt is made by reducing sulphite ions by sulphur in alkaline solution:

$$2SO_3^{2-}(aq) + S(s) \longrightarrow S_2O_3^{2-}(aq)$$

The salt crystallises out from solution very readily.

Thiosulphates of the alkali metals and the alkaline earths are soluble but the heavy metal salts are insoluble. The salts contain the tetrahedral $S_2O_3^{2-}$ ion. They are used as reducing agents and in the formation of complexes.

Chlorine oxidises the $S_2O_3^{2-}$ ion to SO_4^{2-},

$$S_2O_3^{2-}(aq) + 4Cl_2(g) + 5H_2O(l)$$
$$\longrightarrow 2SO_4^{2-}(aq) + 10H^+(aq) + 8Cl^-(aq)$$

and this reaction is used for removing excess chlorine from bleached fabrics. Iodine, which is a weaker oxidising agent than chlorine, only oxidises the thiosulphate ion to the tetrathionate ion, $S_4O_6^{2-}$:

$$2S_2O_3^{2-}(aq) + I_2(aq) \longrightarrow 2I^-(aq) + S_4O_6^{2-}(aq)$$

This reaction takes place quantitatively and is widely used in volumetric analysis.

The formation of complexes with Ag^+ ions, e.g. $[Ag(S_2O_3)_2]^{3-}$, explains the use of sodium thiosulphate in photography.

b Thionic acids, $H_2S_xO_6$ The value of x in the general formula is most commonly 2 or 4, but species in which it is around 20 have been reported. The sulphur atoms are linked together in zig-zag chains.

Solutions containing the various ions are made by oxidising sulphites or thiosulphates, using oxidising agents of different strengths. Oxidation of solutions of sulphite ions or sulphurous acid by manganese(IV) oxide produces dithionate ions, $S_2O_6^{2-}$. Oxidation of sodium thiosulphate solution by iodine produces sodium tetrathionate, $Na_2S_4O_6$; if hydrogen peroxide is used as the oxidising agent, sodium trithionate, $Na_2S_3O_6$, is formed.

c Peroxoacids The two commonest peroxoacids are peroxosulphuric(VI) acid (Caro's acid), H_2SO_5, and peroxodisulphuric(VI) acid, $H_2S_2O_8$.

The structures of the two acids are indicated by the fact that they can be made from chlorosulphonic acid and hydrogen peroxide:

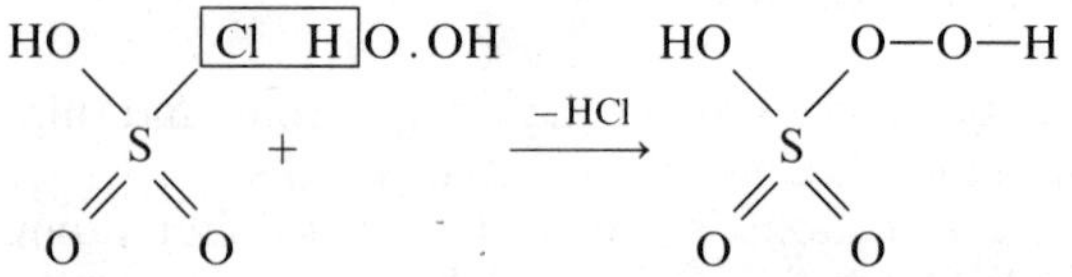

peroxosulphuric(VI) acid

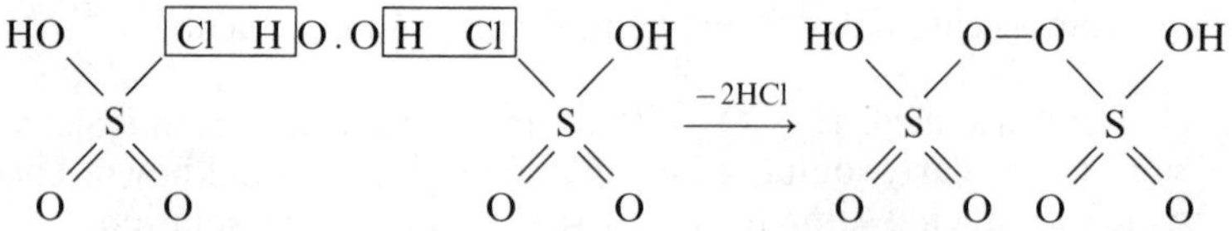

peroxodisulphuric(VI) acid

Both the acids and their salts are strong oxidising agents, e.g.

$$S_2O_8^{2-}(aq) + 2e^- \longrightarrow 2SO_4^{2-}(aq) \qquad E^\ominus = +2.01 \text{ V}$$

d Disulphuric(VI) acid, $H_2S_2O_7$ This condensed acid is present in oleum:

$$H_2SO_4(l) + SO_3(g) \longrightarrow H_2S_2O_7(l)$$

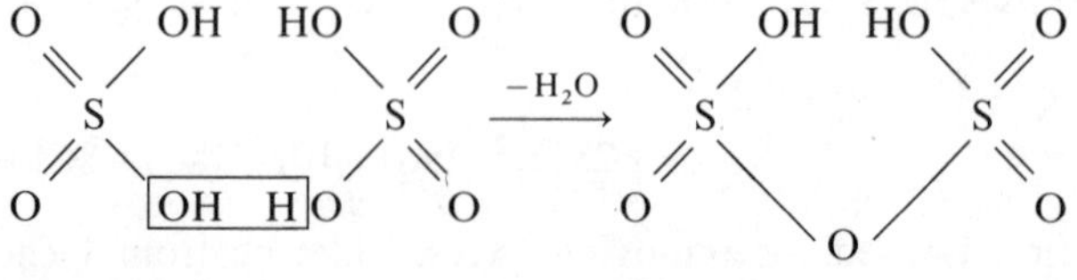

The pure acid cannot be isolated, but salts of it are known. Potassium disulphate(VI), for example, is made by heating potassium hydrogensulphate:

$$2KHSO_4(s) \longrightarrow K_2S_2O_7(s) + H_2O(l)$$

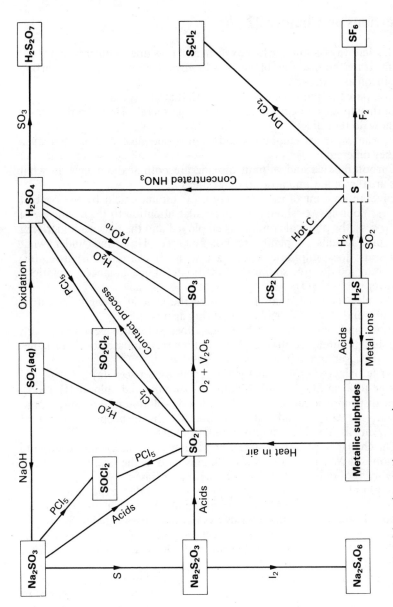

Chart 12 Sulphur and its simple compounds.

Questions on Chapter 22

1 In what ways is sulphur like oxygen, and how does it differ from it?

2 Illustrate the meaning of the word *catenation*, using sulphur, carbon and silicon as examples.

3 Compare the allotropy of sulphur with that of oxygen.

4 Both sulphur and carbon are typical non-metals. How are they alike, and how do they differ?

5 In what ways are sulphur dioxide and ozone alike, and in what ways are they different?

6 Carbon dioxide and sulphur dioxide are both acidic oxides. In what ways are they alike, and how do they differ?

7 Give an account of the production of sulphuric acid by the contact process. Your answer should pay particular attention to the chemistry of the process, the physical conditions employed and the reasons underlying their use. Details of plant are not required. How and under what conditions does sulphuric acid react with (i) charcoal, (ii) iron, (iii) potassium iodide, (iv) sodium ethanedioate? Give balanced equations wherever possible. (O)

8 Use electrode potentials to compare the general strengths of nitric, sulphuric and chloric(V) acids as oxidising agents.

9 How, and under what conditions, does sulphuric acid react with potassium bromide, iron(II) sulphate, sodium methanoate, sodium ethanedioate, zinc, sodium iodide?

10 Write balanced equations for the reactions between (a) $S_2O_3^{2-}$ and chlorine to form SO_4^{2-} and sulphur, (b) sulphite and chlorate(III) ions, (c) the disproportionation of sulphur dioxide solution into (i) SO_4^{2-} and sulphur, (ii) SO_4^{2-} and $S_4O_6^{2-}$, and (d) the disproportionation of HSO_3^- in alkaline solution into sulphur and SO_4^{2-}.

11 The change in oxidation state when iodine solution oxidises a sulphite solution is 2. When cerium(IV) sulphate is used as the oxidising agent, it is effectively about 1.5. How do you account for this?

12 Sulphur forms a chloride, S_2Cl_2. What is the evidence that leads to this formula? Another chloride exists containing a higher proportion of chlorine. In the vapour state it is always partially dissociated to S_2Cl_2 and chlorine. How could its formula be established? S_2Cl_2 hydrolyses completely in water. Suggest possible products. How would you obtain an equation for the reaction? (CS)

13 Discuss the structure of, and the bonding in, the molecules H_2S, SO_2, $SOCl_2$ and H_2SO_4. Give *two* reactions of each of these substances to illustrate their chemical reactivity. State the type of reaction taking place in each case. (JMB)

14 (a) Compare the properties of water and hydrogen sulphide, giving explanations of any similarities or differences that you quote.

(b) Comment on the following observations: (i) Oxygen normally exists in the form of diatomic molecules, whereas sulphur normally exists in the form of S_8 molecules. (ii) Oxygen does not form a hexafluoride analogous to SF_6. (iii) There are no compounds containing the O^- or S^- ions although the electron affinities of oxygen and sulphur are $-142\,kJ\,mol^{-1}$ and $-200\,kJ\,mol^{-1}$ respectively. (iv) Oxygen has a higher electronegativity than sulphur although more energy is released when a sulphur atom accepts an electron than when an oxygen atom accepts an electron. (OC)

15* Neglecting any changes of ΔH and ΔS with temperature, and any phase changes, calculate the temperature at which hydrogen would reduce (a) zinc sulphide, (b) lead sulphide and (c) silver sulphide.

16* Compare the lattice enthalpies of some oxides and sulphides, and comment on the figures.

23
The halogens (Group 7B)

1 Preparation

a Fluorine   The main source of fluorine is calcium fluoride, CaF_2 (fluorspar); an electrolytic method is necessary to obtain fluorine from it because of the difficulty in oxidising F^- ions. Calcium fluoride cannot be melted easily, so it is first converted into hydrogen fluoride (p. 311) and potassium fluoride. A mixture of these two, of composition KF.2HF, melts below 100 °C and is used as the electrolyte.

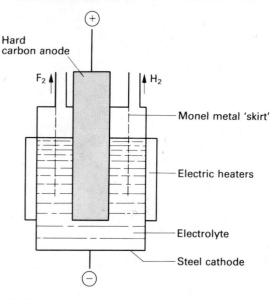

Fig. 91 *Manufacture of fluorine by electrolysis.*

A typical electrolysis cell (Fig. 91) is made of mild steel and contains a central anode of specially hard porous carbon, which is not wetted by the electrolyte, and a steel cathode which surrounds the anode. Between the anode and the cathode there is a perforated diaphragm of monel metal which keeps the fluorine produced at the anode away from the hydrogen at the cathode. The fluorine from the anode compartment contains some hydrogen fluoride, and this is removed by condensation and/or reaction with sodium fluoride. The purified fluorine is liquefied and stored in monel metal containers. The electrolytic process uses up hydrogen fluoride and more is added to the cell periodically.

b Chlorine   Oxidation of the Cl^- ion is much easier than that of F^-, so chlorine can be made by oxidising concentrated hydrochloric acid by, for example, potassium manganate(VII). Hydrogen chloride can also be oxidised by air in the presence of a hot catalyst, and such a reaction is being revived as an industrial process (p. 109).

But the main industrial source of chlorine, at present, is the electrolysis of brine (p. 106). Some chlorine is also available as a by-product from the electrolysis of the molten chlorides of sodium, magnesium and calcium.

c Bromine Bromine is obtained from sea water, from some natural brines and from the solution remaining after the preliminary treatment of the Stassfurt deposits (p. 169). These solutions all contain dissolved bromides, and chlorine can be used to oxidise the Br$^-$ ions to bromine.

The solutions are acidified and chlorine is bubbled through. If sufficient bromine is present it can be removed by steam stripping. In most cases, however, the bromine that is produced is removed by passing air through, and the air–bromine vapour mixture is passed into a solution of sulphur dioxide in water. This reduces the bromine to hydrogen bromide,

$$Br_2(l) + SO_2(g) + 2H_2O(l) \longrightarrow 2HBr(aq) + H_2SO_4(aq)$$

and concentrates it into a small bulk of solution. Bromine is again freed from this solution by passing chlorine in, and there is now enough bromine to remove it by steam stripping. It can be purified if necessary by distillation.

d Iodine Iodine is obtained from caliche, which occurs in Chile. This is the main source of sodium nitrate, or Chile saltpetre, but after the extraction of this salt, considerable amounts of sodium and other iodates(V) remain in the mother liquor. The solution is treated with sodium hydrogensulphite, which reduces the iodates(V) to iodine:

$$IO_3^-(aq) + 3HSO_3^-(aq) \longrightarrow I^-(aq) + 3HSO_4^-(aq)$$
$$5I^-(aq) + IO_3^-(aq) + 6H^+(aq) \longrightarrow 3I_2(s) + 3H_2O(l)$$

The iodine is sparingly soluble in the solution, and can be filtered off and purified by sublimation.

2 Physical properties of the halogens

The elements show a regular gradation of physical properties and pass from the gaseous fluorine and chlorine through liquid bromine (which is the only non-metal that is liquid at room temperature) to solid iodine. Chlorine is easily liquefied at room temperature under a pressure of about 4 atm; iodine sublimes.

The colour changes from pale yellow for fluorine through yellow–green and red to what appears to be a black solid, but a purple vapour, for iodine.

handwritten: $2F_2 + 2H_2O \longrightarrow 4H^+ + 4F^- + O_2$
powerful oxidising agent.

Fluorine reacts vigorously with water; the other halogens react to some extent, but also dissolve, giving solutions of varying composition (p. 309). Aqueous solutions of chlorine and bromine are used as chlorine water and bromine water. Iodine is used in alcoholic solution (tincture of iodine), or dissolved in potassium iodide solution, with which it reacts to form I_3^- ions:

handwritten: down the group, more HX and HOX are produced

$$I_2(s) + I^-(aq) \rightleftharpoons I_3^-(aq)$$

The halogens are soluble in organic solvents such as tetrachloromethane, giving yellow (Cl_2), red (Br_2) and purple (I_2) solutions.

Iodine forms a blue–black complex with starch, and this can be used as a convenient test. The colour is caused by a chain of I_5^- groups being trapped between the helical molecules of amylose.

handwritten: $2Cl_2 + 2H_2O \longrightarrow 4H^+ + 4Cl^- + O_2$

3 Compound formation

a Oxidation states   The halogens are all very reactive and chlorine, bromine and iodine, in particular, exhibit a wide range of oxidation states, as shown in the following summary.

oxidation state

-1	OF_2			
	HF	HCl	HBr	HI
	F^-	Cl^-	Br^-	I^-
0	F_2	Cl_2	Br_2	I_2
$+1$		Cl_2O	Br_2O	
		$HClO$	$HBrO$	HIO
		ClO^-	BrO^-	IO^-
$+3$		$HClO_2$		
		ClO_2^-		
$+4$		ClO_2	BrO_2	
$+5$		$HClO_3$	$HBrO_3$	I_2O_5
		ClO_3^-	BrO_3^-	HIO_3
				IO_3^-
$+6$		Cl_2O_6	BrO_3	
$+7$		Cl_2O_7		H_5IO_6
		$HClO_4$		HIO_4
		ClO_4^-		IO_4^-

There is a decrease in reactivity in passing from fluorine to iodine.

Within the group, it is chlorine and bromine that exhibit the greatest similarities. Fluorine is unique in many ways, as summarized on p. 306, and iodine has some distinctive properties because it is the only member of the group with slight metallic character, and because it has the largest atom. *down group metallic character ↗*

b Formation of X⁻ ions Fluorine, the most electronegative of the halogens, produces X⁻ ions the most easily; iodine the least easily.

c Formation of single X— covalent bonds All the halogens form single covalent bonds, both with themselves and with many other non-metals. The strength of the F—F bond is out of line with the other X—X bonds,

F—F	Cl—Cl	Br—Br	I—I
158	242	193	151 kJ mol^{-1}

HF is extremly strong bond ⇒ weak acid.

and this weakness contributes to the reactivity of fluorine, for it enables fluorine to be atomised easily. It is accounted for by repulsion between the non-bonding electrons on the two combined atoms; they are closer to each other in the F_2 molecule than in the other halogens.

The X_2 molecules all dissociate to some extent at high temperatures. Chlorine and bromine molecules also dissociate under the influence of ultraviolet light, so that they take part in photochemical reactions.

Although the F—F bond is weak, fluorine forms stronger covalent bonds with most other non-metals than the other halogens do. The weakness of F—F bonds and the strength of F—X bonds combine to make fluorine compounds more exothermic, and therefore stabler, than similar compounds of the other halogens. The compounds tend to become endothermic in passing towards iodine, as the following enthalpies of formation (in kJ mol^{-1}) show.

HF(g)	HCl(g)	HBr(g)	HI(g)
−271	−92	−36	+26

$CF_4(g)$	$CCl_4(g)$	$CBr_4(g)$
−925	−100	+50

d Higher covalencies Although the theoretical covalency maximum for fluorine is 4, it is only rarely more than 1-covalent in practice. Chlorine, bromine and iodine can also be 3- or 5-covalent, and chlorine and iodine can be 7-valent, too. This is because of the availability of d orbitals in chlorine, bromine and iodine (p. 36), and it means that these three halogens form many more compounds than fluorine does. There are, for example, many more oxoacids and oxoacid salts (p. 319).

e Formation of cations Iodine is the only halogen to form cations, and this indicates a certain metallic character in iodine which is not apparent

I is enabled to lose e's due to its size, which is greater than the other halogens.

306   Introduction to Inorganic Chemistry

in the other halogens, but which would be expected to develop on passing down a group in the periodic table. The I^+ ion is not stable as a simple ion, but it is stabilised by complexing with pyridine (py), and occurs in salts, such as $[I(py)_2]^+ClO_4^-$ and $[I(py)_2]^+NO_3^-$. Electrolysis of the solutions of these salts produces iodine at the cathode.

The iodine molecule ion, I_2^+, is produced by oxidation of iodine in 60 per cent oleum, and the I^{3+} ion probably exists in $I(ClO_4)_3$ and I_4O_9 (p. 318).

4 The peculiarities of fluorine

Fluorine differs from the other halogens in the following ways.

a It has the highest electronegativity.
b The F—F bond is weak ($158\,kJ\,mol^{-1}$), but bonds between fluorine and most other atoms are strong. These two factors taken together (p. 305) mean that reactions involving fluorine are more highly exothermic than similar reactions involving the other halogens. This makes fluorine more reactive than the other halogens.
c Fluorine forms a monovalent ion, F^-, more easily than the other halogens, and it is the strongest oxidising agent known. It is the fluoride of any element that is most likely to be ionic. due to electronegativ
d The F^- ion is smaller than the other X^- ions. This gives the metallic fluorides high lattice enthalpies and accounts, at least partially, for the odd solubilities of many fluorides. LiF, MgF_2 and CaF_2, for example, are insoluble whereas the corresponding chlorides, bromides and iodides are soluble. On the other hand, AgF is remarkably soluble compared with the insoluble AgCl, AgBr and AgI.
e Fluorine has a theoretical maximum covalency of 4, in common with the other elements in the first short period, but it never actually achieves this (p. 36). It does, however, bring out the highest covalency or the highest positive oxidation state in other elements (p. 134).
f The H—F bond is very polar, and hydrogen fluoride is strongly hydrogen bonded (p. 52).
g Hydrofluoric acid is a weak acid (p. 313).
h Liquid hydrogen fluoride is a good ionising solvent
i Fluorine forms only one oxoacid (fluoric(I) acid, HFO), and this is very unstable, existing only at low temperatures (p. 319).

5 Reactions of the halogens

a Reaction with hydrogen   All the halogens react with hydrogen to form hydrogen halides but the rates of the reactions illustrate the general differences in reactivity.

Fluorine and hydrogen explode in the dark, even at temperatures at which fluorine is solid.

HF weakest acid.
HI strongest acid. (due to weak H-I b

Chlorine will not react at room temperature in the dark, but a photochemical reaction takes place explosively in sunlight or ultraviolet light. The light causes dissociation of Cl_2 molecules into atoms,

$$Cl_2(g) + hv \longrightarrow Cl \cdot + Cl \cdot$$

which initiate a chain reaction:

$$Cl \cdot + H_2(g) \longrightarrow HCl(g) + H \cdot$$
$$H \cdot + Cl_2(g) \longrightarrow HCl(g) + Cl \cdot$$

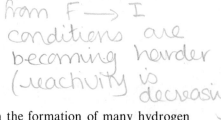 *from F → I conditions are becoming harder (reactivity is decreasing)*

Thus one chlorine atom can result in the formation of many hydrogen chloride molecules. Chlorine and hydrogen react non-explosively in diffused light, on passing over activated charcoal, or on burning a jet of one gas in an atmosphere of the other.

Bromine and hydrogen combine on heating or in the presence of a platinum catalyst at about 200 °C. Hydrogen and iodine react at about 400 °C, particularly in the presence of a platinum catalyst, but the reaction is reversible, resulting in an equilibrium mixture of hydrogen, iodine and hydrogen iodide.

b Reactions with other non-metals Fluorine combines directly with carbon, and even with xenon, krypton and radon, to form fluorides (p. 143); nitrogen and the other noble gases are the only non-metals with which it will not combine.

Chlorine and bromine react with all non-metals except carbon, oxygen, nitrogen and most of the noble gases; with excess halogen, the highest halide is formed. *- the halide with greatest oxidation state.*

Iodine is like chlorine and bromine, but it will not react with boron, and the lower halide is generally formed. The higher halide, e.g. PI_5, may not exist because the iodine atom is too large for five of them to fit round a phosphorus atom. *steric hindrance -*

c Reactions with metals Most metals react with fluorine, particularly if they are hot. Reactive metals catch fire spontaneously and many others become incandescent. The highest fluoride of the metal is formed. If, however, the metallic fluoride is insoluble, reaction is limited to surface corrosion. Thus copper, nickel and mild steel or monel metal containers can be used for storing the gas. *because gas reacts only on surface similar to Al corrosion*

Chlorine, bromine and iodine will also react with most metals. The reactions may cause incandescence or may be explosive, particularly if the metal is powdered; but if the halide formed is insoluble, the reaction may be limited to surface corrosion. Where more than one halide exists, it is generally the higher one that is produced.

However, some higher iodides, e.g. FeI_3 and CuI_2, do not exist, because the I^- ion is sufficiently strong as a reducing agent to reduce the metal to its lower oxidation state, i.e. FeI_2 and Cu_2I_2. *Similar to tin and lead.*

d Oxidising action The following standard electrode potentials show that fluorine is the strongest oxidising agent.

$$\tfrac{1}{2}F_2/F^-(aq) \quad \tfrac{1}{2}Cl_2/Cl^-(aq) \quad \tfrac{1}{2}Br_2/Br^-(aq) \quad \tfrac{1}{2}I_2/I^-(aq)$$
$$+2.87\,V \qquad\quad +1.36\,V \qquad\qquad +1.07\,V \qquad\qquad +0.54\,V$$

It is, indeed, the strongest oxidising agent that is known.

Fluorine will oxidise all the other halide ions,

$$F_2(g) + 2X^-(aq) \longrightarrow 2F^-(aq) + X_2$$

displacement rxns.

chlorine will oxidise Br^- and I^- ions, and iodine will not oxidise any halide ion.

As iodine is the weakest oxidising agent amongst the halogens, so I^- is the strongest reducing agent of the halide ions. It is used as a test for other oxidising agents, the liberated iodine being easily detected by the blue–black colour it gives with starch (p. 304).

Chlorine is the halogen that is most widely used as an oxidising agent. In aqueous solution it oxidises both by conversion into Cl^- ions and by the formation of chloric(I) acid, HClO, which is itself an oxidising agent:

$$Cl_2(g) + H_2O(l) \longrightarrow HCl(aq) + HClO(aq)$$
$$HClO(aq) + H^+(aq) + e^- \longrightarrow \tfrac{1}{2}Cl_2(g) + H_2O(l) \qquad E^\ominus = +1.64\,V$$

most come use for sterilation of water

Dry chlorine will also oxidise by removing hydrogen. Typical reactions are summarised in the following equations:

$$CH_4(g) + Cl_2(g) \longrightarrow CH_3Cl(g) + HCl(g)$$
$$C_{10}H_{16}(l) + 8Cl_2(g) \longrightarrow 10C(s) + 16HCl(g)$$
$$2NH_3(s) + 3Cl_2(g) \longrightarrow N_2(g) + 6HCl(g)$$
$$H_2S(g) + Cl_2(g) \longrightarrow S(s) + 2HCl(g)$$

Bromine reacts in much the same way though less vigorously, and the oxidising action of iodine is very weak. Iodine will, however, oxidise sodium thiosulphate to sodium tetrathionate (p. 91), in an important quantitative reaction.

$$I_2(aq) + 2S_2O_3^{2-}(aq) \longrightarrow 2I^-(aq) + S_4O_6^{2-}(aq)$$

The increase in oxidation number of the sulphur is from 2 to $2\tfrac{1}{2}$ (p. 83). Chlorine and bromine increase the oxidation number from 2 to 6, e.g.

$$4Cl_2(g) + S_2O_3^{2-}(aq) + 5H_2O(l)$$
$$\longrightarrow 2SO_4^{2-}(aq) + 10H^+(aq) + 8Cl^-(aq)$$

e Reactions with organic compounds Fluorine reacts violently with most organic compounds, hydrogen atoms being replaced by fluorine. Even solid methane, for example, reacts explosively with liquid fluorine, and aromatic hydrocarbons are broken down into a mixture of aliphatic and alicyclic products. In fluorination processes using fluorine direct, the reaction has to be controlled by diluting the gas with nitrogen and having excess copper gauze present to absorb the heat.

Chlorine will also chlorinate many organic compounds by direct reaction, i.e. replace hydrogen atoms by chlorine atoms. But direct bromination is more difficult, and iodination even more so.

Chlorine and bromine form addition compounds with many unsaturated organic substances such as the alkenes, and the decolorisation of bromine water is a simple test for unsaturation. Iodine does not readily form addition compounds.

f Reaction with water The standard electrode potentials of fluorine ($+2.87$ V) and chlorine ($+1.36$ V) show that these two halogens might be able to oxidise water to oxygen:.

$$O_2(g) + 4H^+(aq) + 4e^- \longrightarrow 2H_2O(l) \qquad E^\ominus = +1.23 \text{ V}$$
$$2F_2(g) + 2H_2O(l) \longrightarrow O_2(g) + 4HF(aq)$$

Fluorine is, in fact, so strongly oxidising that its reaction with water also produces ozone, hydrogen peroxide and oxygen difluoride, OF_2.

Chlorine, bromine and iodine both dissolve in, and react with, water. Bromine is the most, and iodine the least, soluble. The dissolved molecules can also disproportionate (p. 89):

$$X_2(aq) + H_2O(l) \longrightarrow 2H^+(aq) + X^-(aq) + XO^-(aq)$$

This disproportionation is very limited for bromine, and even more so for iodine, so the concentrations of BrO^- and IO^- in bromine water and iodine water are very low. Chlorine, however, disproportionates more readily, so chlorine water contains a significant concentration of ClO^- ions, and these decompose slowly, to form oxygen:

$$2ClO^-(aq) \longrightarrow 2Cl^-(aq) + O_2(g)$$

The reaction is facilitated by sunlight and by Co^{2+} ions. The overall reaction is an oxidation of water by chlorine:

$$2Cl_2(g) + 2H_2O(l) \longrightarrow 4H^+(aq) + 4Cl^-(aq) + O_2(g)$$

g Reactions with alkalis In hot concentrated solutions, fluorine reacts with alkalis in much the same way as with water, oxygen being the main product:

$$2F_2(g) + 4OH^-(aq) \longrightarrow O_2(g) + 4F^-(aq) + 2H_2O(l)$$

In cold dilute solution, however, oxygen difluoride (the anhydride of the unstable fluoric(I) acid, HFO) is the main product:

$$2F_2(g) + 2OH^-(aq) \longrightarrow OF_2(g) + 2F^-(aq) + H_2O(l)$$

The reactions of chlorine, bromine and iodine with solutions of alkalis depend on a number of possible disproportionation reactions. The disproportionation of X_2 into X^- and XO^-, which occurs in aqueous solution, is much more extensive in the presence of OH^- ions,

$$X_2(aq) + 2OH^-(aq) \longrightarrow X^-(aq) + XO^-(aq) + H_2O(l)$$

giving much higher concentrations of XO^- than in aqueous solution. These ions may then disproportionate into X^- and XO_3^-,

$$3XO^-(aq) \longrightarrow 2X^-(aq) + XO_3^-(aq)$$

the extent of the reaction depending on the temperature and the concentration of XO^-.

For IO^- ions, the reaction takes place even at room temperature, so that the overall reaction of iodine with alkalis is summarised by the equation

$$3I_2(s) + 6OH^-(aq) \longrightarrow 5I^-(aq) + IO_3^-(aq) + 3H_2O(l)$$

Similar reactions take place with bromine (around 50 °C) and chlorine (around 75 °C), so long as the concentration of alkali is high enough to give a high concentration of XO^- ions. The predominant reaction is the formation of XO_3^- ions, e.g.

$$3Cl_2(g) + 6OH^-(aq) \longrightarrow 5Cl^-(aq) + ClO_3^-(aq) + 3H_2O(l)$$

But at low temperatures and low OH^- ion concentrations, the reaction is mainly represented by the equation

$$Cl_2(g) + 2OH^-(aq) \longrightarrow Cl^-(aq) + ClO^-(aq) + H_2O(l)$$

6 The uses of the halogens

Fluorine and chlorine are mainly used in making a wide variety of very useful organic compounds. These include trichloroethane and tetrachloroethane, which are used as solvents; chloroethene (vinyl chloride), which is used in making PVC; chlorinated phenols, which are used in antiseptics; dichlorodifluoromethane, CF_2Cl_2, a typical member of a group of compounds known as Freons or Arctons that are used as refrigerants; Fluothane, $CF_3.CHBrCl$, a very common anaesthetic; and polytetra-

[handwritten annotation: "metal black coating on pans"]

fluoroethene (PTFE, Fluon or Teflon, $(C_2F_4)_n$), which is an extremely inert plastic used as an electrical insulator, as a non-stick agent in treating the surfaces of skis, printing presses and pans, and in making replacement hip joints. Chlorine is also used in sterilising drinking water and swimming pools, in making domestic 'bleaches', and for bleaching materials industrially.

Bromine is used in making dyestuffs, and photographic paper and film. Iodine is also used in photography, in making dyes and pharmaceutical chemicals, and as an antiseptic (in 'tincture of iodine' (p. 304)).

7 Preparation of the hydrogen halides

The hydrogen halides, HX, can all be formed by reaction between the halogen and hydrogen (p. 306), or by treating most metallic halides with sulphuric acid or phosphoric(V) acid. The hydrides of chlorine, bromine and iodine are also formed when most non-metallic halides are hydrolysed by water.

[handwritten: $H_2 + X_2 \longrightarrow 2HX$]

a Hydrogen fluoride Calcium fluoride is heated with concentrated sulphuric acid in steel retorts:

$$CaF_2(s) + H_2SO_4(aq) \longrightarrow CaSO_4(s) + 2HF(g)$$

and the hydrogen fluoride gas evolved is absorbed in water to give an 80 per cent solution of hydrofluoric acid. Subsequent distillation of this solution gives anhydrous hydrogen fluoride.

Hydrogen fluoride can also be made by heating acid fluorides in a copper or platinum tube, e.g.

$$KHF_2(s) \longrightarrow HF(g) + KF(s)$$

b Hydrogen chloride Hydrogen chloride used to be made industrially by heating sodium chloride with concentrated sulphuric acid. At a low temperature, sodium hydrogensulphate and hydrogen chloride are formed; at a higher temperature sodium sulphate is formed, and twice as much hydrogen chloride:

$$NaCl(s) + H_2SO_4(aq) \longrightarrow NaHSO_4(aq) + HCl(g)$$
$$2NaCl(s) + H_2SO_4(aq) \longrightarrow Na_2SO_4(aq) + 2HCl(g)$$

The process is still used as a commercial method for making sodium sulphate, but hydrogen chloride is now made by burning a jet of hydrogen in an atmosphere of chlorine. A lot of hydrogen chloride is also available as a by-product from organic chlorination processes (p. 109).

c Hydrogen bromide Hydrogen bromide can be made by direct combination of hydrogen and bromine vapour using a platinum catalyst, but it cannot be made satisfactorily by reaction between metallic bromides and concentrated sulphuric acid. This is because the concentrated acid is a strong enough oxidising agent to oxidise hydrogen bromide to bromine, but not to oxidise hydrogen chloride to chlorine. If hydrogen bromide is to be made from a metallic bromide, it is necessary to use the non-oxidising phosphoric(V) acid.

The commonest method of making hydrogen bromide is, however, the hydrolysis of phosphorus tribromide:

$$PBr_3(l) + 3H_2O(l) \longrightarrow H_2PHO_3(aq) + 3HBr(g)$$

Bromine is dropped on to a paste of red phosphorus and water and the gas evolved is freed from any unreacted bromine by passing it through a U-tube packed with moist red phosphorus. The hydrogen bromide is collected by downward delivery or is absorbed in water to give hydrobromic acid.

d **Hydrogen iodide** Hydrogen iodide is made, like hydrogen bromide, by the hydrolysis of phosphorus triiodide. Water is dropped on to a mixture of red phosphorus and iodine and the hydrogen iodide evolved is passed through a U-tube containing a paste of moist phosphorus to remove any iodine present. The gas is collected by downward delivery or is absorbed in water to give hydriodic acid.

8 Properties of the hydrogen halides

a **Physical properties** Pure hydrogen fluoride is a colourless, fuming liquid with a boiling temperature of 19.9 °C; the other hydrides are colourless, fuming gases with pungent smells and much lower boiling temperatures.

The relatively high boiling temperature of hydrogen fluoride is due to the strong hydrogen bonding within the substance (p. 53). The hydrogen bonding also causes hydrogen fluoride to be associated, and it normally exists as $(HF)_n$ polymers. In the solid state, the molecules are made up of zig-zag H—F···H chains (p. 52), and there are cyclic $(HF)_6$ groups in the liquid. The associated molecules dissociate on heating, so the gas contains HF molecules, and resembles the other halogen hydrides, at about 100 °C. Hydrogen fluoride also has a high dipole moment (1.9 D or 6.34×10^{-30} C m) and is a good ionising solvent:

$$2HF \rightleftharpoons H_2F^+ + F^-$$

The hydrides are all very soluble in water, one volume of water dissolving about 500 volumes of hydrogen chloride at s.t.p. to give a solution of approximately 12 M hydrochloric acid. Hydrogen bromide is even more soluble; hydrogen iodide less so. All four hydrides give constant boiling mixtures. *(azeotropes)*

b Acidity Pure hydrogen halides are covalent, but polar; they have very low electrical conductivity as liquids, and they are almost insoluble in organic solvents. In aqueous solution, however, hydrogen chloride, bromide and iodide are almost completely dissociated:

$$HX(g) + H_2O(l) \longrightarrow H_3O^+(aq) + X^-(aq)$$

The three acids – hydrochloric, hydrobromic and hydriodic – are all strong, and it is difficult to detect much difference in strength between them in aqueous solution. In other solvents, however, it is clear that the acid strength is HI > HBr > HCl. The decrease in bond enthalpy in passing from HCl to HI must be the predominant factor rather than the greater electronegativity of chlorine as compared with bromine and iodine.

Hydrogen fluoride is a much weaker acid than the other hydrogen halides; it is only about 9 per cent ionised in 0.1 M solution. When hydrogen fluoride is added to water, the hydrogen bonding must first be broken down, and the HF molecules then ionise only slightly, mainly because of the high H—F bond enthalpy:

$$HF + H_2O(l) \rightleftharpoons H_3O^+(aq) + F^-(aq)$$

The strength of hydrogen fluoride as an acid increases at higher concentrations, because the HF molecules react with F^- ions to form HF_2^- ions, and this tips the equilibrium to the right, producing a higher concentration of H_3O^+ ions.

Both normal salts, e.g. NaF, and acid salts, e.g. KHF_2, are known, but there are no stable acid salts for the other hydrohalic acids.

c Thermal stability Hydrogen iodide, with the lowest bond enthalpy, dissociates on heating, being about 30 per cent dissociated at 2000 °C. Hydrogen bromide is only about 1 per cent dissociated at that temperature, and hydrogen chloride and fluoride even less so. The standard enthalpies of formation are -271, -92, -36 and $+26\,\text{kJ}\,\text{mol}^{-1}$, for HF, HCl, HBr and HI respectively. *the more -ve the standard enthalpy the more stable*

d Ease of oxidation Hydrogen iodide is very easily oxidised, even by atmospheric oxygen. It is, therefore, a strong reducing agent, and is used as a test for oxidising agents.

Hydrogen bromide is a less strong reducing agent but is oxidised fairly easily by, for example, concentrated sulphuric acid or hydrogen peroxide. Hydrogen chloride is not oxidised by these two oxidising agents; it needs a stronger agent such as potassium manganate(VII). Hydrogen fluoride can only be oxidised electrically (p. 302).

e Reactions with silica and with glass Anhydrous hydrogen fluoride does not react with these, but hydrofluoric acid reacts with silicon(IV) oxide,

$$SiO_2(s) + 4HF(aq) \longrightarrow SiF_4(g) + 2H_2O(l)$$
$$SiF_4(g) + 2HF(aq) \longrightarrow H_2SiF_6(aq)$$

and with silicates(IV), such as occur in glass:

$$SiO_3^{2-} + 6HF(aq) \longrightarrow [SiF_6]^{2-}(aq) + 3H_2O(l)$$

Therefore, it has to be stored in containers made from copper, monel metal or plastic, and not in glass vessels.

The other acids do not react in this way.

f Summary of properties

	HF	HCl	HBr	HI
state at 15 °C	liquid	gas	gas	gas
melts at	-83.1 °C	-114.2 °C	-86.9 °C	-50.8 °C
boils at	19.9 °C	-85.0 °C	-66.7 °C	-35.4 °C
$\Delta H_{f,m}^{\ominus}/kJ\,mol^{-1}$	-271	-92	-36	$+26$
hydrogen bonding	strong	$\longleftarrow$ much weaker $\longrightarrow$		
dipole moment/D	1.9	1.03	0.78	0.38
ionising solvent	v. good	$\longleftarrow$ much weaker $\longrightarrow$		
acid strength	weak	$\longleftarrow$ strong $\longrightarrow$		
α in 0.1 M soln./%	8.5	92	93	95
bond enthalpy/kJ mol^{-1}	566	431	366	299
thermal stability	$\longrightarrow$ decreases $\longrightarrow$			
ease of oxidation	$\longrightarrow$ increases $\longrightarrow$			

9 Preparation of halides

Non-metallic halides are generally made by reaction between the non-metal and a halogen. If there is more than one possible product, the higher halide is generally formed by using excess halogen, and the lower one by using excess of the non-metal, as in making PCl_3 and PCl_5 (p. 258).

Many *anhydrous* metallic halides can also be made by direct reaction. For chlorides, chlorine gives the higher halide, and hydrogen chloride is used to obtain the lower one, e.g.

$$2Fe(s) + 3Cl_2(g) \longrightarrow 2FeCl_3(s)$$
$$Fe(s) + 2HCl(g) \longrightarrow FeCl_2(s) + H_2(g)$$

Anhydrous metallic chlorides can also be prepared by heating the metallic oxide with carbon and chlorine, as for $TiCl_4$ (p. 357) and $AlCl_3$ (p. 193).

It is easy to make *hydrated* metallic halides by reaction between a metal, or its oxide, hydroxide or carbonate, with the necessary hydrohalic acid. The hydrates cannot, however, always be converted into the anhydrous halides simply by heating, for that would cause hydrolysis if the halide had any covalent character, e.g.

$$AlCl_3.6H_2O(s) \rightleftharpoons Al(OH)_3(s) + 3HCl(g) + 3H_2O(l)$$

Effective dehydration may, however, be brought about by using sulphur dichloride oxide as a dehydrating agent. Anhydrous chromium(III) chloride, for example, can be made from the hydrate by refluxing it with $SOCl_2$:

$$CrCl_3.6H_2O(s) + 6SOCl_2(l) \longrightarrow CrCl_3(s) + 12HCl(g) + 6SO_2(g)$$

The products, other than the one that is required, are gases.

Insoluble metallic halides can readily be made by precipitation reactions, e.g.

$$Ag^+(aq) + Cl^-(aq) \longrightarrow AgCl(s)$$
$$Pb^{2+}(aq) + 2I^-(aq) \longrightarrow PbI_2(s)$$

10 Properties of halides

Halides range from the fully ionic, e.g. NaCl, to the fully covalent, e.g. CCl_4 (p. 26). Non-metallic halides are almost always covalent, but many metallic halides have significant covalent character, e.g. $BeCl_2$, $MgCl_2$, $AlCl_3$, $FeCl_3$ and AgCl. The ionic character is greatest in the fluorides and when the metal is in a low oxidation state. For example, AlF_3 is ionic, with $AlCl_3$ much more covalent; $PbCl_2$ is ionic, with $PbCl_4$ covalent.

Fully ionic halides are not hydrolysed by water, partially ionic ones are partially hydrolysed, and fully covalent ones (with a few exceptions) are fully hydrolysed, e.g.

$$NaCl(s) + H_2O(l) \longrightarrow Na^+(aq) + Cl^-(aq)$$ No hydrolysis
$$MgCl_2(s) + H_2O(l) \rightleftharpoons MgCl(OH)(aq) + HCl(g)$$ partial ''
$$PCl_3(l) + 3H_2O(l) \longrightarrow H_2PHO_3(aq) + 3HCl(g)$$ complete ''

Covalent halides can, however, be hydrolysed in a number of ways (see section **b** below).

a Solubility of metallic halides Most ionic halides dissolve in water to give metal ions and halide ions, but the fluorides of lithium and the alkaline earths, and the chlorides, bromides and iodides of Cu^+, Hg^+, Ag^+ and Pb^{2+} are insoluble.

The trends are very irregular. Sometimes the solubility of metallic halides increases in passing from the fluoride to the iodide; sometimes the trend is in the reverse direction. Sometimes the solubility of a particular halide increases in passing down a group in the periodic table; sometimes the trend is irregular. As with all solubility data, which is often very bewildering, it is only possible to make any sense out of the trends by a detailed consideration of both enthalpy and entropy changes. The latter may be just as important as the former (p. 72).

Silver nitrate solution, in the presence of dilute nitric acid, is commonly used to detect Cl^-, Br^- and I^- ions by precipitation of AgCl (white), AgBr (cream) or AgI (yellowish). It will not detect F^- ions for AgF is, surprisingly, very soluble. The insoluble halides are precipitated almost quantitatively, so they can be used in gravimetric analysis. They will dissolve in solutions of ammonia, sodium thiosulphate or potassium cyanide, owing to the formation of complexes:

$$Ag^+(aq) + 2NH_3(aq) \longrightarrow [Ag(NH_3)_2]^+(aq)$$
$$Ag^+(aq) + 2S_2O_3^{2-}(aq) \longrightarrow [Ag(S_2O_3)_2]^{3-}(aq)$$
$$Ag^+(aq) + 2CN^-(aq) \longrightarrow [Ag(CN)_2]^-(aq)$$

The solubility in ammonia decreases in passing from AgCl to AgI, and this can be used to distinguish the three silver halides. They all dissolve readily in thiosulphate or cyanide solutions and these are used in developing photographs. They remove, from the film or plate, any silver halide that has not been converted into silver by photodecomposition. Amateur photographers use sodium thiosulphate (hypo), but cyanide solutions are used industrially. $[Ag(CN)_2]^-$ has a higher stability constant (p. 335) than $[Ag(S_2O_3)_2]^{3-}$, and that means that cyanide solutions are more effective in complexing Ag^+ ions than thiosulphate solutions are.

b Hydrolysis of halides   Fully ionic halides are not hydrolysed by water, but covalent and partially covalent ones are. The extent of the hydrolysis is greater for the iodides than for the fluorides. The nature of the process in the hydrolysis of, for example, MCl_x depends on M.

Covalent chlorides in which M can act as an acceptor, i.e. in which M has a maximum covalency (p. 133) greater than 4, are 'attacked' by water, by the donation of a lone pair and the subsequent elimination of HCl, e.g.

$$SiCl_4(l) + 2H_2O(l) \longrightarrow SiO_2(s) + 4HCl(g)$$

$$PCl_3(l) + 3H_2O(l) \longrightarrow H_2PHO_3(aq) + 3HCl(g)$$

When M can act as a donor, but cannot act as an acceptor because it is in the first short period and has a maximum covalency of 4, it is the lone pair of M that attacks the water molecule, with the subsequent elimination of HClO, e.g.

$$NCl_3(g) + 3H_2O(l) \longrightarrow NH_3(g) + 3HClO(aq)$$
$$FCl(g) + H_2O(l) \longrightarrow HF(g) + HClO(aq)$$

When M can act neither as a donor nor as an acceptor, as in CCl_4 and SF_6, there is no reaction with water, even though the reaction may be thermodynamically feasible (p. 216).

11 Uses of metallic halides

All four halide ions play a part in metabolic processes in humans. F^- ions occur in bones and, particularly, in tooth enamel. Therefore, a shortage of F^- ions in a diet may lead to poor teeth. To prevent this, a carefully controlled concentration of sodium fluoride (for the F^- ion can be poisonous) is added to many water supplies, and most toothpastes contain added fluorides.

Sodium chloride is also an essential component of a healthy diet, for Cl^- ions are present in quite high concentrations in blood plasma, saliva and sweat. A deficiency of Cl^- ions may cause muscular cramp, which is why athletes take salt tablets after prolonged activity. People who live in hot climates or who sweat extensively at their work, may also require extra salt.

Potassium bromide is used as a sedative. Iodide ions are required for the correct functioning of the thyroid gland, which is just below the Adam's apple. This gland produces the hormone thyroxin, which contains iodine atoms. Lack of I^- ions in a diet may lead to malfunctioning of the gland, with a subsequent swelling of the neck, in a disease called goitre.

Nowadays, the disease is virtually non-existent in the UK because potassium iodide is commonly added to supplies of table salt.

Other important uses of halides include the use of sodium chloride to prevent icing of roads; the use of uranium(VI) fluoride, UF_6, in the separation of the isotopes of uranium; and the use of silver halides in photography.

12 Oxides of the halogens

a Summary The main halogen oxides (or oxygen fluorides in the case of fluorine) are summarised below.

oxidation state	oxide	associated oxoacid(s)	nature of compound
−1	O_2F_2		unstable orange–red solid
	OF_2	(HFO)	colourless gas
+1	Cl_2O	HClO	explosive yellow gas
	Br_2O	HBrO	explosive brown liquid
+4	ClO_2	$HClO_2$, $HClO_3$	explosive orange gas
	BrO_2	$HBrO_2$, $HBrO_3$	unstable yellow solid
+5	I_2O_5	HIO_3	white crystalline solid
+6	Cl_2O_6		unstable red liquid
	BrO_3		unstable white solid
+7	Cl_2O_7	$HClO_4$	explosive colourless oil

Many of the oxides are very unstable, and they do not have much in common. However, most of them can be regarded, at least formally, as acid anhydrides or mixed acid anhydrides.

b Iodine(V) oxide, I_2O_5 This is the only true oxide of iodine: two other 'oxides', I_2O_4 and I_4O_9, are probably iodates, $IO^+IO_3^-$ and $I^{3+}(IO_3^-)_3$. Iodine(V) oxide is made by heating iodic(V) acid at 200 °C:

$$2HIO_3(s) \longrightarrow I_2O_5(s) + H_2O(l)$$

On heating to 300 °C, the oxide decomposes into its elements. It can be used as an oxidising agent, and it will convert carbon monoxide quantitatively into carbon dioxide:

$$5CO(g) + I_2O_5(s) \longrightarrow I_2(s) + 5CO_2(g)$$

The carbon dioxide or iodine can be estimated, so that the reaction provides a method of measuring the carbon monoxide content of a gas sample.

13 Oxoacids of the halogens

Fluorine forms no stable oxoacids (although HFO can exist at low temperatures) but the other halogens form the following main acids and corresponding salts:

name of acid	formula of acid			name of salt
halic(I)	HClO	HBrO	HIO	halate(I)
halic(III)	HClO$_2$			halate(III)
halic(V)	HClO$_3$	HBrO$_3$	HIO$_3$	halate(V)
halic(VII)	HClO$_4$	HBrO$_4$	HIO$_4$	halate(VII)
			(H$_5$IO$_6$)	

handwritten margin notes: hypohalite, halite, halate, perhalate

Apart from HIO$_3$, HClO$_4$ and HIO$_4$, the oxoacids are only stable in solution.

a Structures of oxoacid ions The structures and shapes of typical oxoacid ions are shown in Fig. 92.

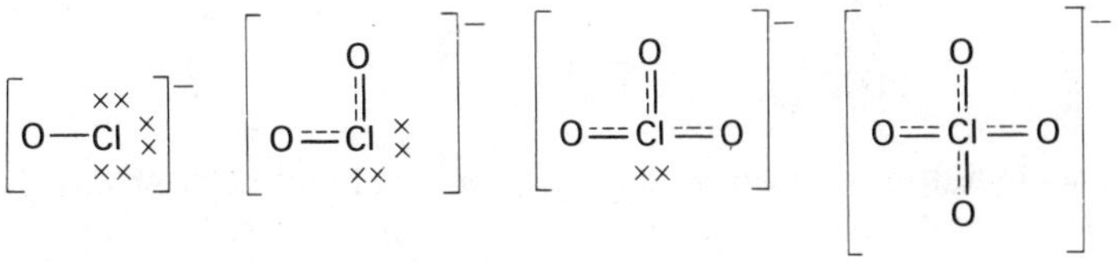

| Chlorate(I) ion (collinear) | Chlorate(III) ion (triangular) | Chlorate(V) ion (pyramidal) | Chlorate(VII) ion (tetrahedral) |

Fig. 92 *Oxoacid ions of chlorine.*

As iodine is large enough to coordinate six oxygen atoms around it, it can form the octahedral IO$_6^{5-}$ ion, as well as the IO$_4^-$ ion.

b Strengths of oxoacids There is a marked increase in strength in passing from the +1 acid to the +7 acid (p. 279) and in passing from the iodine acid to the chlorine acid:

increase in strength →

	HClO	HClO$_2$	HClO$_3$	HClO$_4$
increase in strength ↑	HBrO		HBrO$_3$	HBrO$_4$
	HIO		HIO$_3$	HIO$_4$

Chloric(VII) acid is one of the strongest known acids, whereas iodic(I) acid is extremely weak.

c Halic(I) acids Solutions containing these acids are formed when the halogens react with water,

$$X_2 + H_2O(l) \rightleftharpoons HX(aq) + HXO(aq)$$

disproportionation rxn.

but the concentration of halic(I) acid is never very high at room temperature, because the equilibrium constant is too low and/or because the acid is not stable. The conditions are most favourable for chloric(I) acid.

A very low temperature is required in order to make fluoric(I) acid, and the highest concentrations of the other acids are obtained in the presence of freshly precipitated mercury(II) oxide. This forms an insoluble basic mercury(II) halide, which shifts the equilibrium to the right:

$$2X_2(g) + HgO(s) + H_2O(l) \longrightarrow HgX_2(s) + 2HXO(aq)$$

All the solutions are unstable, either because the halic(I) acid decomposes,

$$2HXO(aq) \rightleftharpoons 2HX(aq) + O_2(g)$$

or because it disproportionates:

$$3HXO(aq) \rightleftharpoons 2HX(aq) + HXO_3(aq)$$

Fluoric(I) acid solutions decompose just above $0\,°C$, and both bromic(I) and iodic(I) acid solutions disproportionate at room temperature so that they do not store well. Chloric(I) acid decomposes and disproportionates only slowly at room temperature, but decomposition is speeded up by sunlight and by Co^{2+} ions, and disproportionation by heating. This acid must therefore be stored in a cold, dark place.

All the acids are oxidising agents, e.g.

$$2HClO(aq) + 2H^+(aq) + 2e^- \longrightarrow 2H_2O(l) + Cl_2(g) \qquad E^\ominus = +1.64\,V$$

The corresponding electrode potentials for HBrO and HIO are $+1.59\,V$ and $+1.45\,V$ respectively, so that iodic(I) acid is the weakest oxidising agent.

d Halic(V) acids Fluoric(V) acid has not been made, but solutions of the other acids can be made by treating the barium halates(V), obtained from the halogens and hot concentrated barium hydroxide solution (p. 310), with concentrated sulphuric acid. Barium sulphate is precipitated and a solution containing the halic(V) acid is formed. Chloric(V) and bromic(V) acids are known only in solution, but iodic(V) acid can be crystallised out. It can also be made by oxidising iodine with concentrated nitric acid.

The acids are all strong oxidising agents, particularly in concentrated solution. Bromic(V) acid is the strongest:

$$2BrO_3^-(aq) + 12H^+(aq) + 10e^- \longrightarrow Br_2(aq) + 6H_2O(l)$$
$$E^\ominus = +1.52\,V$$

The corresponding figures for $ClO_3^-(aq)$ and $IO_3^-(aq)$ are $+1.47\,V$ and $+1.195\,V$. Under standard conditions, chloric(V) acid is not able to oxidise Cl^- ions to chlorine, but the other two halic(V) acids will oxidise Br^- or I^-:

$$IO_3^-(aq) + 5I^-(aq) + 6H^+(aq) \longrightarrow 3I_2(aq) + 3H_2O(l)$$
$$BrO_3^-(aq) + 5Br^-(aq) + 6H^+(aq) \longrightarrow 3Br_2(aq) + 3H_2O(l)$$

Under alkaline conditions, the electrode potentials change, and these reactions are not possible.

e Halic(VII) acids Only chloric(VII), bromic(VII) and iodic(VII) acids are known. They can be obtained in quite concentrated solution and, although it is not very stable, pure chloric(VII) acid can also be made.

Chloric(VII) acid is a strong acid, and it is also a very strong oxidising agent when pure or in hot, concentrated solution. Under these conditions, it reacts vigorously, or even explosively, with many reducing agents and many organic materials. It is, however, only weakly oxidising in cold, dilute solution.

14 Halates(I)

These can be made by treating a cold, dilute solution of an alkali with a halogen, e.g.

$$Cl_2(g) + 2NaOH(aq) \longrightarrow NaCl(aq) + NaClO(aq) + H_2O(l)$$

Fluorates(I) and iodates(I) rapidly decompose or disproportionate, so that they are not useful. Solutions of bromates(I) will keep only at a low temperature; when they are needed in organic chemistry, it is best to use a freshly made solution. Chlorates(I) are the most stable and useful.

a Sodium chlorate(I) This can be crystallised as a hydrate, $NaClO.6H_2O$, from the solution formed by bubbling chlorine through cold, dilute sodium hydroxide solution, but it is usually used in solution. The solution can be made by electrolysing cold brine in such a way that the chlorine and sodium hydroxide produced can interact.

Sodium chlorate(I) is a strong oxidising agent, particularly in acidic solution:

$$ClO^-(aq) + 2H^+(aq) + 2e^- \longrightarrow Cl^-(aq) + H_2O(l) \qquad E^\ominus = +1.64\,V$$

It is a very common bleach, being sold under a number of trade names, such as Parazone and Domestos.

The solution decomposes slowly into oxygen and sodium chloride (p. 310) and the decomposition is accelerated by sunlight or by Co^{2+} ions. Disproportionation into the chloride and chlorate(V) also occurs, particularly on heating (p. 320).

b Bleaching powder This is manufactured by passing chlorine up a rotating inclined cylinder, down which slightly moist slaked lime is dropping. The product is an off-white powder of variable composition. It is, predominantly, a mixture of basic calcium chlorates(I), e.g. $3Ca(ClO)_2.2Ca(OH)_2$ and $Ca(ClO)_2.2Ca(OH)_2$, with basic calcium chloride, $CaCl_2.Ca(OH)_2.H_2O$. As a simplification, it is often regarded as a mixture of calcium chlorate(I), $Ca(ClO)_2$, and basic calcium chloride. In more elementary terms, the composition can be expressed reasonably well by the 'formula' $CaOCl_2$.

It is used as an oxidising and bleaching agent, as it reacts with acids, even with carbonic acid, to liberate chlorine.

15 Halates(V)

a Sodium and potassium chlorates(V) These salts are made by reaction between chlorine and hot, concentrated sodium hydroxide or potassium hydroxide solution:

$$3Cl_2(g) + 6OH^-(aq) \longrightarrow ClO_3^-(aq) + 5Cl^-(aq) + 3H_2O(l)$$

Industrially, the reaction is brought about by the electrolysis of a hot solution of brine in such a way that the chlorine formed at the anode mixes, and reacts, with the hydroxide formed at the cathode. The resulting solution contains a mixture of chlorides and chlorates(V). For the sodium salts, the chloride crystallises out first, followed by the chlorate(V); for the potassium salts, the chlorate(V) crystallises out first.

Both chlorates(V) decompose on heating, e.g.

$$4KClO_3(s) \xrightarrow[\text{temperature}]{\text{low}} 3KClO_4(s) + KCl(s)$$

$$KClO_4(s) \xrightarrow[\text{temperature}]{\text{high}} KCl(s) + 2O_2(g)$$

They are strong oxidising agents, forming dangerously explosive mixtures with easily oxidisable substances such as charcoal, sulphur, phosphorus, sugar and other organic materials.

Potassium chlorate(V) is used in making matches, fireworks and potassium chlorate(VII); sodium chlorate(V) is used as a weedkiller, and in making sodium chlorate(VII) and chlorine dioxide.

b Potassium bromate(V) This is made, like potassium chlorate(V), from bromine and hot, concentrated potassium hydroxide solution. Potassium bromate(V) crystallises out first from the mixture of bromate(V) and bromide.

Solid potassium bromate(V) decomposes into potassium bromide and oxygen on heating. It is used as an oxidising agent in volumetric analysis:

$$BrO_3^-(aq) + 6H^+(aq) + 6e^- \longrightarrow Br^-(aq) + 3H_2O(l)$$
$$E^\ominus = +1.52\,V$$

As soon as all the reducing agent present has been oxidised, excess bromate(V) ions react with the bromide ions formed, giving bromine:

$$BrO_3^-(aq) + 5Br^-(aq) + 6H^+(aq) \longrightarrow 3Br_2(aq) + 3H_2O(l)$$

The end-point can be detected by methyl orange indicator, whose colour is bleached by the bromine formed.

c Sodium and potassium iodates(V) Sodium iodate(V) occurs naturally in caliche and is the main source of iodine (p. 303), but potassium iodate(V) is more commonly used as a reagent in volumetric analysis. It is made from iodine and hot concentrated potassium hydroxide, the potassium iodate(V) crystallising out before the potassium iodide. The salt can be obtained in a very pure state by crystallisation.

In dilute acid solutions it oxidises iodide ions,

$$IO_3^-(aq) + 5I^-(aq) + 6H^+(aq) \longrightarrow 3I_2(aq) + 3H_2O(l)$$

and the resulting iodine can be titrated with sodium thiosulphate. The process, using a definite amount of iodate(V) with an excess of iodide, can be used to standardise thiosulphate solutions. Alternatively, iodates(V) or iodides or acids can be estimated by having any two in excess.

16 Halates(VII)

Potassium and sodium chlorates(VII) can be made by careful heating of the chlorates(V), but the latter compound is made industrially by the electrolysis of sodium chlorate(V) solution using an iron cathode and a platinum anode. The platinum has a high oxygen over-voltage, so anodic oxidation is favoured:

$$ClO_3^-(aq) + H_2O(l) \longrightarrow ClO_4^-(aq) + 2H^+(aq) + 2e^-$$

Potassium and ammonium chlorates(VII) are only slightly soluble, so they can be made from a solution of sodium chlorate(VII) by adding K^+ or

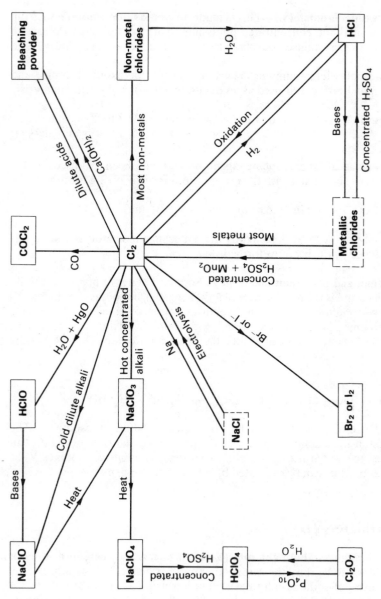

Chart 13 Chlorine and its simple compounds. NaCl occurs naturally.

NH_4^+ ions; the sodium salt, $NaClO_4$, can therefore be used as a test for K^+ and NH_4^+ ions.

The chlorates(VII) decompose on heating to give the chloride and oxygen. They are strong oxidising agents in hot, concentrated solution, but only weak ones in cold, dilute solution. Potassium chlorate(VII) is used in making detonators and explosives, and ammonium chlorate(VII) in making solid propellants.

Bromates(VII) have only recently been made, by oxidising bromates(V) with fluorine in strongly alkaline solution. They are stronger oxidising agents than either chlorates(VII) or iodates(VII).

Both normal and acid salts of the different iodic(VII) acids are known. They are strong oxidising agents, and will oxidise manganese(II) salts to manganates(VII).

17 Interhalogen compounds

Thirteen interhalogen compounds are known, as summarised below; they are generally made by reaction between the two halogens concerned.

type	compound	m.p./°C	b.p./°C	nature of compound
AX	ClF	−154	−101	colourless gas
	BrF	−33	20	red–brown gas
	BrCl		5	red gas
	ICl	27	97	red solid
	IBr	36		black solid
AX_3	ClF_3	−83	12	colourless gas
	BrF_3	9	127	yellow liquid
	IF_3	dec. into I_2 and IF_5		yellow solid
	ICl_3	dec. into ICl and Cl_2		yellow solid
AX_5	ClF_5	−103	−14	colourless gas
	BrF_5	−61	40	colourless liquid
	IF_5	10	101	colourless liquid
AX_7	IF_7		5	colourless gas

In the AX_n compound, n is always an odd number, and when n is 3, 5 or 7, X is always a lighter halogen than A. Fluorine, which forms the stablest halides (p. 304), forms the greatest number of interhalogen compounds.

The valency of fluorine in these compounds is, however, never greater than 1; that of chlorine reaches 5 in ClF_5; bromine has a valency of 5 in BrF_5; and iodine is 7-valent in IF_7. The compounds are covalent with the X atoms arranged round A on the basis of electron pair repulsion ideas (p. 27). The covalent bonds in the compounds are, however, significantly polar, depending on the electronegativities of A and X, e.g. $Cl^{\delta+}$—$F^{\delta-}$, $I^{\delta+}$—$Cl^{\delta-}$.

The interhalogen compounds are highly reactive, like the halogens. They react with most elements to produce mixtures of halides.

Questions on Chapter 23

1 Describe the manufacture of chlorine by the electrolysis of brine. Give two industrial uses for the element. How may chlorine be converted to (a) potassium chlorate(V) crystals, (b) nitrogen trichloride, (c) disulphur dichloride, S_2Cl_2. (C)

2 Discuss the statement: 'The halogens show the most regular gradations in properties of all the groups in the periodic table.' (L)

3 'The fact that iodine is the least electronegative of the halogens shows up in its properties and those of its compounds.' Illustrate.

4 'The electrolysis of aqueous sodium chloride involves oxidation, reduction and the displacement of ionic equilibrium.' Carefully explain this statement. Outline how this process under varying conditions has been adapted for the manufacture of (a) sodium hydroxide, (b) sodium chlorate(I), (c) sodium chlorate(V). (L)

5 How does potassium iodide react with each of the following: (a) mercury(II) chloride, (b) sulphuric acid, (c) copper(II) sulphate, (d) nitrous acid, (e) iodine? Give observations, conditions and equations. How and why is potassium iodide used in the purification of commercial iodine? (L)

6 Give a comparative account of the oxoacids of the halogens from the viewpoint of (a) their acid properties or the thermal stability of their alkali salts, and (b) their properties as oxidants. (L)

7 What factors contribute to the extremely high reactivity of fluorine? Explain the inertness in chemical reactivity of sulphur hexafluoride. Why, in the halogen fluorides, is the ratio of fluorine to other halogen atoms always odd? Explain why phosphorus trifluoride boils at a higher temperature than phosphorus pentafluoride. (SUJB)

8 'When a metal exhibits more than one oxidation state, the halides of the lower state are the more ionic.' Illustrate that statement.

9 Comment on the formation of hydrates by halides.

10 What shapes would you expect the interhalogen molecules to have?

11 The ionic radii of F^- and O^{2-} are almost the same. Are fluorides in any way like oxides?

12 What physical and chemical properties would you expect for astatine?

13 In what ways do the halogens occur naturally? Compare their methods of extraction.

14 Compare the solubilities of chlorides and fluorides.

15 Why is (a) fluorine so very reactive, and (b) hydrogen fluoride such a weak acid?

16 Discuss the reactions of chlorine with (a) non-metals, (b) metals, and (c) sodium hydroxide solution.

17 Write short notes on (a) the reaction of glass with hydrofluoric acid, (b) Chile saltpetre, (c) iodine(V) oxide, (d) bleaching powder.

18 Describe some uses of chlorine and bromine in organic chemistry.

19 Briefly describe the preparation of (a) chlorine, and (b) iodine. Suggest a reason why the same method is not used for the preparation of both halogens.

When an excess of liquid chlorine is added to iodine, a reaction occurs. When the unreacted chlorine is allowed to evaporate, an orange solid of formula ICl_x remains. One mole of ICl_x reacts with excess potassium iodide solution to liberate two moles of iodine (I_2). Write an equation for the reaction between ICl_x and iodide ions. What is the oxidation state of the iodine in ICl_x? (OC)

20 (a) Use the halogens to illustrate how the physical and chemical properties of elements and their compounds change in moving down a group in the periodic table. (b) Predict the shapes of the following species, giving reasons for your answers: (i) H_5IO_6, (ii) Cl_2O_7, (iii) I_3^-. (O)

21 Compare the chemical and physical properties of the two compounds in any *three* of the following pairs of chlorides, explaining the role of the electronic structures of the elements involved, where appropriate: (a) CCl_4 and $SnCl_4$, (b) $AlCl_3$ and PCl_3, (c) $CaCl_2$ and $FeCl_2$, (d) HCl and NaCl. (OC)

22 Metallic character increases from right to left and from top to bottom in the periodic table. Select any *two* properties of elements which determine metallic character and discuss the trends which occur in (a) the halogens, and (b) the period from sodium to chlorine. Illustrate your answer with suitable examples. How far do you agree with the view that a classification of elements into metals and non-metals is unrealistic? (L)

23 Give an account of the chlorides of the elements from sodium to chlorine in the periodic table, paying particular attention to (a) methods of preparation, including brief experimental details, (b) the nature of the bonding, (c) reactions with water.

0.800 g of a chloride of sulphur was hydrolysed with water and the solution diluted to $100\,cm^3$. $25.00\,cm^3$ of this solution was titrated with 0.100 M silver nitrate solution, of which $29.60\,cm^3$ were required. Determine the empirical formula of the chloride. [S = 32.0; Cl = 35.5] (L)

24 For *one* chloride of *each* of the elements sodium, magnesium, aluminium, silicon, phosphorus and sulphur, state and illustrate the following characteristic properties: (a) bonding, (b) structure, (c) their reaction with water. (AEB)

25 (a) Compare the elements chlorine and iodine with respect to (i) the appearance of, and bonding in (1) the elements at room temperature and (2) their vapours, (ii) the solubility in water and in aqueous ammonia of their silver compounds, (iii) the preparation, ease of oxidation and reaction with water of their compounds with hydrogen. (b) State how the reaction between iodine and aqueous thiosulphate can be used to determine the concentration of certain aqueous oxidising agents, e.g. iodate(V). (Details of apparatus and manipulations are not required.) (AEB)

26* Summarise the bond enthalpy values for the A—X bond when A represents different non-metals and X different halogens. What general conclusions can you draw?

27* Compare and contrast the values of the standard enthalpies of formation of the hydrogen halides.

28* Compare the ionic radii values for F^-, Cl^-, Br^-, I^-, O^{2-}, S^{2-}, Se^{2-}

and Te^{2-}. Comment on any significant points.

29* Compare the boiling temperatures of the AX interhalogen compounds with those of the halogens. What conclusions can you draw?

30* Compare the melting and boiling temperatures of some common hydrides and halides. Can you draw any general conclusions?

31* Investigate the enthalpy of formation of the chlorides of sodium, magnesium, aluminium, silicon and phosphorus on a 'per bond' basis. What do you find?

32* Compare the enthalpies of formation of the HX, OX_2 and NX_3 compounds of the halogens. Comment on the figures.

33* Use the enthalpies of formation of the oxides and the bond enthalpies of chlorine and oxygen to calculate the bond enthalpies in Cl_2O and ClO_2. If the bonding in Cl_2O is a single covalent bond of length 0.171 nm, what can you deduce about the bonding in ClO_2?

Complex or coordination compounds

ions of d-block elements are hydrated in sol'n.

1 Complex ions

An ion that is formed from a single atom, e.g. Cl^- and Na^+, is known as a simple ion, whereas ions containing more than one atom are known as complex ions. Such complex ions consist of a simple ion linked to other ions, atoms or molecules. Even very common ions such as sulphate, nitrate and hydroxide are, strictly, complex ions. They can be regarded as $S^{6+}(O^{2-})_4$, $N^{5+}(O^{2-})_3$ and $O^{2-}H^+$, but they are so stable (i.e. they do not dissociate into their component simple ions) that they function in much the same way as simple ions do, and they are not typical complex ions.

In a typical complex ion, there is always some measure of dissociation into its components. The following are common examples:

hexacyanoferrate(II) $[Fe(CN)_6]^{4-}$ hexaaquacopper(II) $[Cu(H_2O)_6]^{2+}$
hexachlorostannate(IV) $[SnCl_6]^{2-}$ diamminesilver(I) $[Ag(NH_3)_2]^+$
tetrahydridoaluminate(III) $[AlH_4]^-$ hexaamminenickel(II) $[Ni(NH_3)_6]^{2+}$

Salts containing ions such as these are known as complex salts and the groups attached to the central simple ion are known as *ligands*. Complex salts were first studied by Werner (1866–1919), who called them coordination compounds. He used the term coordination number to indicate the number of ligands round the central simple ion. Coordination numbers from 2 to 12 are known, with 4 and 6 being much the most common; coordination numbers from 9 to 12 are very rare.

Some typical examples, together with their shapes, are summarised below.

Coordination number = 2 $[Ag(CN)_2]^-$
   $[Cu(NH_3)_2]^+$
   linear

Coordination number = 4 $[FeCl_4]^-$ $[PdCl_4]^{2-}$
   $[Zn(NH_3)_4]^{2+}$ $[Pt(NH_3)_4]^{2+}$
   $[Ni(CO)_4]$ $[Ni(CN)_4]^{2-}$
   tetrahedral   square

Coordination number = 6 $[FeF_6]^{3-}$
   $[Cr(H_2O)_6]^{2+}$
   $[Fe(CN)_6]^{3-}$
   octahedral

2 Types of ligand

Simple, or monodentate, ligands are linked to the central ion by only one atom of the ligand; these are generally anions or neutral molecules, the commonest examples being listed below.

cyano	CN^-	fluoro	F^-	aqua	H_2O
hydroxo	OH^-	chloro	Cl^-	ammine	NH_3
nitrito	ONO^-	bromo	Br^-	carbonyl	CO
nitro	$NO_2{}^-$	iodo	I^-	nitrosyl	NO

Other ligands make use of two or more atoms to form more than one bond to the central ion. Such ligands are called chelate groups, and the compounds formed, *chelate compounds* (from the Greek $\chi\eta\lambda\eta$ = a crab's claw). A group capable of forming two bonds is called a bidentate group, one forming three groups a tridentate group, and so on.

Ethane-1,2-diamine (usually abbreviated to en) is a bidentate group that gives typical chelate compounds, as shown below.

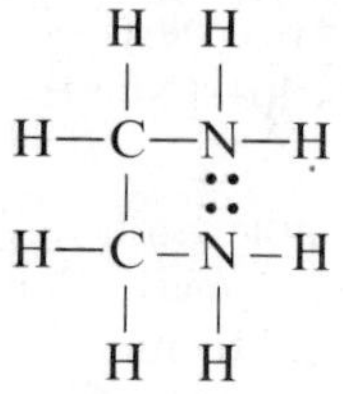

| ethane-1,2-diamine | dichlorobis(ethane-1,2-diamine)chromium(III) chloride | tris(ethane-1,2-diamine)cobalt(III) chloride |

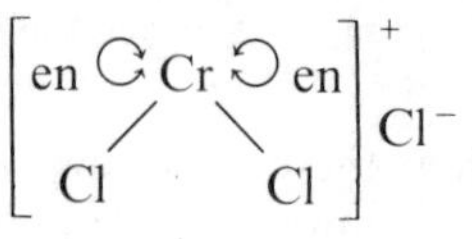

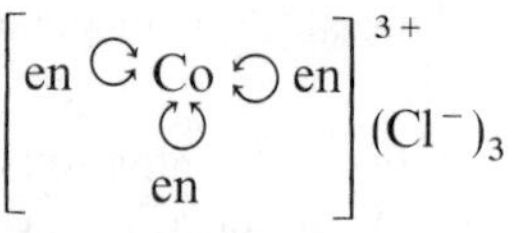

Other common bidentate groups are acetylacetone (pentane-2,4-dione), dipyridyl, dimethylglyoxime (butanedione dioxime) and 1,2-dihydroxybenzene.

Ethylenediaminetetraacetic acid, which is often simplified to $H_4[edta]$, (1,2-bis[bis(carboxymethyl)amino]ethane) is an important hexadentate group (p. 338).

3 Types, nomenclature and formulae of complexes

a Complex cations   The name used begins with the numbers and names of the ligands, followed by the name of the central atom with its oxidation number (p. 81) indicated by Roman numerals in parentheses. If different ligands are present, they are listed with the names in alphabetical order (disregarding any numerical prefixes such as di-, tri-...). In the formulae, which are enclosed in square brackets, anion ligands are written before neutral ones, the groups being placed in alphabetical order of the symbol for the donor atom of the ligands, e.g.

$[Co(NH_3)_6]^{3+}$ $[CrCl_2(H_2O)_4]^+$
hexaamminecobalt(III) ion   tetraaquadichlorochromium(III) ion

$[PtCl(NO_2)(NH_3)_4]^{2+}$
tetraamminechloronitroplatinum(IV) ion

b Complex anions The name used gives the number and names of the ligands, followed by the name of the central atom with the ending -ate (or -ic, for an acid) and then its oxidation number in parentheses. The other rules, as given in section **a**, also apply, e.g.

$[SiF_6]^{2-}$ $[Fe(CN)_5NO]^{2-}$
hexafluorosilicate(IV) ion   pentacyanonitrosylferrate(III) ion

$H_2[SiF_6]$
hexafluorosilicic(IV) acid

c Neutral complexes If the charge on the central atom is equal and opposite to the total charge on the ligands, a neutral complex results, e.g. $[CrCl_3(NH_3)_3]$ and $[CoCl_2(NH_3)_4]$.

Neutral complexes also occur when the central species is an atom rather than an ion and the ligands are uncharged. Transition metals form many carbonyls of this type, e.g.

$[Cr(CO)_6]$ $[Mn_2(CO)_{10}]$ $[Fe(CO)_5]$ $[Co_2(CO)_8]$ $[Ni(CO)_4]$
$[Fe_2(CO)_9]$

Transition metals also form interesting complexes involving hydrocarbon rings as ligands. These are known as *sandwich compounds* because the central atom is sometimes 'sandwiched' between C_xH_y rings. Examples of these are $[Fe(C_5H_5)_2]$, $[Cr(C_6H_6)_2]$ and $[Ni(C_5H_5)NO]$.

4 Isomerism in complexes

Complexes with coordination number 6 are generally octahedral and those with coordination number 4 are tetrahedral or square planar depending on the nature of the bonding (p. 333). With certain ligand arrangements, isomerism is possible.

a Geometric isomerism This does not occur in tetrahedral complexes but it is found in Ma_4b_2 octahedral complexes, e.g. tetraamminedichlorocobalt(III) chloride (Fig. 93), and Ma_2b_2 square planar complexes, e.g. diamminedichloroplatinum(II) (Fig. 94).

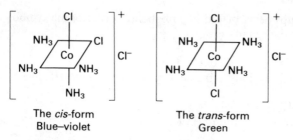

The *cis*-form
Blue–violet

The *trans*-form
Green

Fig. 93 *Geometric isomers of tetraamminedichlorocobalt(III) chloride.*

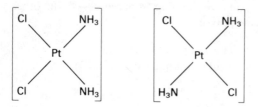

Fig. 94 *Geometric isomers of diamminedichloroplatinum(II).*

b Optical (stereo) isomerism As in organic compounds, this arises when a compound can be represented by two asymmetrical structures, one being the mirror image of the other. $[Cr(en)_3]^{3+}$, for example, has two optically active isomers (Fig. 95), and $[CoCl_2(en)_2]^+$ has two optically active *cis*-forms and an optically inactive *trans*-form (Fig. 96).

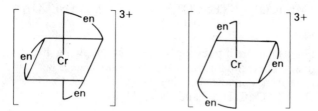

Fig. 95 *Optical isomers of* $[Cr(en)_3]^{3+}$ *('en' representing ethane-1,2-diamine). The two ions have structures which are mirror images.*

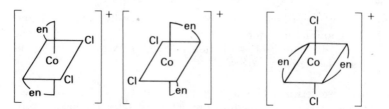

Fig. 96 *Isomers of* $[CoCl_2(en)_2]^+$. *The two left-hand isomers are mirror image (cis) optically active isomers. The right-hand structure is an optically inactive (trans) form.*

c Ionisation isomerism Isomers also occur when an ion can occupy positions inside or outside a complex ion, e.g.

$[Co(SO_4)(NH_3)_5]^+Br^-$ $[CoBr(NH_3)_5]^{2+}SO_4^{2-}$
pentaamminesulphato-   pentaamminebromo-
cobalt(III) bromide   cobalt(III) sulphate

The first compound will give a precipitate with silver nitrate solution but not with barium chloride solution; the second compound will behave in the opposite way.

There are three isomers of $CrCl_3.6H_2O$:

$[Cr(H_2O)_6]^{3+}(Cl^-)_3$ violet
$[CrCl(H_2O)_5]^{2+}(Cl^-)_2.H_2O$ green
$[CrCl_2(H_2O)_4]^+Cl^-.2H_2O$ green

They give different amounts of silver chloride on adding Ag^+ ions, and they lose different amounts of water on standing over concentrated sulphuric acid.

d Coordination isomerism This occurs when both cation and anion are complex, e.g.

$[Co(NH_3)_6]^{3+}[Cr(CN)_6]^{3-}$ and $[Cr(NH_3)_6]^{3+}[Co(CN)_6]^{3-}$

$[Pt^{II}(NH_3)_4]^{2+}[Pt^{IV}Cl_6]^{2-}$ and $[Pt^{IV}Cl_2(NH_3)_4]^{2+}[Pt^{II}Cl_4]^{2-}$

e Linkage isomerism This may be found when the ligand itself has isomeric forms, e.g.

$[Co(NO_2)(NH_3)_5]^{2+}$ and $[Co(NH_3)_5(ONO)]^{2+}$
a nitro-complex   a nitrito-complex

5 Bonding in complexes (dative bonds

The ligands in complexes are ions or molecules that have lone pairs of electrons, and the earliest ideas of bonding in the complexes were developed by Sidgwick (1873–1952). He envisaged dative bonds forming between the ligands (acting as the donors) and the central ion (acting as the acceptor). Such an idea is now known to be an over-simplification, for it cannot provide an explanation of the colours and the variety of geometrical shapes found in the complexes.

These can be explained by using *crystal field theory* or *ligand field theory*, as developed by Bethe, van Vleck, Orgel and others. On these theories, a complex is regarded as an agglomeration, consisting of a central ion surrounded by other ions or molecules with electrical fields. The electrical field of the central ion will affect the surrounding ligands, whilst the combined field of the ligands will influence the arrangement of electrons in the central ion.

In crystal field theory, the bonding between the ligands and the central ion, or the interaction between them, is regarded as fully ionic so that the forces concerned can be limited to electrostatic forces. Ligand field theory develops the idea further by taking into account the molecular orbitals that can be formed from the orbitals of the central ion and from those of the ligands. These molecular orbitals introduce the possibilities of covalent bonding within the complex.

In both theories, the splitting of the five d orbitals of the central ion into two groups is one of the fundamental ideas. In a free ion the five d orbitals are *degenerate*, i.e. energetically alike, and electrons that occupy them do so according to the rule of maximum multiplicity (pp. 4 and 341).

The five d orbitals are not, however, all alike in shape and that is why they split up into two groups of different energy under the influence of the ligand field. The two groups consist of three orbitals (the t_{2g} group) and of two orbitals (the e_g group). The extent and nature of the difference in energy between the two groups depends on the field strength of the ligands and on their geometrical arrangement around the central ion (see Fig. 105, p. 351). And the arrangement of electrons in the central ion is decided by the energy difference between the t_{2g} and e_g groups of orbitals.

In ligand field theory, the molecular orbitals which can be formed from these t_{2g} and e_g orbitals together with the ligand orbitals are taken into account.

6 Stability constants

When ammonia is added to an aqueous solution of, say, nickel(II) sulphate, there is a noticeable change of colour, because some of the $[Ni(H_2O)_6]^{2+}$ ions are converted into $[Ni(NH_3)_6]^{2+}$ ions:

$$[Ni(H_2O)_6]^{2+}(aq) + 6NH_3(g) \rightleftharpoons [Ni(NH_3)_6]^{2+}(aq) + 6H_2O(l) \qquad i$$

The equilibrium constant for the reaction, taking the concentration of water as being constant, is

$$\frac{[Ni(NH_3)_6{}^{2+}]_{eq}}{[Ni(H_2O)_6{}^{2+}]_{eq}[NH_3]_{eq}^6} = K$$

and the value of K is known as the stability constant for the $[Ni(NH_3)_6]^{2+}$ ion; the numerical value is $4.8 \times 10^7 \, mol^{-6} \, dm^{18}$. The higher the value of the stability constant, the more stable the $[Ni(NH_3)_6]^{2+}$ ion would be, i.e. the more the equilibrium would shift to the right. The stability constant, K, is commonly quoted as a lg K value, which in this case would be 7.7; a high lg K value means a stable complex.

The reaction represented in equation i actually takes place in six stages, the first two being represented by

$$[Ni(H_2O)_6]^{2+}(aq) + NH_3(g) \rightleftharpoons [Ni(NH_3)(H_2O)_5]^{2+}(aq) + H_2O(l)$$
$$[Ni(NH_3)(H_2O)_5]^{2+}(aq) + NH_3(g)$$
$$\rightleftharpoons [Ni(NH_3)_2(H_2O)_4]^{2+}(aq) + H_2O(l)$$

Each stage has its own equilibrium constant, known as a *step-wise* stability constant. The *overall* stability constant for the overall process (equation *i*) is the product of the five step-wise stability constants. Alternatively, the logarithm of the overall stability constant is the sum of the logarithms of the step-wise stability constants.

a Use of stability constants   If the values of the necessary stability constants are known, it is possible to calculate the concentrations of the various species present in a solution of the complex. Chemical facts can also be accounted for and, indeed, predicted.

The lg K values for $[Ag(NH_3)_2]^+$, $[Ag(S_2O_3)_2]^{3-}$ and $[Ag(CN)_2]^-$ for example, are 7.23, 13 and 21 respectively. That is why CN^- ions are more effective in dissolving silver chloride than $S_2O_3^{2-}$ ions, whilst $S_2O_3^{2-}$ ions are better than ammonia solution.

Addition of ammonia to copper(II) sulphate solution produces a deep blue solution that mainly contains $[Cu(NH_3)_4(H_2O)_2]^{2+}$ ions, as this complex is more stable than $[Cu(H_2O)_6]^{2+}$. Addition of edta^{4-} ions to this solution converts some of the ammine complex into the lighter coloured, more stable, $[Cu(edta)]^{2-}$ complex.

Similarly, iron(III) chloride solution gives a deep red colour with SCN^- ions, due to the formation of $[Fe(SCN)(H_2O)_5]^{2+}$ ions (p. 380). Addition of F^- or $C_2O_4^{2-}$ ions removes the colour, owing to the formation of the colourless and very stable complexes, $[FeF_6]^{3-}$ and $[Fe(C_2O_4)_3]^{3-}$.

b Factors affecting the stability of complexes   Considerable work has been done on trying to relate the stability of a complex to its structure, but has only led to some broad, and controversial, generalisations.

The greater the electrical fields of the central ion and the ligand, i.e. the smaller the size and the higher the charge, the more stable a complex tends to be. $[Fe(CN)_6]^{3-}$, for example, with a lg K of 31, is much more stable than $[Fe(CN)_6]^{4-}$, which has a lg K of 8.3; and fluoro-complexes tend to be more stable than the corresponding chloro-complexes.

The electron donating (basic) properties of the ligand are also very important, and many common ligands which form stable complexes are strong bases. The formation of a complex and the formation of a Brønsted–Lowry acid are, indeed, similar processes, i.e.

$$H^+ + base \rightleftharpoons acid \qquad M^+ + ligand \rightleftharpoons complex$$

As a broad generalisation, too, multidentate ligands form more stable complexes than monodentate ones. The edta complex, for example, is invariably much more stable than the corresponding ammine.

c Labile and inert complexes   The stability constant of a complex gives no indication of the rate at which it might react. That is a matter of chemical kinetics, and is not related to thermodynamic stability. A complex in which ligands can rapidly be replaced by others is called a *labile* complex; one in which ligand substitution is slow is called *inert*.

Chromium(III) complexes are particularly kinetically inert, and this explains why so many of them have been made and studied.

7 Applications of complex formation

a Sequestration   The concentration of a simple ion in a solution can be lowered by converting it into a complex ion; the more stable the complex, the greater the concentration lowering. The process is known as *sequestration*, and the chemical used for it is called a sequestrating or complexing agent, or a complexone.

Rochelle salt, sodium potassium 2, 3-dihydroxybutanedioate (tartrate), acts as a sequestrating agent in making Fehling's solution. Addition of sodium hydroxide solution to a solution of copper(II) sulphate gives a precipitate of basic copper(II) sulphate (p. 386). But the precipitate is not formed if Rochelle salt is present, for this salt lowers the Cu^{2+} ion concentration by forming a complex with it. Fehling's solution, then, is an alkaline solution of a copper(II) complex with Rochelle salt. It is used in making copper(I) oxide (p. 385), and in the detection and volumetric estimation of reducing sugars.

Fehling's sol'n is a test for reducing sugar.

b Formation of soluble complexes   There are many examples in which a compound that is insoluble in water is 'dissolved' by conversion into a complex. Typical examples are summarised below.

substance insoluble in water	'soluble in'	complex formed	use
AgCl, AgBr, AgI	NH_3 soln. KCN soln. $Na_2S_2O_3$ soln.	$[Ag(NH_3)_2]^+$ $[Ag(CN)_2]^{2-}$ $[Ag(S_2O_3)_2]^{3-}$	in photography
I_2	KI soln.	$[I_3]^-$	to make 'iodine' solutions
HgI_2	KI soln.	$[HgI_4]^{2-}$	Nessler's soln. (p. 338)
Au	aqua regia (conc. HNO_3 + conc. HCl)	$[AuCl_4]^-$	dissolving Au and Pt
Pt		$[PtCl_6]^{2-}$	
Au	NaCN soln. and air	$[Au(CN)_4]^-$	extraction of Au and Ag
Ag		$[Ag(CN)_2]^-$	
insoluble Fe^{3+} compounds	ethanedioate soln.	$[Fe(C_2O_4)_3]^{3-}$	removal of rust spots and some ink spots

c Formation of insoluble complexes   Insoluble complexes are useful in analysis. If a simple ion in a solution can be converted into an insoluble complex a precipitate will be formed, and weighing the precipitate provides a method for estimating the original simple ion if the complex formation takes place quantitatively. Many complexing agents which form specific, coloured complexes (ideally, with just one simple ion) have been developed. Butanedione dioxime (dimethyl glyoxime) is typical. It forms a red precipitate with Ni^{2+} ions in slightly alkaline solution, and can be used for the qualitative or quantitative estimation of Ni^{2+} ions.

$$
\begin{array}{c}
\overset{\displaystyle OH}{|} \qquad \overset{\displaystyle O}{\uparrow} \\
H_3C.C{=}N \qquad N{=}C.CH_3 \\
\diagdown \qquad \diagup \\
Ni \\
\diagup \qquad \diagdown \\
H_3C.C{=}N \qquad N{=}C.CH_3 \\
\underset{\displaystyle O}{\downarrow} \qquad \underset{\displaystyle HO}{|}
\end{array}
$$

bis(butanedione dioximato)nickel(II)

d Stabilisation of oxidation states   Replacement of H_2O ligands in an aqua-complex by other ligands can cause large changes in electrode potential values.

The standard electrode potential for the $Fe^{3+}(aq)/Fe^{2+}(aq)$ couple, for example, is $+0.77\,V$, but that for the $[Fe(CN)_6]^{3-}/[Fe(CN)_6]^{4-}$ couple is $+0.36\,V$. $[Fe(CN)_6]^{3-}$ is, therefore, a weaker oxidising agent than $[Fe(H_2O)_6]^{3+}$; it will not, for instance, oxidise I^- ions. Alternatively, it is easier to oxidise $[Fe(CN)_6]^{4-}$ ions than $[Fe(H_2O)_6]^{2+}$ ions. In other words, the $+3$ oxidation state of iron is stabilised with respect to the $+2$ state by cyanide ligands.

Similar values for Co^{3+} complexes show that the $+3$ state is stabilised by NH_3 and CN^- ligands:

$$
\begin{array}{ll}
[Co(CN)_6]^{3-}(aq)+e^- \longrightarrow [Co(CN)_6]^{4-}(aq) & E^{\ominus} = -0.80\,V \\
[Co(NH_3)_6]^{3+}(aq)+e^- \longrightarrow [Co(NH_3)_6]^{2+}(aq) & E^{\ominus} = +0.10\,V \\
[Co(H_2O)_6]^{3+}(aq)+e^- \longrightarrow [Co(H_2O)_6]^{2+}(aq) & E^{\ominus} = +1.82\,V
\end{array}
$$

$[Co(H_2O)_6]^{3+}$ ions are so strongly oxidising that they react with water to produce oxygen. $[Co(NH_3)_6]^{2+}$ and $[Co(CN)_6]^{4-}$ ions are so strongly reducing that they are oxidised by air.

The nature of redox processes can therefore be controlled by changing the ligand. An example of this is the use of CN^- ions in electroplating, as described in section **e** below.

e Control of electrode potential values   Electrodeposition of a metal is more effective if carried out from a solution containing the metallic ion in low concentration. Silver plating, for example, can be done using silver nitrate(V) solution but the layer of metal deposited adheres better if cyanide ions are added,

$$Ag^+(aq)+2CN^-(aq) \rightleftharpoons [Ag(CN)_2]^-(aq)$$

and cyanides are widely used in the electroplating industry. Addition of CN^- ions lowers the Ag^+ ion concentration, and this affects the value of the Ag^+/Ag electrode potential. The standard electrode potential in aqueous solution is $+0.80 V$ but it is lowered, and eventually becomes negative, as more and more CN^- ions are added.

Similarly, addition of CN^- ions to a mixture containing Cu^{2+} and Zn^{2+} ions lowers the Cu^{2+}/Cu and Zn^{2+}/Zn electrode potentials. As the cyanide complexes do not have equal stability, the lowering for the copper complex is greater than for the zinc complex. Concentrations can be chosen which give about equal electrode potentials for the two couples, and electrolysis of such a mixture deposits both copper and zinc, i.e. it can be used for brass plating.

f Coloured complexes   Coloured complexes can be used as dyes or as pigments, e.g. Prussian blue (p. 378).

The formation of a coloured complex can also be used in colorimetric analysis. Nessler's solution, for example, consists of an alkaline solution of mercury(II) iodide in potassium iodide. It contains $[HgI_4]^{2-}$ ions and these react with ammonia to form an orange–brown colour. The solution is used, colorimetrically, to determine the concentration of ammonia in water supplies.

Similarly, thiocyanate solutions can be used to estimate Fe^{3+} ions by the formation of the blood-red $[Fe(SCN)(H_2O)_5]^{2+}$ ions.

g Complexometric titrations   In a complexometric titration, the concentration of a metallic ion is estimated by direct titration with a standard solution of a complexing agent. The determination of the hardness of water provides a typical example.

Both Ca^{2+} and Mg^{2+} ions form stable complexes with $edta^{4-}$, one mole of the simple ions combining with one mole of $edta^{4-}$ anions:

$$Mg^{2+}(aq) + edta^{4-}(aq) \longrightarrow [Mg(edta)]^{2-}(aq)$$

The calcium complex is more stable than the magnesium one, so that the calcium complex is formed first, when $edta^{4-}$ solution is added to water, followed by the magnesium complex.

An indicator is necessary to denote the point at which all the Ca^{2+} ions and/or all the Mg^{2+} ions are used up. The indicator is itself a complexing agent: one which forms coloured complexes with both Ca^{2+} and/or Mg^{2+} ions, the complexes being less stable than those formed with $edta^{4-}$. The indicator forms complexes with both Ca^{2+} and Mg^{2+}, but as soon as enough $edta^{4-}$ is added to react with all the Ca^{2+} and Mg^{2+}, the colour changes, the complexes with the indicator being replaced by the more stable complexes with $edta^{4-}$. By using an indicator that forms a complex only with Ca^{2+} and not with Mg^{2+}, it is possible to determine the Ca^{2+} ion concentration.

The titrations are carried out in an alkaline buffer solution which favours the quantitative formation of the complexes, and a solution of the sodium salt of $H_4[edta]$ is used, as it is more soluble than the acid itself.

Questions on Chapter 24

1 Account for the following facts:

(a) Silver chloride dissolves in ammonia solution but not in sodium hydroxide.

(b) A white precipitate is formed when silver nitrate is added to sodium chloride solution. The precipitate disappears on adding ammonia solution but reappears on adding nitric acid.

(c) Copper(I) chloride and lead(II) chloride are both insoluble in water, but dissolve in concentrated hydrochloric acid.

(d) Mercury(II) iodide is more soluble in organic solvents than in water, and is also soluble in potassium iodide solution.

(e) Iodine is only very slightly soluble in water but dissolves in potassium iodide solution.

(f) Aluminium hydroxide dissolves in sodium hydroxide solution, but magnesium hydroxide does not.

(g) Silver chloride will dissolve in concentrated ammonia solution but silver iodide is less soluble.

(h) Silver bromide is precipitated from an ammoniacal solution of silver chloride by the addition of potassium bromide.

(i) Copper(II) hydroxide will dissolve both in sulphuric acid, which is acidic, and in ammonium hydroxide, which is basic.

(j) A blue solution of copper(II) chloride turns green on the addition of Cl^- ions.

(k) Glass is attacked by hydrofluoric acid but not by hydrochloric acid.

2 Give the names, the oxidation numbers of the central atoms, and the probable geometrical shapes of the following: $[Cu(CN)_2]^-$, $[FeCl_4]^-$, $[Zn(NH_3)_4]^{2+}$, $[Co(NH_3)_6]^{2+}$, $[TaF_7]^{2-}$, $[TaF_8]^{3-}$, $[Mo(CN)_8]^{4-}$.

3 Write short notes on the following: (a) chelation, (b) labile complexes, (c) geometrical isomerism in inorganic compounds.

4 Give formulae for the following: (a) sodium tetrahydridoborate(III), (b) ammonium hexathiocyanatochromate(III), (c) hexaaquairon(II) sulphate, (d) sodium tetrahydroxoaluminate(III), (e) pentaamminechlorocobalt(III) chloride.

5 Name the following: (a) $K_2[Co(CN)(CO)_2(NO)]$, (b) $H[AuCl_4]$, (c) $K_3[Al(C_2O_4)_3]$, (d) $(NH_4)_2(PtCl_6)$, (e) $NH_4[Cr(SCN)_4(NH_3)_2]$.

6 What is a complex ion? What experimental observations suggest that such ions exist in certain solutions?

7 Account for the following: (a) $[PtCl_2(NH_3)_2]$ can exist in two forms, (b) $[Pt(NO_2)(NH_3)(NH_2OH)py]^+$ can exist in three forms (py is an abbreviation for pyridine), (c) $[CoCl_2(NH_3)_4]^+$ can exist in two forms.

8 Pentaamminesulphatocobalt(III) bromide and pentaamminebromocobalt(III) sulphate are isomeric. Write down the formula of each isomer and explain how you would distinguish between them.

9 How is the solubility of metallic hydroxides in (a) sodium hydroxide solution and (b) ammonium hydroxide solution, accounted for by complex ion formation?

10 Compare and contrast the structure and properties of ammonium copper(II) sulphate and tetraamminecopper(II) sulphate.

11 Describe, in detail, the laboratory preparation of any one complex salt.

12 Describe the types of isomerism found in octahedral transition metal complexes.

Two compounds of empirical formula $CoBr(SO_4)(NH_3)_5$ can be prepared. Describe, giving brief experimental details, how you would confirm that the $Br : NH_3$ ratio is $1 : 5$ in these compounds. Give the possible structural formulae for these compounds and describe how you would distinguish between them. (JMB)

25
General properties of transition elements

1 Arrangement of d electrons

The term transition element was first used to denote the three groups of elements Fe, Co, Ni; Ru, Rh, Pd, and Os, Ir, Pt. They were placed in what was called Group 8 in older arrangements of the periodic table (see the front end-paper), and they were called transitional because they linked up the A and B sub-groups.

Nowadays, the term is most commonly used to include any d-block element that has unfilled d orbitals, either in its atom or its ions. That gives three series of transition elements: scandium to copper, yttrium to silver, and lanthanum to gold, as summarised below. Zinc, cadmium and mercury are also d-block elements but they do not have atoms or ions with unfilled d orbitals, so they are not regarded as transition elements, and they have little in common with them.

periodic table group	3A	4A	5A	6A	7A		8		1B	2B
first transition series	Sc	Ti	V	Cr	Mn	Fe	Co	Ni	Cu	Zn
second transition series	Y	Zr	Nb	Mo	Tc	Ru	Rh	Pd	Ag	Cd
third transition series	La	Hf	Ta	W	Re	Os	Ir	Pt	Au	Hg
number of 3d electrons	1	2	3	5	5	6	7	8	10	10

The number of 3d electrons in the ground states of the atoms of the first transition series are as follows:

	1s	2s	2p	3s	3p	3d					4s
Sc	2	2	6	2	6	↓					2
Ti	2	2	6	2	6	↓	↓				2
V	2	2	6	2	6	↓	↓	↓			2
Cr	2	2	6	2	6	↓	↓	↓	↓	↓	1
Mn	2	2	6	2	6	↓	↓	↓	↓	↓	2
Fe	2	2	6	2	6	↓↑	↓	↓	↓	↓	2
Co	2	2	6	2	6	↓↑	↓↑	↓	↓	↓	2
Ni	2	2	6	2	6	↓↑	↓↑	↓↑	↓	↓	2
Cu	2	2	6	2	6	↓↑	↓↑	↓↑	↓↑	↓↑	1
Zn	2	2	6	2	6	↓↑	↓↑	↓↑	↓↑	↓↑	2

All the atoms have one or two 4s electrons, and the same pattern is repeated for 4d and 5s, or 5d and 6s, electrons in the second and third series. It is the availability of the d electrons, as well as the s electrons, that gives the transition elements their peculiar characteristics. They are all metals, with general properties summarised as follows:

a Similarity of physical properties.
b Wide variation of oxidation states.
c Formation of coloured ions.
d Paramagnetism of ions. → lone electron
e Catalytic activity. ie. Pt, Ni, dissassohian of H_2O_2 → O_2 + H_2
f Formation of interstitial 'compounds'. (compounds that are not stoichometric
g Ready formation of stable complexes.   ie. you find Fe^{2+}
                                                and Fe^{3+}) because
2 Physical properties of the metals
they are small in size so they stick
in vacancies

The physical properties of the metals do not vary greatly and many of
them show a regular trend.

mallible + ductile due to electron when hammered the plane separate

a Metallic character   The transition metals are true metals and many of
them are widely used industrially. They are hard, strong and lustrous, and
have high melting and boiling temperatures (Fig. 97) and high enthalpies
of atomisation (Fig. 98). These properties suggest strong metallic bonding,
and this is also indicated by low compressibilities of the metals. The strong
bonding is probably due to the availability of d electrons. Sodium, for
example, which is soft and highly compressible, has only one 1s electron
available for bonding (p. 47); iron has two 4s electrons, and a maximum
of six 3d electrons.

On the other hand, the electrical and thermal conductivities of the
metals are unexpectedly low as compared with those of copper, gold, silver
and aluminium.

b Crystal structure   The metals have face-centred cubic, body-centred
cubic or hexagonal close-packed structures (p. 50), and most of them exist

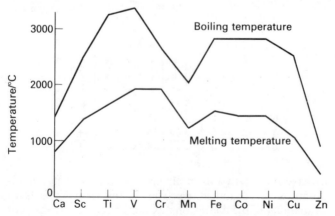

Fig. 97 Melting and boiling temperatures of elements in the first transition series.

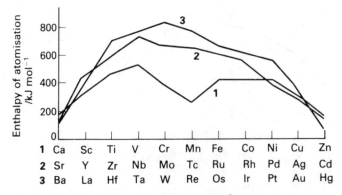

Fig. 98 Enthalpies of atomisation of the transition elements.

in allotropic modifications. They readily form alloys with each other, partially because they do not differ greatly in metallic radius (p. 131).

c Metallic radius, density and atomic volume The metallic radius decreases in passing from scandium to nickel (Fig. 99). Addition of electrons might be expected to result in an increase in radius, but the electrons are being added to an inner orbital and it is the increase in nuclear charge in passing from scandium to nickel that causes the slight decrease in radius.

The decrease in radius, coupled with an increasing relative atomic mass, causes an increase in density and a decrease in atomic volume in passing from scandium to nickel (Figs. 99 and 100).

d Ionic radius and isomorphism There is a small, fairly regular decrease in ionic radius, for ions of like charge, in passing from scandium to nickel

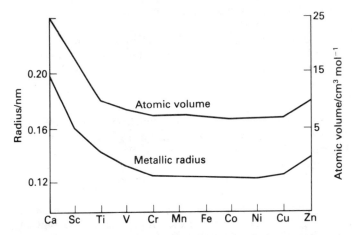

Fig. 99 Metallic radii and atomic volumes of elements in the first transition series. Compare Fig. 58, p. 119.

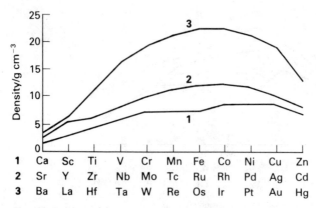

Fig. 100 *Densities of the transition elements. Compare Fig. 59, p. 120.*

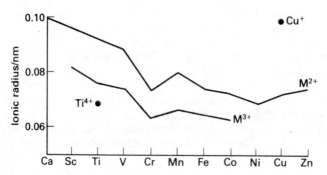

Fig. 101 *Ionic radii of elements in the first transition series.*

(Fig. 101). Nevertheless, many of the ions are close enough in size for their compounds to be isomorphous.

The Mg^{2+} ion in schönite, $K_2SO_4.MgSO_4.6H_2O$, for example, can be replaced by V^{2+}, Mn^{2+}, Fe^{2+}, Co^{2+} or Ni^{2+}, whilst the Al^{3+} ion in the alums, $M^+Al^{3+}(SO_4)_2.12H_2O$, can be replaced by Ti^{3+}, V^{3+}, Cr^{3+}, Mn^{3+}, Fe^{3+} or Co^{3+}.

e Ionisation energy The energy required to remove the first 4s electron, the second 4s electron and the first 3d electron are given by the first, second and third ionisation energies (p. 9). There is a fairly general increase in the values in passing from scandium to nickel (Fig. 102) and this is due to the increasing nuclear charge holding the electrons more strongly.

The sum of the first and second ionisation energies gives the energy required to form M^{2+} gaseous ions, and the sum of the first, second and third ionisation energies gives the energy required to form M^{3+} gaseous ions. It will be seen (Fig. 102) that the trends are almost regular.

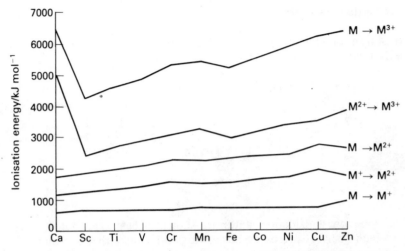

Fig. 102 *Various ionisation energy values for elements and ions in the first transition series.*
Compare Fig. 63, p. 123.

Metals are expected to have low Electronegativity

f Electronegativity The ionisation energy values (Fig. 102) show that the metals towards scandium form M^{2+} or M^{3+} ions most easily. In other words, the electronegativity increases in passing from scandium to nickel (Fig. 103). However, the differences in electronegativity between any of the transition metals are not very great:

Electronegativities of d-block elements

Sc	Ti	V	Cr	Mn	Fe	Co	Ni	Cu	Zn
1.3	1.5	1.6	1.6	1.5	1.8	1.8	1.8	1.9	1.6
Y	Zr	Nb	Mo	Tc	Ru	Rh	Pd	Ag	Cd
1.3	1.4	1.6	1.8	1.9	2.2	2.2	2.2	1.9	1.7
La	Hf	Ta	W	Re	Os	Ir	Pt	Au	Hg
1.1	1.3	1.5	1.7	1.9	2.2	2.2	2.2	2.4	1.9

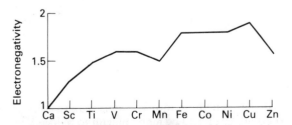

Fig. 103 *Electronegativities of elements in the first transition series.*

3 Oxidation states

Most transition metals exhibit variable positive oxidation states, as shown in the following summary and in the table on p. 347.

K	Ca	Sc	Ti	V	Cr	Mn	Fe	Co	Ni	Cu	Zn
1						1			1	1	
	2		2	2	2	2	2	2	2	2	2
		3	3	3	3	3	3	3	3	3	
			4	4	4	4					
				5	5	5					
					6	6	6				
						7					

(marginal note: memorize oxid states of V and Mn with example on each)

The commonest oxidation states are shown in **bold** type, but it is not possible to be very precise because an oxidation state that is common in simple compounds may be uncommon in complexes. There are, moreover, some rather 'freak' chemicals that show very unusual oxidation states. Many of the transition metals can also show an oxidation state of zero in their carbonyls (p. 206).

The variability of the oxidation state is due to the availability of the 3d electrons. The energy level of these electrons is not unlike that of the 4s or 4p electrons so that they participate in bond formation, which may be ionic or covalent.

(marginal note: due to multiple oxidation states they are very good conductors)

Manganese, with the highest number of unpaired 3d electrons, shows the widest range of oxidation states and is the only metal to exhibit an oxidation state of $+7$. Copper forms many compounds with an oxidation state of $+1$, but this is unusual amongst transition metals; calcium and zinc, at the extreme ends of the series, generally have fixed oxidation states of $+2$.

a Change of oxidation state Because of the wide variety of oxidation states, the transition metals and their compounds participate in many redox reactions. These are best considered from the point of view of the standard electrode potentials involved. In predicting whether or not a particular reaction will take place, or in comparing the likelihood of one reaction against another, it is a matter of deciding the relative strengths of the oxidising or reducing agents involved.

Vanadium, for example, is a better reducing agent than cobalt when conversion into M^{2+} ions is concerned: *(marginal note: because it is more negative.)*

$$V^{2+}(aq) + e^- \longrightarrow V(s) \qquad E^{\ominus} = -1.2\,V$$
$$Co^{2+}(aq) + e^- \longrightarrow Co(s) \qquad E^{\ominus} = -0.28\,V$$

It will, then, be easier to convert vanadium into V^{2+} ions than to convert cobalt into Co^{2+} ions. Alternatively, it will be more difficult to reduce V^{2+} ions than to reduce Co^{2+} ions. This means that a reaction between

Typical compounds of the first transition series metals, showing the variation in oxidation state. The highest oxidation state is generally shown by the oxide or fluoride.

+1	+2	+3	+4	+5	+6	+7
		Sc_2O_3 ScF_3 Sc^{3+}				
	TiO $TiCl_2$	Ti_2O_3 $TiCl_3$ Ti^{3+}	TiO_2 $TiCl_4$ TiO^{2+}			
	VO VCl_2 V^{2+}	V_2O_3 VCl_3 V^{3+}	VO_2 VCl_5 VO^{2+}	V_2O_5 VF_5 VO_3^- VO_2^+		
	CrO $CrCl_2$ Cr^{2+}	Cr_2O_3 $CrCl_3$ Cr^{3+}	CrO_2 $CrCl_4$	CrF_5	CrO_3 CrO_4^{2-} $Cr_2O_7^{2-}$	
$[Mn(CN)_6]^{5-}$	MnO $MnCl_2$ Mn^{2+}	Mn_2O_3 $MnCl_3$	MnO_2 MnO_3^{2-}	MnO_4^{3-} MnO_3^-	MnO_4^{2-}	Mn_2O_7 MnO_4^-
	FeO $FeCl_2$ Fe^{2+}	Fe_2O_3 $FeCl_3$ Fe^{3+}			FeO_4^{2-}	
	CoO $CoCl_2$ Co^{2+}	CoF_3 $[Co(NH_3)_6]^{3+}$				
$NiCN$ $[Ni(CN)_3]^{2-}$	NiO $NiCl_2$ Ni^{2+}	$Ni_2O_3 . xH_2O$				
Cu_2O $CuCl$ Cu^+	CuO $CuCl_2$ Cu^{2+}	$[CuF_6]^{3-}$				
	ZnO $ZnCl_2$ Zn^{2+}					

vanadium and Co^{2+} ions would be expected, but not one between cobalt and V^{2+} ions. Similarly, V^{2+} ions are much stronger reducing agents than Co^{2+} ions when conversion into M^{3+} ions is concerned,

$$V^{3+}(aq) + e^- \longrightarrow V^{2+}(aq) \qquad E^\ominus = -0.255 \, V$$
$$Co^{3+}(aq) + e^- \longrightarrow Co^{2+}(aq) \qquad E^\ominus = +1.82 \, V$$

Read

so that the change from V^{2+} to V^{3+} is easier than the change from Co^{2+} to Co^{3+}. Alternatively, it is more difficult to reduce V^{3+} than Co^{3+}, and a reaction would be expected between V^{2+} and Co^{3+}, but not between Co^{2+} and V^{3+}.

A reaction which will not take place under standard conditions may, however, be made to take place by altering the conditions. The following standard electrode potentials, under acid conditions, indicate that it will not be possible to make MnO_4^{2-} ions from MnO_4^- ions and MnO_2:

$$2MnO_4^-(aq) + 2e^- \longrightarrow 2MnO_4^{2-}(aq) \qquad E^\ominus = +0.56 \text{ V}$$
$$MnO_4^{2-}(aq) + 4H^+(aq) + 2e^- \longrightarrow MnO_2(s) + 2H_2O(l)$$
$$E^\ominus = +2.26 \text{ V}$$

Indeed, the figures show that MnO_4^{2-} ions would be expected to disproportionate under these conditions:

$$\underset{+6}{3MnO_4^{2-}(aq)} + 4H^+(aq) \longrightarrow \underset{+7}{2MnO_4^-(aq)} + \underset{+4}{MnO_2(s)} + 2H_2O(l)$$

However, the $E^\ominus$ value for the process that involves H^+ ions is different in alkaline solution (p. 94). Thus, in the presence of OH^- ions,

$$2MnO_4^-(aq) + 2e^- \longrightarrow 2MnO_4^{2-}(aq) \qquad E^\ominus = +0.56 \text{ V}$$
$$MnO_4^{2-}(aq) + 2H_2O(l) + 2e^- \longrightarrow MnO_2(s) + 4OH^-(aq)$$
$$E^\ominus = +0.60 \text{ V}$$

Reaction between MnO_4^- and MnO_2 is now much more feasible. It will, in fact, occur when the concentration of OH^- ions is sufficiently high, and a green solution containing MnO_4^{2-} ions is formed:

$$2MnO_4^-(aq) + MnO_2(s) + 4OH^-(aq) \longrightarrow 3MnO_4^{2-}(aq) + 2H_2O(l)$$

On acidification, the mixture reverts to MnO_4^- and MnO_2, through disproportionation.

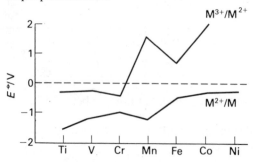

Fig. 104 *Variation in the $M^{2+}(aq)/M(s)$ and $M^{3+}(aq)/M^{2+}(aq)$ standard electrode potentials in passing from titanium to nickel.*

b The +2 oxidation state Fig. 104 shows that, with the exception of manganese, the change from M to M^{2+}(aq) and the further oxidation of M^{2+}(aq) to M^{3+}(aq) both become more difficult in passing from titanium to nickel. Manganese is out of line, for the Mn^{2+}(aq) ion is both formed from the metal more easily, and oxidised to Mn^{3+}(aq) less easily, than might be expected. This extra stability of the +2 state of manganese is attributed to the fact that the 3d orbital is half full in manganese. For this reason, manganese always tends to be the odd one out. But in general, the +2 oxidation state becomes more stable in passing from titanium to nickel.

Ti^{2+}(aq) does not exist, V^{2+}(aq) is such a strong reducing agent that it reduces water to hydrogen (p. 359), and Cr^{2+}(aq) and Fe^{2+}(aq) are both oxidised by air (the former particularly easily). On the other hand, it is very difficult to oxidise Ni^{2+}(aq) to Ni^{3+}(aq).

c The +3 oxidation state Scandium, at the beginning of the first transition series, exists only in an oxidation state of +3, but this state becomes less stable in passing from titanium to copper.

Ti(III) and V(III) compounds are not uncommon, but they can be oxidised to +4 (TiO^{2+} and VO^{2+}) or +5 (VO_2^+) fairly easily. Iron(III) and chromium(III) compounds are very common and quite stable; but the +3 state is not very stable for manganese, cobalt or, particularly, nickel. The Mn^{3+}(aq) ion disproportionates (Fig. 49, p. 94) into Mn^{2+}(aq) and MnO_2; the Co^{3+}(aq) ion is such a strong oxidising agent that it liberates oxygen from water; nickel only exists in the +3 state in unusual circumstances, such as in the nickel–cadmium cell (p. 383), and in some complexes. Copper forms a few copper(III) compounds, e.g. $K_3[CuF_6]$.

d Other oxidation states Some of the transition metals exist in the +1 state but, apart from those of copper, such compounds are rare.

Oxidation states higher than +3 occur mainly in the metals in the middle of the series. Compounds with oxidation state +4 for titanium, +5 for vanadium, +6 for chromium, and +7 for manganese are particularly common, and the elements achieve a noble gas electronic structure in these states. The compounds are usually strong oxidising agents.

Manganates(VII) and dichromates(VI) are particularly well known as oxidising agents. The former are stronger than the latter; they are capable of oxidising water very slowly and, for this reason, do not store well in aqueous solution. That is why standard solutions of potassium manganate(VII) must be freshly made. Ferrates(VI) are stronger oxidising agents than manganates(VII) and will liberate oxygen in aqueous solution.

4 The colour of transition metal ions

Most complexes of transition metal ions are coloured, with the actual colour depending on the oxidation state of the central ion and on the nature of the ligands surrounding it.

[Handwritten margin notes: "As ligands approach metal degeneracy is lost forming two separate energy levels. As e⁻ moves to fill vacancy it is unstable and it falls enough emitting visible light."]

Typical colours for some common hydrated ions (i.e. with H_2O as the ligand) are summarised below.

[Handwritten margin note: "Memo colours"]

$[V(H_2O)_6]^{2+}$ violet	$[Co(H_2O)_6]^{2+}$ pink	$[Sc(H_2O)_6]^{3+}$ colourless
$[Cr(H_2O)_6]^{2+}$ blue	$[Ni(H_2O)_6]^{2+}$ green	$[Ti(H_2O)_6]^{3+}$ purple
$[Mn(H_2O)_6]^{2+}$ pale pink	$[Cu(H_2O)_6]^{2+}$ blue	$[V(H_2O)_6]^{3+}$ green
$[Fe(H_2O)_6]^{2+}$ pale green	$[Zn(H_2O)_6]^{2+}$ colourless	$[Cr(H_2O)_6]^{3+}$ violet

The colours cannot always be described very precisely for they may change significantly, depending on the concentration of the ion. They may also be affected by the nature of the anion present.

a Oxidation state and colour Marked colour changes may take place when the transition metal involved in a complex changes its oxidation state, as summarised below. These colour changes can be seen in many common redox reactions.

+2	+3	+4	+5	+6	+7
$V^{2+}(aq)$ violet	$V^{3+}(aq)$ green	$VO^{2+}(aq)$ blue	$VO_2^{+}(aq)$ yellow		
$Cr^{2+}(aq)$ blue	$Cr^{3+}(aq)$ violet			$CrO_4^{2-}(aq)$ yellow	
$Mn^{2+}(aq)$ pale pink	$Mn^{3+}(aq)$ red		$MnO_4^{3-}(aq)$ blue	$MnO_4^{2-}(aq)$ green	$MnO_4^{-}(aq)$ purple
$Fe^{2+}(aq)$ pale green	$Fe^{3+}(aq)$ pale purple				

b Ligand field splitting To account for the particular colour of a transition metal complex, it is necessary to understand how the complex is able to undergo an energy change that enables it to absorb visible light of a certain frequency. If, for example, an ion absorbs blue light, it will appear to be red, for that will be the colour of the reflected light; if it absorbs red light, it will look blue. Therefore, transition metal ions of different colours must be able to absorb different frequencies from the visible region. How this comes about can be explained, particularly for common oxidation states, by the ideas of ligand field splitting and charge transfer (p. 352).

In a *free* ion of a transition metal of the first series, the five 3d orbitals are degenerate, i.e. energetically alike, and the electrons occupy the orbitals according to the rule of maximum multiplicity (p. 341). The five d orbitals are not, however, all the same shape, and the electrical field of a ligand around a central atom affects the energy of three of the orbitals

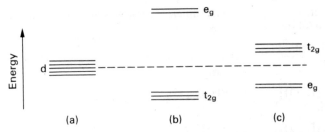

Fig. 105 *Crystal field splitting. Five degenerate orbitals are shown in (a), unaffected by any ligand field. These orbitals split into two groups when surrounded octahedrally by six ligands; the e_g group has the higher energy, as in (b). With four ligands arranged tetrahedrally, the energy splitting is lower, and it is the t_{2g} group that has the higher energy, as in (c).*

(known as the t_{2g} group) differently from the other two (known as the e_g group). Under the influence of a ligand, then, the 3d orbitals split into two groups with slightly different energy levels. The effect is known as crystal field or ligand field splitting, and the energy difference that it causes depends on the nature of the ligand and the shape of the complex, as summarised in Fig. 105. Some ligands, e.g. NH_3, exert stronger fields than others, e.g. H_2O, and cause a larger energy difference between the two groups of d orbitals. The actual value of this energy difference is significant, for it is the promotion of electrons from the lower to the higher level, by absorption of light, that accounts for the colours of transition metal complexes.

The energy difference between the two groups of d orbitals for the $[Cu(H_2O)_6]^{2+}$ ion, for example, is of just the right magnitude for promotion of an electron to be brought about by the absorption of red light. That is why the ion appears to be blue. As ammonia is added, the $[Cu(H_2O)_6]^{2+}$ ion is converted into ions such as $[Cu(NH_3)_4(H_2O)_2]^{2+}$, in which the energy difference between the two groups of d orbitals is larger. The absorbed light is orange, and the ion appears violet in colour.

For still larger energy differences between the two groups of d orbitals, the light absorption will be in the ultraviolet region. As no visible frequencies are absorbed, the complex will be colourless.

The spectrochemical series The colour of a complex of any transition metal ion is dependent on the field strength of the ligand involved, and it is possible to list the common ligands in the order of their field strength. The list is known as the spectrochemical series.

$$I^- \quad Br^- \quad Cl^- \quad F^- \quad OH^- \quad H_2O \quad SCN^- \quad NH_3 \quad en \quad NO_2^- \quad CN^-$$

— increasing field strength →

If the ligands in a complex ion are replaced by others with stronger fields, there will be a larger energy difference between the two groups of d orbitals and the light absorbed by the complex will shift from the red to the blue end of the spectrum. That is why different complexes have different colours, as in the following examples.

$[Cr(H_2O)_6]^{3+}$ Cu^{2+}
violet   colourless

$[Cr(NH_3)_5(H_2O)]^{3+}$ $[Cu(H_2O)_6]^{2+}$
orange–yellow   blue

$[Cr(NH_3)_3(H_2O)_3]^{3+}$ $[CuCl_4]^{2-}$
red   yellow

$[Cr(NH_3)_6]^{3+}$ $[Cu(NH_3)_4(H_2O)_2]^{2+}$
yellow   violet

d Charge transfer   Ligand field splitting does not account for the colour of all complexes. With a ligand that is strongly oxidising and a central metal ion that is strongly reducing (or vice versa), strong absorption of light may take place due to the transfer of an electron from the reducing agent to the oxidising agent; there is an *internal* redox process, brought about by the absorption of radiation. The charge transfer may be from the ligand to the metal (L→M) or from the metal to the ligand (M→L). The absorption is usually in the blue or ultraviolet region, causing yellow or red colours.

The colours of the MnO_4^- and CrO_4^{2-} ions are mainly due to charge transfer, as are the colours of many complexes used in testing for metallic ions and in colorimetry, e.g. thiocyanate complexes of Fe^{3+} (p. 380) and butanedione dioxime complexes of Ni^{2+} (p. 337).

Similar charge transfer may account for the colour in compounds that contain a metal in two oxidation states, and in many inorganic compounds. In Prussian blue (p. 378), for example, the transfer is from Fe^{2+} to Fe^{3+}, and it is from the negative to the positive ion in compounds such as AgI, PbI_2, HgI_2, HgO, HgS and Fe_2O_3. In general, the more oxidising the metallic ion and the more reducing the non-metallic ion, the stronger the charge transfer. That explains why, for instance, AgCl is white whilst AgI is yellow; why KI and CaI_2 are white whereas PbI_2 is yellow and HgI_2 red; and why Al_2O_3 is white but Fe_2O_3 is yellow or red.

5 Paramagnetism   unpaired e⁻.

Substances or ions may be classified as paramagnetic or diamagnetic. Paramagnetic substances are drawn into a strong magnetic field, i.e. a rod of such a substance (or a tube containing the substance) takes up a position parallel to the field (Fig. 106(a)). This is because a paramagnetic material is more permeable to magnetic lines of force than a vacuum. In other words, the permeability is greater than 1. Diamagnetic materials tend to be drawn out of a magnetic field and to set themselves at right angles to the field (Fig. 106(b)). They have permeabilities less than 1.

Substances that are normally regarded as magnetic, e.g. steel, iron, cobalt, nickel and magnetic alloys, are actually paramagnetic, but the

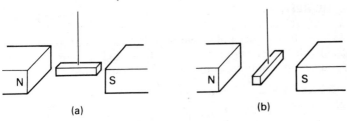

Fig. 106 (*a*) *Paramagnetic substance.* (*b*) *Diamagnetic substance.*

degree of magnetism possessed by such substances is much greater than that of any others; they are said to be *ferromagnetic*.

A substance or an ion in which all the electrons are paired is diamagnetic, whilst substances or ions with unpaired electrons are paramagnetic. This is because a spinning electron is equivalent to an electric current in a circular conductor and, as such, it acts as a tiny magnet. When an orbit contains two paired electrons, the magnetic moment of one is compensated by the equal but opposite moment of the other.

The magnetic moment of a substance or ion depends, then, on the number of unpaired electrons, and it can be shown that the magnetic moment is equal to $\sqrt{n(n+2)}$ where n is the number of unpaired electrons. The structures of the simple ions of the transition metals are summarised below.

ion	arrangement of 3d electrons	number of unpaired electrons	calculated magnetic moment
K^+, Ca^{2+}, Sc^{3+}, Ti^{4+}	none	0	0
V^{4+}	1	1	1.73
V^{3+}	1 1	2	2.83
V^{2+}, Cr^{3+}	1 1 1	3	3.87
Mn^{3+}, Cr^{2+}	1 1 1 1	4	4.90
Mn^{2+}, Fe^{3+}	1 1 1 1 1	5	5.92
Fe^{2+}	2 1 1 1 1	4	4.90
Co^{2+}	2 2 1 1 1	3	3.87
Ni^{2+}	2 2 2 1 1	2	2.83
Cu^{2+}	2 2 2 2 1	1	1.73
Cu^+, Zn^{2+}	2 2 2 2 2	0	0

Experimental measurement of magnetic moments gives values close to the calculated values, and shows that the ions with the greatest number of unpaired electrons, i.e. Mn^{2+} and Fe^{3+}, have the highest moments. This provides good evidence in favour of the rule of maximum multiplicity (p. 4). Measurement of magnetic moments provides a method of discovering the number of unpaired electrons in any substance or ion, and this helps enormously in elucidating electronic structures (pp. 36 and 241).

6 Catalytic activity

Transition metals have high catalytic activity, both as metals (particularly when finely divided) in heterogeneous catalysis, and as ions in homogeneous catalysis in solution. Some typical examples are summarised below.

Memo underlined reaction

catalyst	reaction catalysed	conditions
Ti compounds	$xC_2H_4 \longrightarrow (C_2H_4)_x$	Zeigler catalysts
V_2O_5	$2SO_2 + O_2 \longrightarrow 2SO_3$	450 °C
$Cr_2O_3 + ZnO$	$CO + 2H_2 \longrightarrow CH_3OH$	400 °C; 200 atm
MnO_2	$2KClO_3 \longrightarrow 2KCl + 3O_2$	300 °C
$Mn(CH_3COO)_2$	$2CH_3CHO + O_2 \longrightarrow 2CH_3COOH$	room temp.
Fe	$N_2 + 3H_2 \longrightarrow 2NH_3$	550 °C; 200 atm
$FeCl_3$	$C_6H_6 + Cl_2 \longrightarrow C_6H_5Cl + HCl$	room temp.
CoO	$CO + H_2 \longrightarrow$ hydrocarbons	400 °C; 200 atm
Co^{2+}	$2ClO^- \longrightarrow 2Cl^- + O_2$	room temp.
$Pd^{2+} + Cu^{2+}$	$C=C \xrightarrow[O_2]{H_2O}$ aldehydes or ketones	in solution
Ni	$C=C \xrightarrow{H_2} HC-CH$	150 °C

a Heterogeneous catalysis In heterogeneous catalysis, the metals seem to provide an active surface on which reactants can be chemisorbed. In the catalytic hydrogenation of an alkene in the presence of nickel, for example, it is thought that both the hydrogen and the alkene are chemisorbed on the metal surface. The bonds in both molecules are, therefore, weakened; so much so in the hydrogen molecule that it splits into atoms, which can then react with adjacent alkene molecules. The process is summarised in Fig. 107.

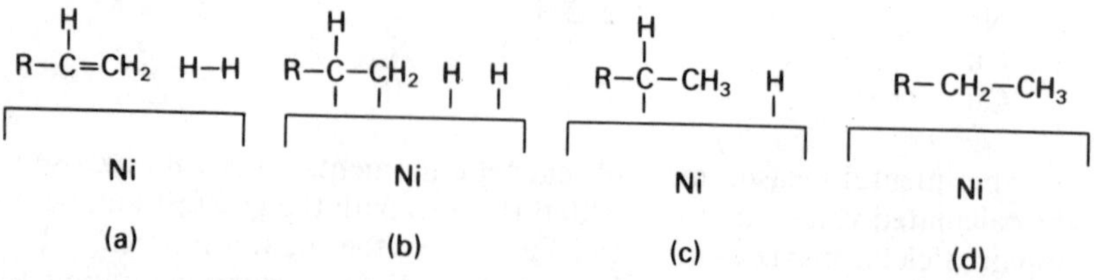

*Fig. 107 The mechanism of hydrogenation of an alkene at a nickel surface. (**a**) Shows the approach of an alkene molecule and a hydrogen molecule to the surface. (**b**) The molecules are chemisorbed and the H_2 bond has broken. (**c**) and (**d**) show the two-step addition of H atoms to the alkene molecule.*

b Homogeneous catalysis Homogeneous catalysts generally function by enabling a reaction to take place via a quicker, two-stage process. The reaction

$$A + B \longrightarrow X$$

for example, might take place in the presence of a catalyst, Ct, via the quicker stages

$$A + Ct \longrightarrow A.Ct$$
$$A.Ct + B \longrightarrow X + Ct$$

The precise nature of the intermediate compounds formed is not always known, but when carbon compounds are involved, the intermediate contains bonds between carbon atoms and the transition metal atoms.

The ease with which transition metals can be converted from one oxidation state to another also plays a large part in their catalytic activity. The reaction

$$A + B \longrightarrow R + S$$

for example, may take place in the presence of an M^{x+} catalyst via the quicker stages

$$A + M^{x+} \longrightarrow M^{y+} + R$$
$$M^{y+} + B \longrightarrow M^{x+} + S$$

Cu^{2+} ions catalyse the oxidation of vanadium from $+3$ to $+4$ by Fe^{3+} in this way:

$$V^{3+}(aq) + Cu^{2+}(aq) + H_2O(l) \longrightarrow Cu^{+}(aq) + VO^{2+}(aq) + 2H^{+}(aq)$$
$$Cu^{+}(aq) + Fe^{3+}(aq) \longrightarrow Fe^{2+}(aq) + Cu^{2+}(aq)$$

A possible mechanism for the reaction in which $I^{-}(aq)$ ions are oxidised by $S_2O_8^{2-}(aq)$ ions, in the presence of Fe^{3+} (or Fe^{2+}) as a catalyst, involves a similar two-stage process, i.e.

$$2I^{-}(aq) + 2Fe^{3+}(aq) \longrightarrow 2Fe^{2+}(aq) + I_2(aq)$$
$$2Fe^{2+}(aq) + S_2O_8^{2-}(aq) \longrightarrow 2Fe^{3+}(aq) + 2SO_4^{2-}(aq)$$

In the Wacker process for oxidising alkenes to aldehydes or ketones, the catalyst of $PdCl_2$ and $CuCl_2$ undergoes the oxidation state changes summarised in the following equations.

$$C_2H_4 + PdCl_2 + H_2O \longrightarrow CH_3CHO + Pd + 2HCl$$
$$Pd + 2CuCl_2 \longrightarrow PdCl_2 + 2CuCl$$
$$2CuCl + 2HCl + \tfrac{1}{2}O_2 \longrightarrow 2CuCl_2 + H_2O$$

This gives an overall reaction of

$$C_2H_4 + \tfrac{1}{2}O_2 \longrightarrow CH_3CHO$$

The processes involved are, however, more complicated than these simple equations suggest, for species such as $[PdCl_3(C_2H_4)]^-$ and $[PdCl_2(C_2H_4)(H_2O)]$, in which there are bonds between palladium and carbon atoms, are involved.

Homogeneous catalysis by transition metals, notably iron, also plays a very large part in important biochemical processes.

7 Interstitial compounds

The sizes of the transition metal atoms, and their crystal structures, are such that small atoms like hydrogen, boron, carbon and nitrogen can occupy the spaces within the crystal and form interstitial compounds.

These are generally made by heating the metal to a high temperature with hydrogen, boron, carbon (or a hydrocarbon) and nitrogen (or ammonia). The compounds maintain the crystal structure of the original metal, though with some distortion, and they have metallic properties such as high melting temperatures, high electrical conductivity and hardness. Some of the borides, carbides and nitrides find use as refractory and hard materials.

Some of the 'compounds' are exactly stoicheiometric, with a fixed, whole-number ratio of atoms, e.g. TiB_2, Fe_3C and VN. Others have odd atom ratios, e.g. $TiH_{1.73}$ and $VH_{0.56}$, and some can exist over a range of compositions, e.g. $VC_{0.37-0.47}$.

(Questions on this chapter will be found on p. 391.)

26
Titanium to zinc

1 Titanium and its compounds

a The metal Titanium occurs very widely and abundantly as rutile, TiO_2, and ilmenite, $FeO.TiO_2$. However, the manufacture of a usable metal is difficult and expensive because hot titanium reacts readily with carbon, oxygen and nitrogen, forming interstitial compounds, which make the metal sample brittle.

Titanium(IV) oxide (rutile) is first heated with carbon and chlorine, at around $900\,°C$, to form titanium(IV) chloride:

$$TiO_2(s) + 2C(s) + 2Cl_2(g) \longrightarrow TiCl_4(l) + 2CO(g)$$

The chloride is then reduced by heating it with sodium or magnesium in an atmosphere of helium or argon. The reaction is highly exothermic, e.g.

$$TiCl_4(l) + 2Mg(l) \longrightarrow Ti(s) + 2MgCl_2(l)$$
$$\Delta H_m^{\ominus}(298\,K) = -534\,kJ\,mol^{-1}$$

and the temperature has to be carefully controlled at about $850\,°C$. The magnesium chloride is tapped off from time to time and electrolysed to provide fresh supplies of magnesium and chlorine. The crude titanium is refined by vacuum distillation.

The metal is much less dense than steel but just as strong, it has a high melting temperature ($1675\,°C$), and it is resistant to corrosion unless it gets too hot. It is used in building aircraft, space vehicles, chemical plant and nuclear reactors, and it would be an excellent replacement for steel in many other uses if it could be produced more cheaply.

b Titanium(IV) compounds Titanium(IV) oxide (rutile) is an amphoteric oxide, forming titanates(IV), e.g. $BaTiO_3$, and basic salts, e.g. $TiOSO_4.xH_2O$. The Ti^{4+} ion exists in rutile (p. 16), but it is a polymerised oxo-cation, $(TiO^{2+})_x$, that exists in aerated aqueous solution. Titanium(IV) oxide is a brilliant white powder that is widely used in paints and for coating papers.

Titanium(IV) chloride is a covalent liquid that fumes in moist air, due to hydrolysis:

$$TiCl_4(l) + 2H_2O(l) \longrightarrow TiO_2(s) + 4HCl(g)$$

The formation of the titanium(IV) oxide makes it suitable for use in smoke grenades.

c Titanium(III) compounds These can be obtained by reduction of the corresponding titanium(IV) compound by gaseous hydrogen or by zinc and a dilute acid. The Ti^{3+} ion is also formed when titanium reacts with dilute acids in the absence of air.

If air is present, the Ti^{3+} ions are oxidised to 'TiO^{2+}':

$$Ti^{3+}(aq) + H_2O(l) \longrightarrow 'TiO^{2+}(aq)' + 2H^+(aq) + e^-$$

Solutions of titanium(III) chloride are used as reducing agents in volumetric analysis.

d Titanium(II) compounds Anhydrous titanium(II) chloride and iodide can be made, because the titanium(III) halides disproportionate on heating, e.g.

$$2TiCl_3(s) \longrightarrow TiCl_4(s) + TiCl_2(s)$$

but the Ti^{2+} ion cannot exist in solution. In acidic solution in the absence of air, it is oxidised to Ti^{3+}; if air is present, the product is $(TiO^{2+})_x$.

2 Vanadium and its compounds

a The metal Vanadium occurs as vanadinite, $3Pb_3(VO_4)_2.PbCl_2$, and carnotite, $KUO_2VO_4.1.5H_2O$, which also serves as a source of uranium. The extraction process is complex, but pure vanadium is rarely made, for the main use of the metal is in making alloys. Vanadium steels, for example, are very strong and tough; they are used for such articles as crankshafts, exhaust valves and high-speed tools.

b Vanadium(V) compounds Vanadates(V) can be extracted from the naturally occurring ores; the commonest is generally called ammonium vanadate(V) and is formulated in an over-simplified way as NH_4VO_3. However, the VO_3^- ion, poly-trioxovanadate(V), is probably polymerised, e.g. $V_3O_9^{3-}$. The tetraoxovanadate(V) (orthovanadate) ion, VO_4^{3-}, can also exist in strongly alkaline solution.

Ammonium vanadate(V) is a yellow solid which is converted into vanadium(V) oxide on heating:

$$2NH_4VO_3(s) \longrightarrow V_2O_5(s) + 2NH_3(g) + H_2O(l)$$

This oxide is an orange solid, and is familiar as the catalyst used in the contact process (p. 104). It is an amphoteric oxide, giving vanadates(V) with alkalis and solutions of salts with concentrated acids. The salts contain the VO_2^+ ion, and their solutions are yellow.

c Oxidation states and colours of vanadium ions The VO_2^+ ion, in which the vanadium has an oxidation state of $+5$, can be reduced to give distinctly coloured ions, in which the vanadium has oxidation states of $+4$, $+3$ and $+2$, as follows:

+5	+4	+3	+2
$VO_2^+(aq)$	$VO^{2+}(aq)$	$V^{3+}(aq)$	$V^{2+}(aq)$
yellow	blue	green	violet

The standard electrode potentials of the couples involved, together with those of some other redox couples, are summarised below.

$Zn^{2+}(aq)$	$Zn(s)$	$-0.763\,V$
$V^{3+}(aq)$	$V^{2+}(aq)$	$-0.255\,V$
$Cu^{2+}(aq)$	$Cu(s)$	$+0.337\,V$
$VO_2{}^+(aq)$	$V^{3+}(aq)$	$+0.361\,V$
$I_2(aq)$	$2I^-(aq)$	$+0.54\,V$
$Fe^{3+}(aq)$	$Fe^{2+}(aq)$	$+0.771\,V$
$VO_2{}^+(aq)$	$VO^{2+}(aq)$	$+1.00\,V$
$Br_2(aq)$	$2Br^-(aq)$	$+1.07\,V$

Any of the oxidising agents in the left-hand column will, theoretically and under standard conditions, be reduced by any of the reducing agents higher up in the right-hand column (p. 88).

Thus, zinc will reduce a yellow, acidic solution of ammonium vanadate(V) through the blue ($+4$) and green ($+3$) to the violet ($+2$) oxidation state. The resulting V^{2+} solution is such a strong reducing agent that it will reduce water to hydrogen, or copper(II) sulphate solution to copper.

Weaker reducing agents than zinc will reduce $+5$ vanadium to $+4$ or $+3$. Copper will reduce $VO_2{}^+$ to V^{3+}; Fe^{2+} or I^- ions will reduce it to VO^{2+}; Br^- ions will not reduce it at all.

CHROMIUM AND ITS COMPOUNDS

3 The metal

Chromium occurs naturally as *chromite* or *chrome ironstone* (iron(II) chromate(III), $FeCr_2O_4$). A useful alloy, ferrochrome (70% Cr : 30% Fe), is obtained from the ore by direct reduction with carbon in an electric furnace:

$$FeCr_2O_4(s) + 4C(s) \longrightarrow Fe(s) + 2Cr(s) + 4CO(g)$$

Pure chromium is obtained in the thermit process reduction of chromium(III) oxide with aluminium (p. 114), the chromium(III) oxide being obtained from chromite in three stages:

$$FeCr_2O_4 \xrightarrow[K_2CO_3 + air]{heat\ with} K_2CrO_4 \xrightarrow{H_2SO_4} K_2Cr_2O_7 \xrightarrow[with\ carbon]{heat} Cr_2O_3$$

$$Cr_2O_3(s) + 2Al(s) \longrightarrow 2Cr(s) + Al_2O_3(s)$$

Pure chromium is also made by electrolysis of chromic(VI) acid solution.

Chromium is a hard, malleable, silvery-white metal. It reacts with non-oxidising acids (in the absence of air) to form chromium(II) salts, and with oxidising acids to form chromium(III) salts, but it is rendered passive by concentrated nitric acid. It will form a sulphide, a nitride, a carbide and halides by direct combination.

Chromium is used in making stainless steels (85% Fe : 13% Cr : 2% Ni), and *nichrome* (60% Ni : 15% Cr : 25% Fe), which is used as resistance wiring. Chromium is also widely used in plating because it is very resistant to atmospheric corrosion.

4 Oxidation states of chromium

Chromium can exist in the following oxidation states, the main ones being shown in bold type.

$+6$
- CrO_3 (chromium(VI) oxide)
- CrO_2Cl_2 (chromium(VI) dichloride dioxide)

$Cr_2O_7^{2-}$ (dichromate(VI) ion) orange

CrO_4^- (chromate(VI) ion) yellow

$+5$ CrF_5

$+4$
- $CrCl_4$
- CrO_2 (chromium(IV) oxide)

$+3$ Cr_2O_3 (chromium(III) oxide) Cr^{3+} (chromium(III) ion) green

$+2$ Cr^{2+} (chromium(II) ion) blue

5 Chromium(VI) compounds

a Chromium (VI) oxide, CrO_3   This oxide is precipitated when concentrated sulphuric acid is added to solutions of chromates(VI) or dichromates(VI):

$$K_2Cr_2O_7 \xrightarrow{\text{conc. } H_2SO_4} H_2Cr_2O_7 \xrightarrow{-H_2O} 2CrO_3$$

$$K_2CrO_4 \xrightarrow{\text{conc. } H_2SO_4} H_2CrO_4 \xrightarrow{-H_2O} CrO_3$$

It is a red crystalline solid, not unlike sulphur(VI) oxide, SO_3. It is the acid anhydride of chromic(VI) acid, H_2CrO_4, which it forms with water but which is only known in solution.

The solution of the acid is a powerful oxidising agent; it is used in chromium plating, in treating metal surfaces to resist corrosion, in making dyestuffs, and in cleaning laboratory glassware.

Chromium(VI) oxide is used in making chromium(IV) oxide, CrO_2, by heating with hydrogen. The chromium(IV) oxide is a black, conducting, ferromagnetic solid that is used in making magnetic tapes.

b Chromates(VI)   These are salts of chromic(VI) acid. They contain the tetrahedral CrO_4^{2-} ion (Fig. 108), which is yellow in colour, and they are generally isomorphous with the corresponding sulphates.

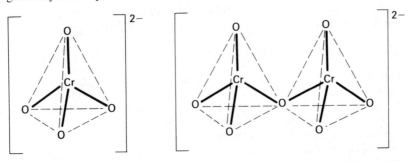

Fig. 108 *The shapes of the CrO_4^{2-} ion (left) and the $Cr_2O_7^{2-}$ ion (right). The ions have delocalised bonding, not shown here.*

Chromates(VI) only exist in neutral or alkaline solutions, for they are converted into dichromates(VI) in acidic solution:

$$2CrO_4^{2-}(aq) + 2H^+(aq) \longrightarrow Cr_2O_7^{2-}(aq) + H_2O(l)$$

Soluble sodium chromate(VI), which is obtained by heating a mixture of chromite and sodium carbonate in contact with air at 1100 °C, is the most important chromate(VI):

$$4FeCr_2O_4(s) + 8Na_2CO_3(s) + 7O_2$$
$$\longrightarrow 2Fe_2O_3(s) + 8Na_2CrO_4(s) + 8CO_2(g)$$

It is used in making sodium dichromate(VI), and in the leather, photographic and printing industries. Insoluble chromates(VI) include silver chromate(VI), which is used as an indicator in silver nitrate titrations, and barium and lead chromates(VI), which are used as pigments and as tests for barium and lead ions.

c Dichromates(VI)   These are orange-coloured salts containing the $Cr_2O_7^{2-}$ ion, which consists of two CrO_4^- tetrahedra linked through an oxygen atom (Fig. 108). They are made by acidifying solutions of chromates(VI) as above, but form chromates(VI) in alkaline solution:

$$Cr_2O_7^{2-}(aq) + 2OH^-(aq) \underset{acid}{\overset{alkali}{\rightleftharpoons}} 2CrO_4^{2-}(aq) + H_2O(l)$$

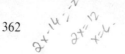

Dichromates(VI) are strong oxidising agents in acid solution:

$$Cr_2O_7^{2-}(aq) + 14H^+(aq) + 6e^- \longrightarrow 2Cr^{3+}(aq) + 7H_2O(l)$$
(orange)   (green)

but they are not strong enough to oxidise chlorides to chlorine, so they can be used in volumetric analysis in the presence of hydrochloric acid, unlike potassium manganate(VII).

In dilute acidified solutions, dichromates(VI) react with hydrogen peroxide solution to form a blue chromium peroxide, $CrO(O_2)_2$:

$$Cr_2O_7^{2-}(aq) + 2H^+(aq) + 4H_2O_2(aq) \longrightarrow 2CrO(O_2)_2 + 5H_2O(l)$$

This is unstable in aqueous solution, rapidly changing into green chromium(III) salt solution, but if the reaction is carried out in the presence of ethoxyethane (ether), the blue peroxide is stabilised in the organic layer. The reaction provides a sensitive test.

Sodium dichromate(VI) is the most important, and cheapest, chromium compound. It is made from sodium chromate(VI) and is used as a source of other chromium compounds, in tanning leather, in treating metals for corrosion resistance, and as an oxidising agent in organic chemistry.

Potassium dichromate(VI) is made from sodium dichromate(VI) and potassium chloride solutions; it crystallises out because it is less soluble than sodium dichromate(VI). Potassium dichromate(VI) is not deliquescent, whereas the sodium salt is; this explains the use of the potassium, rather than the sodium, salt in volumetric analysis.

d Chromium(VI) dichloride dioxide, CrO_2Cl_2   This is made by heating a mixture of a chloride, potassium dichromate(VI) and concentrated sulphuric acid. The red fumes which form can be condensed into a red liquid. As there is no corresponding reaction with bromides or iodides, the reaction can be used as a test for chlorides.

Chromium(VI) dichloride dioxide (chromyl chloride) is the acid chloride of chromic(VI) acid. It fumes in moist air because it forms the acid and hydrogen chloride:

$$CrO_2Cl_2(l) + 2H_2O(l) \longrightarrow H_2CrO_4(aq) + 2HCl(g)$$

6 Chromium(III) compounds not required

a Chromium(III) oxide, Cr_2O_3   This green insoluble solid is made by heating 'chromium(III) hydroxide' or ammonium dichromate(VI). It is an amphoteric oxide, giving chromium(III) salts with acids (but only slowly if previously ignited), and chromates(III) with alkalis:

$$Cr_2O_3(s) + 6H^+(aq) \longrightarrow 2Cr^{3+}(aq) + 3H_2O(l)$$
$$Cr_2O_3(s) + 6OH^-(aq) + 3H_2O(l) \longrightarrow 2[Cr(OH)_6]^{3-}(aq)$$

Fusion with sodium peroxide oxidises chromium(III) oxide to sodium chromate(VI):

$$Cr_2O_3(s) + 3Na_2O_2(s) \longrightarrow Na_2O(s) + 2Na_2CrO_4(aq)$$

Chromium(III) oxide is a refractory material used in making chromium, as an abrasive, as a pigment, and as a catalyst in organic reactions.

b 'Chromium(III) hydroxide,' $Cr(OH)_3$ This is formed as a green precipitate on adding hydroxide ions to a solution of a chromium(III) salt. The composition of the precipitate varies, and it is probably a hydrated oxide, $Cr_2O_3.xH_2O$.

Like the oxide, it is amphoteric, and it can be oxidised to a chromate(VI) by heating with hydrogen peroxide in alkaline solution.

$$2Cr(OH)_3(s) + 3H_2O_2(aq) + 4OH^-(aq) \longrightarrow 2CrO_4^{2-}(aq) + 8H_2O(l)$$

c Chromium(III) salts   These are made by reaction between chromium(III) oxide or chromium(III) 'hydroxide' and acids. They are purplish in colour, e.g. chromium(III) sulphate(VI), $Cr_2(SO_4)_3.18H_2O$ and chromium(III) chloride, $CrCl_3.6H_2O$.

Solutions of chromium(III) salts give a green precipitate of the 'hydroxide' on adding sodium hydroxide or ammonia solution.

Alums (p. 195) in which the trivalent metal is chromium are known as chrome alums. Common chrome alum is hydrated chromium(III) potassium sulphate, $K_2SO_4.Cr_2(SO_4)_3.24H_2O$ or $KCr(SO_4)_2.12H_2O$. It is made by reduction of potassium dichromate(VI) in sulphuric acid solution by ethanol or sulphur dioxide. The resulting mixture contains an equimolar mixture of potassium sulphate and chromium(III) sulphate, which deposits crystals of chrome alum on cooling:

$$K_2Cr_2O_7(aq) + H_2SO_4(aq) + 3SO_2(g)$$
$$\longrightarrow K_2SO_4(aq) + Cr_2(SO_4)_3(aq) + H_2O(l)$$

7 Chromium(II) compounds

Chromium(II) salts can be made by reducing solutions of chromium(III) salts, using amalgamated zinc and acid, or by treating chromium with non-oxidising acids. The reactions must be carried out in the absence of air, for the Cr^{2+} ion is such a strong reducing agent that it is readily converted into Cr^{3+} by the air:

$$Cr^{3+}(aq) + e^- \longrightarrow Cr^{2+}(aq) \qquad E^\ominus = -0.41 \text{ V}$$

The Cr^{2+} ion will also slowly reduce water to hydrogen.

Chromium(II) ethanoate is a red, insoluble compound, which can be prepared by precipitation between Cr^{2+} and ethanoate ions.

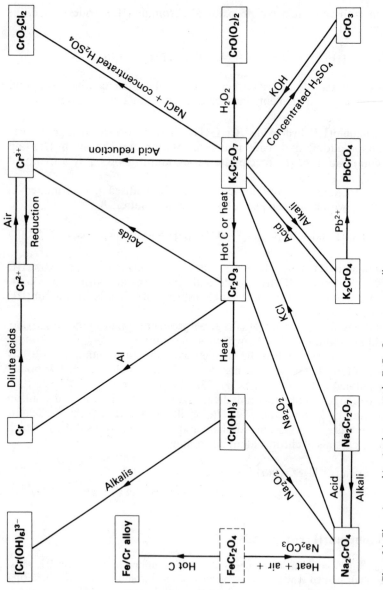

Chart 14 Chromium and its simple compounds. FeCr₂O₄ occurs naturally.

MANGANESE AND ITS COMPOUNDS

8 The metal

This is a hard, grey, brittle metal, occurring naturally as pyrolusite, MnO_2, and manufactured in the pure form by electrolysis of manganese(II) sulphate, or by the thermit process (p. 114) from manganese(II) dimanganese(III) oxide (obtained by heating pyrolusite) and aluminium:

$$3MnO_2(s) \longrightarrow Mn_3O_4(s) + O_2(g)$$
$$3Mn_3O_4(s) + 8Al(s) \longrightarrow 4Al_2O_3(s) + 9Mn(s)$$

However, it is more commonly made, and used, in the form of *ferro-manganese* (50% Fe and 50% Mn), or *spiegeleisen* (75% Fe, 20% Mn and 5% C). These are made by heating a mixture of iron and manganese oxides with carbon in a blast furnace.

Manganese is fairly reactive, combining with hot oxygen, nitrogen, sulphur, carbon and halogens to form the oxide (Mn_3O_4), the nitride (Mn_3N_2), the sulphide (MnS), the carbide (Mn_3C), and halides, respectively. It also reacts with most acids.

It is used in making alloys. Manganese steels are very tough and are used in making railway points and safes; *manganin* (80–85% Cu, 10–15% Mn and 2–5% Ni) is used for resistance wiring; and *manganese bronze* (55% Cu, 44% Zn, 1% Mn) is used in making propellors.

9 Oxidation states of manganese

Manganese can exist in the following oxidation states, the main ones being shown in bold type.

+7	Mn_2O_7 (manganese(VII) oxide)	MnO_4^- (manganate(VII) ion) purple
+6		MnO_4^{2-} (manganate(VI) ion) green
+5		MnO_4^{3-} (manganate(V) ion) blue
+4	MnO_2 (manganese(IV) oxide)	black
+3	Mn_2O_3 (manganese(III) oxide)	Mn^{3+} (manganese(III) ion) red
+2	MnO (manganese(II) oxide)	Mn^{2+} (manganese(II) ion) pale pink

10 Manganese(VII) compounds

a Manganese(VII) oxide, Mn_2O_7   This can be obtained as a dark, oily liquid by treating potassium manganate(VII) with concentrated sulphuric acid. The manganic(VII) acid first formed is dehydrated:

$$2KMnO_4 \xrightarrow{\text{conc. } H_2SO_4} 2HMnO_4 \xrightarrow{-H_2O} Mn_2O_7$$

The reaction is highly dangerous as the oxide is very explosive, liberating oxygen and giving manganese(IV) oxide. Chlorine(VII) oxide, Cl_2O_7, is similarly explosive, but this is one of only very few likenesses between manganese and the halogens.

b Preparation of potassium manganate(VII)   Fusion of manganese(IV) oxide and potassium hydroxide with an oxidising agent such as potassium chlorate(VII) produces green potassium manganate(VI). The manganese(IV) oxide acts as an acidic oxide and forms, first, potassium manganate(IV), K_2MnO_3, but this is readily oxidised to the manganate(VI):

$$MnO_2(s) + 2OH^-(aq) \longrightarrow MnO_3^{2-}(aq) + H_2O(l)$$

$$MnO_3^{2-}(aq) + 2OH^-(aq) \longrightarrow MnO_4^{2-}(aq) + H_2O(l) + 2e^-$$

The potassium manganate(VI) is extracted from the mix with water and the resulting solution is converted into potassium manganate(VII) by oxidation. This is usually brought about by chlorine in the laboratory, and electrolytically in the industrial process:

$$2MnO_4^{2-}(aq) + Cl_2(g) \longrightarrow 2MnO_4^-(aq) + 2Cl^-(aq)$$
$$MnO_4^{2-}(aq) \longrightarrow MnO_4^-(aq) + e^-$$

Alternatively, the manganate(VI) solution is boiled and the disproportionation which begins to take place,

$$3MnO_4^{2-}(aq) + 2H_2O(l) \rightleftharpoons 2MnO_4^-(aq) + MnO_2(s) + 4OH^-(aq)$$

is brought to completion by passing in carbon dioxide to lower the OH^- ion concentration.

Potassium manganate(VII) can be crystallised out after filtering off the manganese(IV) oxide.

c Properties of potassium manganate(VII)   Potassium manganate(VII) crystals appear to be almost black but they give a highly-coloured purple solution. The crystals contain the tetrahedral MnO_4^- ion and are isomorphous with potassium chlorate(VII) crystals.

On heating, potassium manganate(VII) decomposes to give oxygen, but its major use is as an oxidising agent. Conditions can be found in which the MnO_4^- ion can be reduced to MnO_4^{2-}, MnO_4^{3-}, MnO_2, Mn^{3+} or Mn^{2+}, but MnO_4^{2-}, MnO_2 and Mn^{2+} are the most usual products, depending on the pH of the solution used and on the strength of the reducing agent.

Potassium manganate(VII) is usually used in the presence of excess dilute sulphuric acid. Under these conditions, it is generally reduced to Mn^{2+} ions, and the purple colour disappears:

$$MnO_4^-(aq) + 8H^+(aq) + 5e^- \longrightarrow Mn^{2+}(aq) + 4H_2O(l)$$
$$E^\ominus = +1.52\,V$$

The solution becomes less strong as an oxidising agent as the pH increases, the redox potential falling from 1.52 V at pH 1 to 0.96 V at pH 6 (p. 88).

In neutral or alkaline solution the manganate(VII) is reduced to manganese(IV) oxide

$$MnO_4^-(aq) + 4H^+(aq) + 3e^- \longrightarrow MnO_2(s) + 2H_2O(l)$$
$$E^\ominus = +1.67\,V$$

The manganate(VII) is still a strong oxidising agent under these conditions but it has a lower capacity than in acid solution. The formation of brown, insoluble manganese(IV) oxide also makes its use difficult in volumetric analysis.

In more strongly alkaline solution, manganate(VII) can be reduced by MnO_2 (p. 348) to green manganate(VI)

$$MnO_4^-(aq) + e^- \longrightarrow MnO_4^{2-}(aq) \qquad E^\ominus = +0.56\,V$$

11 Manganese(VI) compounds

The manganates(VI), containing the tetrahedral MnO_4^{2-} ion, are the only common examples of manganese(VI) compounds. They are salts of the hypothetical MnO_3 or H_2MnO_4 and are generally isomorphous with the corresponding sulphates or chromates(VI).

Sodium and potassium manganates(VI) are dark green solids which can be made as in the first stage of the preparation of potassium manganate(VII), or by the reduction of potassium manganate(VII) in alkaline solution. This reduction can be brought about by manganese(IV) oxide acting as the reducing agent, in alkaline solution (p. 348):

$$2MnO_4^-(aq) + MnO_2(s) + 4OH^-(aq) \rightleftharpoons 3MnO_4^{2-}(aq) + 2H_2O(l)$$

The reverse process (the disproportionation of the manganate(VI) ion) takes place very readily in acidic solution (p. 94):

$$3MnO_4^{2-}(aq) + 4H^+(aq) \rightleftharpoons 2MnO_4^-(aq) + MnO_2(s) + 2H_2O(l)$$

12 Manganese(V) compounds

These are very rare and unstable, but a blue solution containing MnO_4^{3-} ions can be obtained by adding a few crystals of potassium manganate(VII) to a very concentrated solution of potassium hydroxide. The MnO_4^- ion is reduced to MnO_4^{3-}, with the liberation of oxygen:

$$MnO_4^-(aq) + 2e^- \longrightarrow MnO_4^{3-}(aq)$$
$$2OH^-(aq) \longrightarrow H_2O(l) + \tfrac{1}{2}O_2(g) + 2e^-$$

13 Manganese(IV) compounds

Manganese(IV) oxide, MnO_2, is the main manganese(IV) compound. It is a black, insoluble powder with a rutile structure. It occurs naturally as pyrolusite and is made by heating manganese(II) nitrate or, as a hydrate, by oxidising Mn^{2+} solutions by bleaching powder or peroxodisulphates(VI).

It is converted into manganese(II) dimanganese(III) oxide, Mn_3O_4, on heating in air (p. 365). It is a strong oxidising agent, being reduced, mainly, to Mn^{2+},

$$MnO_2(s) + 4H^+(aq) + 2e^- \longrightarrow Mn^{2+}(aq) + 2H_2O(l) \qquad E^\ominus = +1.23\,V$$

Manganese(IV) oxide is amphoteric. The Mn^{4+} salts that it forms with acids are unstable and revert to Mn^{2+} salts. With alkalis it gives manganates(IV), which are of doubtful composition.

Manganese(IV) oxide is used as a catalyst, as a depolariser in batteries, as an oxidising agent, and in making glass and pottery.

14 Manganese(III) compounds

Red solutions, containing Mn^{3+} ions, can be obtained by reducing potassium manganate(VII) solution with a solution of manganese(II) sulphate in concentrated sulphuric acid,

$$MnO_4^-(aq) + 4Mn^{2+}(aq) + 8H^+(aq) \longrightarrow 4H_2O(l) + 5Mn^{3+}(aq)$$

or by electrolytic oxidation of Mn^{2+} ions.

The Mn^{3+} ion is, however, readily converted into Mn^{2+}, both by oxidation by water,

$$2Mn^{3+}(aq) + H_2O(l) \longrightarrow 2Mn^{2+}(aq) + 2H^+(aq) + \tfrac{1}{2}O_2(g)$$

and through disproportionation (p. 94):

$$2Mn^{3+}(aq) + 2H_2O(l) \longrightarrow Mn^{2+}(aq) + MnO_2(s) + 4H^+(aq)$$

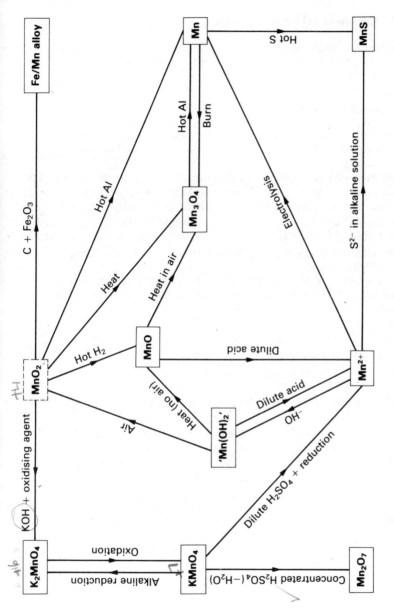

Chart 15 Manganese and its simple compounds. MnO$_2$ occurs naturally.

15 Manganese(II) compounds

a Manganese(II) oxide, MnO   This is a grey–green solid with a sodium chloride structure. It is made by reduction of any higher oxide of manganese with hydrogen or carbon monoxide, or by heating manganese(II) hydroxide or carbonate in the absence of air. It is a basic oxide, giving manganese(II) salts with acids.

b Manganese(II) hydroxide   This is a white hydrated powder, of uncertain composition, that is precipitated by sodium hydroxide solution from solutions of manganese(II) salts. It forms manganese(II) oxide on heating in the absence of air, but is readily oxidised by air to a brown, hydrated manganese(IV) oxide.

c Manganese(II) salts   These are generally pale pink in colour. The soluble chloride, $MnCl_2.4H_2O$, and sulphate, $MnSO_4.5H_2O$, are made from the necessary acid and base. The insoluble sulphide, MnS, and carbonate, $MnCO_3$, are made by precipitation methods.

Solutions of Mn^{2+} salts can be oxidised to manganates(VII) by sodium bismuthate(V) in the presence of dilute nitric acid, and the formation of the purple colour is a useful test.

Traces of Mn^{2+} are essential in fertilisers, and manganese deficiency in animals causes deformation of the bones. Manganese(II) sulphate is generally used as the source of Mn^{2+} ions.

IRON AND ITS COMPOUNDS

16 Pig or cast iron and the blast furnace

Pig or cast iron is manufactured from oxide ores, e.g. haematite, Fe_2O_3, or carbonate ores, e.g. siderite, $FeCO_3$, after a preliminary roasting. The ore is mixed with coke and limestone and heated in a blast furnace (Fig. 109). This consists of a steel, cylindrical shell, 25–30 m high and 4–9 m in diameter, which is thickly lined with firebrick. A blast of hot, dry air, enriched with oxygen and fuel oil, is passed into the bottom of the furnace through small pipes known as tuyères, which lead off from a circulating bustle pipe.

The temperature inside the furnace varies from about 2000 °C close to the tuyères to about 200 °C at the top of the furnace. In the lower part, where the temperature is high, the iron oxide is reduced by the carbon:

$$2Fe_2O_3(s) + 3C(s) \longrightarrow 4Fe(l) + 3CO_2(g)$$

This overall reaction, however, probably takes place in stages, using carbon monoxide as the intermediary, because reactions between two solids are not rapid. Thus

$$Fe_2O_3(s) + 3CO(g) \longrightarrow 2Fe(l) + 3CO_2(g)$$
$$3CO_2(g) + 3C(s) \longrightarrow 6CO(g)$$

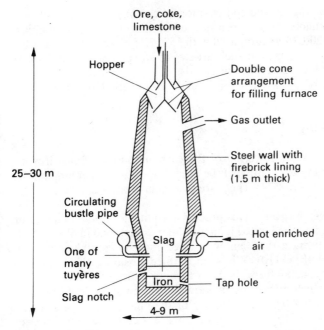

Ore, coke, limestone

Hopper

Double cone arrangement for filling furnace

Gas outlet

Steel wall with firebrick lining (1.5 m thick)

25–30 m

Circulating bustle pipe

Slag

Hot enriched air

One of many tuyères

Iron

Tap hole

Slag notch

4–9 m

Fig. 109 A blast furnace.

It is, then, the carbon that is used up in the constant regeneration of carbon monoxide. Carbon will, however, only reduce carbon dioxide at temperatures above about 710 °C (p. 113),

$$C(s) + CO_2(g) \longrightarrow 2CO(g)$$

so that, at the lower temperatures higher up in the furnace, the reduction process is

$$Fe_2O_3(s) + 3CO(g) \longrightarrow 2Fe(l) + 3CO_2(g)$$

Overall, the oxide ore is reduced to iron, and the silica impurities form a slag of calcium silicate(IV) with the limestone. Both the molten iron and the molten slag sink to the bottom (the hearth) of the furnace, with the less dense slag floating on top of the molten iron, which is known in the industry as 'hot metal'. The whole process is continuous; it may go on for over two years until, eventually, the furnace needs re-lining with new heat-resistant brick. As the slag and 'hot metal' are periodically tapped off, through the slag notch and the tap hole respectively, raw materials are fed in through the hopper at the top. The hot gases passing out of the furnace are freed from dust in dust catchers and are used, in large heat exchangers known as Cowper stoves, to heat the air before it is blown into the bottom of the blast furnace.

Cast iron contains 90–95 per cent of iron. The main impurity is carbon (3–5 per cent) but silicon, manganese, sulphur and phosphorus may also be present, depending on the composition of the original coke and ore. These impurities make the cast iron brittle but they also lower its melting temperature to 1200 °C, as compared with 1535 °C for pure iron. Cast iron is used for making castings in such articles as stoves, cookers, man-hole covers, cylinder blocks and machine stands.

80–90 per cent of the cast iron that is made is converted into steel.

17 Steel making

Steel making involves a lowering of the carbon content of cast iron to below 1.5 per cent and a control of the other impurities, depending on the type of steel required (p. 373). There are two main processes.

a The basic oxygen furnace This process accounts for about 60 per cent of steel production in the United Kingdom; it was first used in about 1962.

A cylindrical furnace that can be rotated into horizontal or vertical positions is used (Fig. 110). Whilst in a tilted position, the furnace is charged with 'hot metal' from the blast furnace and up to 30 per cent of scrap metal. It is then moved into a vertical position and a high-speed jet of oxygen is directed on to the surface of the charge through a water-cooled lance. Some of the carbon in the charge is converted into gaseous oxides, which escape from the furnace through a hood. Other impurities also form oxides, and these react with added lime to form a slag. After the 'blow', the furnace is tipped into a horizontal position so that the molten steel can be run off through the tap-hole. It is then inverted to pour away the slag. A large furnace can produce 250 tonnes of steel in 40 minutes.

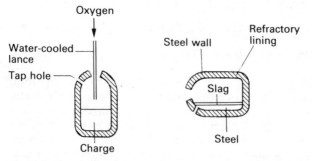

Fig. 110 *A basic oxygen furnace, showing (left) the oxygen lance in position, and (right) the position for tapping off the steel.*

b The electric arc process This furnace (Fig. 111) consists of a cylindrical vessel with a removable lid carrying three graphite electrodes that can be lowered into the vessel when the lid is in place. The charge is made up almost entirely of scrap steel, and it is heated by forming arcs between the ends of the electrodes and the surface of the charge. Lime, fluorspar and

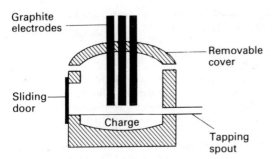

Fig. 111 *An electric arc furnace.*

iron oxide are added to form a slag with the impurities and to give a steel of the required composition. The slag is raked or poured from the surface and the steel is run out through the tapping spout by tilting the furnace.

Because the process can be closely controlled, due to the absence of oxidising gases, it was originally used solely for making high quality steels; but it now accounts for about 40 per cent of steel production in the United Kingdom.

c Casting steel The molten steel produced from any type of furnace can be cast in different ways.

In continuous casting, a stream of molten metal is fed through a water-cooled mould, which causes the outside metal to solidify. The stream is further cooled by spraying it with water, and is then fed through rollers to form steel plates, which are known as 'slabs' or 'blooms'.

In ingot casting, molten steel is poured into specially shaped moulds and allowed to solidify. The resulting ingots can be rolled, whilst red-hot, into slabs or blooms or sections of different shapes. The ingots can also be cold-rolled or forged (i.e. hammered into shape).

In special casting, the molten steel is poured into sand moulds which can be broken away when the metal has solidified. Quite complicated shapes, from the very small to the very large, can be made in this way.

18 Types and uses of steel

Mild steels, containing 0.1–0.25 per cent of carbon, are soft, malleable and ductile. They are used in making tin-plate, corrugated sheets, ships' plates, car bodies, rolled steel joists and steel pipes, but they cannot be heat treated successfully. Medium steels (0.25–0.6 per cent carbon) and hard steels (0.6–1.5 per cent carbon) are harder than mild steel and can be heat treated; they are used for making tools and most alloy steels.

These alloy steels vary widely. Incorporation of chromium and nickel gives stainless steel; cobalt gives a highly magnetic steel; tungsten gives very hard steels for making cutting tools; manganese steels are hard and tough and are used in rock drills; silicon gives spring steels; and molybdenum steels resist acid corrosion.

Medium or hard steels can also be heat treated. On heating to red heat and quenching in oil or water they become brittle, but this brittleness can be removed by *tempering.* This is done by reheating the steel and then recooling; different degrees and rates of heating and cooling give different products to meet different requirements.

In *case hardening*, steel can be hardened on the surface by heating in a pack of charcoal (*carburising*); carbon diffuses into the steel surface and hardens it. Alternatively, some alloy steels can be heated in an atmosphere of ammonia (*nitriding*); nitrogen from the decomposition of the ammonia diffuses into the steel surface, forming very hard nitrides. A mixture of carburising and nitriding can be achieved by treating steels in baths of molten cyanides. Case-hardened steels are used in such articles as gear wheels.

Annealing is a process in which steels are maintained at a temperature of about 1000 °C and then allowed to cool very slowly; a soft steel which can be machined very easily results.

Hundreds of different sorts of steel are in everyday use.

19 Properties of iron

Iron is a soft, ductile and malleable metal that is strongly ferromagnetic. It combines with oxygen or air on heating, and will catch fire if in wire or powdered form, sometimes even burning spontaneously if it is dry and very finely powdered (pyrophoric iron); iron(II) diiron(III) oxide and iron(III) oxide are formed:

$$3Fe(s) + 2O_2(g) \longrightarrow Fe_3O_4(s)$$
$$4Fe(s) + 3O_2(g) \longrightarrow 2Fe_2O_3(s)$$

Iron(II) diiron(III) oxide is also formed on passing steam over red-hot iron:

$$3Fe(s) + 4H_2O(g) \rightleftharpoons Fe_3O_4(s) + 4H_2(g)$$

This reaction is reversible but a steady supply of steam ensures that it is always in excess and removes any hydrogen formed. Rusting in moist air produces hydrated iron(III) oxide, $Fe_2O_3.xH_2O$ (rust).

Iron also reacts on heating with sulphur, chlorine, hydrogen chloride, carbon and carbon monoxide to form iron(II) sulphide (FeS), iron(III) chloride ($FeCl_3$ or Fe_2Cl_6), iron(II) chloride ($FeCl_2$), iron carbide (Fe_3C) and pentacarbonyl iron(0) ($[Fe(CO)_5]$) respectively.

Iron displaces hydrogen from dilute acids to form iron(II) salts, but concentrated sulphuric acid gives iron(III) sulphate and sulphur dioxide. Concentrated nitric acid renders iron passive (p. 243), due to the formation of an impermeable oxide layer.

20 Oxides of iron

a Iron(II) oxide, FeO This is a basic oxide and is made by heating iron(II) ethanedioate (oxalate) in an inert atmosphere or by reduction of higher oxides by hydrogen. It is easily oxidised by air, can be pyrophoric and reacts with acids to give iron(II) salts.

Iron(II) oxide is non-stoicheiometric, with a deficiency of iron and a composition generally best represented as $Fe_{0.94}O$.

b Iron(III) oxide, Fe_2O_3 This is made by heating iron(II) sulphate, iron(II) oxide or iron(III) 'hydroxide', and by roasting iron(II) disulphide (pyrites) in air or oxygen.

The oxide can range in colour from purple to yellow; the form made by heating iron(II) sulphate is dark red. It is used, as jeweller's rouge, as a mild abrasive and in pigments. Hydrated iron(III) oxides also occur naturally as haematites; they are used in making iron and as pigments, e.g. ochres, umbers and siennas.

Iron(III) oxide is essentially basic but it reacts only slowly with acids, particularly if it has been strongly ignited; iron(III) salts are formed. It also reacts with fused alkalis to give ferrates(III), which contain $[Fe(OH)_6]^{3-}$ ions.

c Iron(II) diiron(III) oxide, Fe_3O_4 This is made by the action of steam on red-hot iron, or by heating iron(III) oxide *in vacuo* at 1000 °C. It also occurs naturally as *magnetite* or *lodestone*, and it is sometimes referred to as magnetic oxide of iron because it is ferromagnetic. It is a black solid which functions as a mixed oxide ($FeO + Fe_2O_3$ or $Fe^{II}Fe^{III}_2O_4$), reacting with acids to give a mixture of iron(II) and iron(III) salts, e.g.

$$Fe_3O_4(s) + 8HCl(aq) \longrightarrow FeCl_2(aq) + 2FeCl_3(aq) + 4H_2O(l)$$

21 'Hydroxides' of iron

a 'Iron(II) hydroxide' This is formed as a white precipitate if caustic alkalis are added to solutions of iron(II) salts in the absence of any air, but it is very easily oxidised by dissolved air to a green compound that can be regarded as a mixed oxide–hydroxide, i.e. $Fe_2O_3.Fe(OH)_3$ or $FeO(OH)$:

$$FeSO_4(aq) + 2NaOH(aq) \longrightarrow Fe(OH)_2(s) + Na_2SO_4(aq)$$
$$\text{(white)}$$

$$4Fe(OH)_2(s) + O_2(g) \longrightarrow 4FeO(OH)(s) + 2H_2O(l)$$
$$\text{(green)}$$

The green compound rapidly oxidises into 'iron(III) hydroxide' on further exposure to air.

Iron(II) hydroxide gives iron(II) salts with acids and also shows weakly acidic properties with bases.

b 'Iron(III) hydroxide' The precipitate obtained on adding hydroxide ions to a solution of an iron(III) salt is reddish-brown and gelatinous. It is not iron(III) hydroxide, $Fe(OH)_3$ but a varying mixture of hydrated iron(III) oxides, $Fe_2O_3.xH_2O$, with iron(III) oxyhydroxide, $FeO(OH)$. It reacts with acids to give iron(III) salts and is also weakly acidic.

22 Chlorides of iron

a Iron(II) chloride, $FeCl_2$ Anhydrous iron(II) chloride is made by passing hydrogen chloride over hot iron; it is a white crystalline solid. Treatment of iron(II) oxide, or iron, with dilute hydrochloric acid gives a hydrate, $FeCl_2.4H_2O$, which is pale green.

b Iron(III) chloride, $FeCl_3$ or Fe_2Cl_6 Anhydrous iron(III) chloride is made by passing chlorine over hot iron; it is very dark green. Iron(III) oxide and iron(III) hydroxide give the hydrate, $FeCl_3.6H_2O$, with dilute hydrochloric acid; this compound is mustard yellow in colour and very deliquescent.

Iron(III) chloride exists as covalent $FeCl_3$ molecules in organic solvents, but relative molecular mass measurements show that it is considerably associated into Fe_2Cl_6 molecules in the vapour at low temperatures. At higher temperatures, it decomposes into iron(II) chloride and chlorine.

As iron(III) chloride can be reduced to iron(II) chloride, it is an oxidising agent and is sometimes used as such in organic chemistry. It is also a halogen-carrier catalyst.

23 Sulphides of iron

a Iron(II) sulphide, FeS This is made by heating powdered sulphur and iron (the reaction produces much heat) or by precipitation from alkaline solutions of iron(II) salts.

It is a black, insoluble solid which reacts with acids to give hydrogen sulphide. Like iron(II) oxide it is non-stoicheiometric, ranging in composition from $Fe_{0.86}S$ to $Fe_{0.89}S$.

b Iron(III) sulphide, Fe_2S_3 Since iron(III) ions oxidise hydrogen sulphide to sulphur, the sulphide cannot be prepared by precipitation methods. Dry hydrogen sulphide will, however, react with hydrated iron(III) oxide to form the sulphide as a black solid. The sulphide readily decomposes into iron(II) sulphide and iron(II) disulphide:

$$Fe_2O_3(s) + 3H_2S(g) \longrightarrow Fe_2S_3(s) + 3H_2O(l)$$
$$Fe_2S_3(s) \longrightarrow FeS_2(s) + FeS(s)$$

c Iron(II) disulphide, FeS_2 This occurs naturally as golden yellow pyrites (fools' gold) and as white marcasite. The latter is converted into the former on heating at 450 °C. Both pyrites and marcasite produce sulphur dioxide on roasting in air (p. 103).

24 Iron(II) salts of oxoacids

a General properties These salts are generally green or blue–green crystalline solids that give $[Fe(H_2O)_6]^{2+}$ ions in solution. The solutions give a green precipitate (p. 375) with sodium hydroxide (unless air is completely absent, in which case the precipitate is white); a dark blue precipitate with potassium hexacyanoferrate(III); a white precipitate (which turns pale blue on oxidation) with potassium hexacyanoferrate(II); and a brown complex, $[Fe(NO)(H_2O)_5]^{2+}$, as in the brown ring test, with nitrogen oxide. They give no colour or precipitate with potassium thiocyanate.

Acidified solutions of iron(II) salts are readily oxidised to iron(III) salts by strong oxidising agents and, slowly, by air:

$$[Fe(H_2O)_6]^{3+}(aq)+e^- \longrightarrow [Fe(H_2O)_6]^{2+}(aq) \qquad E^{\ominus} = +0.77\,V$$
$$\tfrac{1}{2}O_2(g)+2H^+(aq)+2e^- \longrightarrow H_2O(l) \qquad E^{\ominus} = +1.23\,V$$

The oxidation is much easier in alkaline solution:

$$\tfrac{1}{2}(Fe_2O_3.3H_2O)(s)+e^- \longrightarrow Fe(OH)_2(s)+OH^-(aq)$$
$$E^{\ominus} = -0.56\,V$$

b Iron(II) sulphate, $FeSO_4$ This is made as the hydrate, $FeSO_4.7H_2O$ (*green vitriol*), by reaction between iron and dilute sulphuric acid. It is obtained commercially from the solutions remaining after sulphuric acid has been used to clean steel, and is very cheap.

It is used medicinally as a source of iron, as a trace component of fertilisers, and in making inks, iron(III) oxides and Prussian blue (p. 378).

c Ammonium iron(II) sulphate Iron(II) sulphate forms a series of double salts, $M_2SO_4.FeSO_4.6H_2O$, with M being potassium, rubidium, caesium or ammonium. Ammonium iron(II) sulphate-6-water, $FeSO_4.(NH_4)_2SO_4.6H_2O$, is pale green. It is used as a cheap primary standard in volumetric analysis, having the advantage over iron(II) sulphate that it is much more resistant to atmospheric oxidation.

25 Iron(III) salts of oxoacids

a General properties These are generally yellow or yellow–brown crystalline solids that give $[Fe(H_2O)_6]^{3+}$ ions in solution. They give a red–brown precipitate of iron(III) hydroxide with sodium hydroxide; a dark blue precipitate with potassium hexacyanoferrate(II); a brown or green coloration with potassium hexacyanoferrate(III); a yellow precipitate of iron(III) phosphate(V) with sodium phosphate(V); and a blood red coloration with potassium thiocyanate.

Iron(III) salts and their solutions can be reduced to iron(II) salts by many reducing agents. If an anion, e.g. I^-, can do this, it is not possible to make the iron(III) salt of that ion. Attempts to make iron(III) iodide, for example, yield iron(II) iodide and iodine.

b Iron(III) sulphate, $Fe_2(SO_4)_3$ The hydrate, $Fe_2(SO_4)_3.9H_2O$ is made by oxidation of iron(II) sulphate with concentrated sulphuric acid. Careful heating gives a white anhydrous salt but this decomposes at a higher temperature into iron(III) oxide and sulphur trioxide.

The salt is a component of iron(III) alums (p. 195), and iron(III) alum, $FeNH_4(SO_4)_2.12H_2O$, is made by crystallising an equimolar mixture of iron(III) sulphate and ammonium sulphate.

26 Complexes of iron

a Potassium hexacyanoferrate(II), $K_4[Fe(CN)_6]$ This is a yellow crystalline solid that is obtained in the laboratory by treating iron(II) salts with excess potassium cyanide. The iron(II) cyanide first formed as a precipitate reacts with more cyanide to form the hexacyanoferrate(II):

$$Fe^{2+}(aq) + 2CN^-(aq) \longrightarrow Fe(CN)_2(s)$$
$$Fe(CN)_2(s) + 4CN^-(aq) \longrightarrow [Fe(CN)_6]^{4-}(aq)$$

With solutions of iron(III) salts it gives a blue precipitate of Prussian blue, and this is used as a test for Fe^{3+} ions. The formation of brown copper(II) hexacyanoferrate(II) with Cu^{2+} ions also serves as a test for these ions.

b Potassium hexacyanoferrate(III), $K_3[Fe(CN)_6]$ This is a dark red solid that is obtained by oxidation of potassium hexacyanoferrate(II), either with chlorine or electrolytically:

$$2[Fe(CN)_6]^{4-}(aq) + Cl_2(g) \longrightarrow 2[Fe(CN)_6]^{3-}(aq) + 2Cl^-(aq)$$
$$[Fe(CN)_6]^{4-}(aq) \longrightarrow [Fe(CN)_6]^{3-}(aq) + e^-$$

A solution of potassium hexacyanoferrate(III) gives a deep blue precipitate of Turnbull's blue with Fe^{2+} ions, and this is used as a test (p. 396). The $[Fe(CN)_6]^{3-}$ ion is a weaker oxidising agent than $[Fe(H_2O)_6]^{3+}$:

$$[Fe(CN)_6]^{3-}(aq) + e^- \longrightarrow [Fe(CN)_6]^{4-}(aq) \qquad E^\ominus = +0.36\,V$$
$$[Fe(H_2O)_6]^{3+}(aq) + e^- \longrightarrow [Fe(H_2O)_6]^{2+}(aq) \qquad E^\ominus = +0.77\,V$$

In other words, $[Fe(CN)_6]^{4-}$ ions are stronger reducing agents than $[Fe(H_2O)_6]^{2+}$ ions, or the $+3$ oxidation state is stabilised by cyanide ligands. It is, therefore, easier to oxidise $[Fe(CN)_6]^{4-}$ ions to the $+3$ oxidation state than $[Fe(H_2O)_6]^{2+}$ ions (p. 337).

c Prussian blues   The Prussian blues are a series of complexes of general formula $MFe[Fe(CN)_6]$, where M is an alkali metal (usually potassium). It is not possible to distinguish between potassium iron(III) hexacyanoferrate(II), i.e. Prussian blue, $KFe^{III}[Fe^{II}(CN)_6]$, and potassium iron(II) hexacyanoferrate(III), i.e. Turnbull's blue, $KFe^{II}[Fe^{III}(CN)_6]$. The complexes are used as pigments.

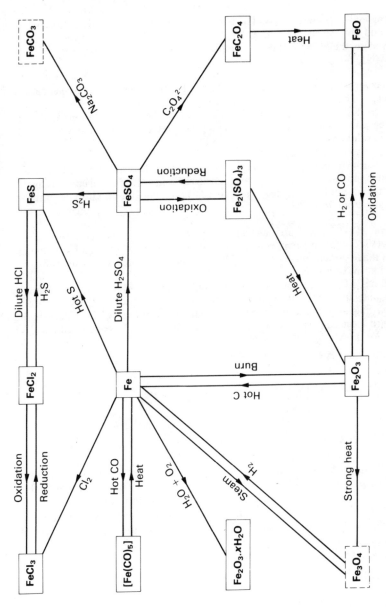

Chart 16 Iron and its simple compounds. Fe_2O_3, Fe_3O_4 and $FeCO_3$ occur naturally.

d Complex thiocyanates Iron(III) ions give a blood red coloration with SCN^- ions, and this serves as a test. The colour is due to a mixture of iron(III) thiocyanate, $Fe(SCN)_3$, and to complexes such as $[Fe(SCN)(H_2O)_5]^{2+}$ and $[Fe(SCN)_6]^{3-}$.

e Disodium pentacyanonitrosylferrate(III), $Na_2[Fe(CN)_5NO]2H_2O$ (sodium nitroprusside) This is a red crystalline solid that is made by boiling potassium hexacyanoferrate(II) with concentrated nitric acid and then, after dilution, making it alkaline with sodium hydroxide. The complex gives an intense violet colour with alkaline solutions of sulphides, and serves as a test for S^{2-} ions.

f Pentacarbonyl iron(0), $[Fe(CO)_5]$ This is a pale yellow liquid made by passing carbon monoxide under pressure over powdered iron at 150 °C and condensing the vapour. It decomposes on heating to give carbon monoxide and iron, or burns in air to give iron(III) oxide.

g Ferrates(VI) These are salts of the hypothetical acid H_2FeO_4. Potassium ferrate(VI) is the most important; it is made as a deep red solution by treating a freshly made suspension of iron(II) hydroxide in concentrated potassium hydroxide with chlorine or ozone.

Ferrates(VI) contain the tetrahedral FeO_4^{2-} ion, and they are the only simple compounds in which iron exhibits an oxidation number of $+6$. They are strong oxidising agents, similar to, but stronger than, the manganates(VII). Compare the standard electrode potential of 2.2 V for the FeO_4^{2-}/Fe^{3+} couple with that of 1.52 V for the MnO_4^-/Mn^{2+} couple.

COBALT AND ITS COMPOUNDS

27 The metal

The main ores of cobalt are smaltite, $CoAs_2$, and cobaltite, $CoAsS$. Various purification and roasting processes yield cobalt(II) dicobalt(III) oxide, Co_3O_4, which is reduced by heating with aluminium (p. 114).

Cobalt is a hard, lustrous metal, less chemically reactive than iron. It is not affected by moist air, but oxidises slowly on heating, to give cobalt(II) oxide or, at a higher temperature, cobalt(II) dicobalt(III) oxide.

Cobalt is used in making high-speed cutting steels and magnetic steels; it is also a useful catalyst.

28 Simple compounds of cobalt

Iron(II) and iron(III) compounds have similar stabilities and are both well known, but the simple compounds of cobalt(III) are very unstable.

The electrode potential for $Co^{3+}(aq)/Co^{2+}(aq)$ is $+1.82$ V as compared with $+0.77$ V for $Fe^{3+}(aq)/Fe^{2+}(aq)$. This means that $Co^{3+}(aq)$ ions are much stronger oxidising agents than Fe^{3+}(aq) ions; they are, in fact, strong enough to oxidise water to oxygen, so they cannot exist in aqueous solution.

Most cobalt(II) compounds are blue when anhydrous, but the hydrates and aqueous solutions contain pink $[Co(H_2O)_6]^{2+}$ ions.

Cobalt(II) chloride is typical. It is formed as a red hydrate, $CoCl_2.6H_2O$, by reaction between dilute hydrochloric acid and cobalt(II) oxide or hydroxide. The anhydrous salt, which is blue, can be obtained by careful heating of the hydrate or by direct reaction between the metal and chlorine. A dilute solution of the chloride is pink but the colour changes to blue on concentration or on adding hydrochloric acid. This is due to the conversion of pink $[Co(H_2O)_6]^{2+}$ ions into blue $[CoCl_4]^{2-}$ ions:

$$[Co(H_2O)_6]^{2+}(aq) + 4Cl^-(aq) \rightleftharpoons [CoCl_4]^{2-}(aq) + 6H_2O(l)$$
(pink)   (blue)

The change is reversible, and cobalt(II) chloride is used as a test for the presence of water as, for example, in self-indicating silica gel (p. 215).

Similarly, pink $[Co(H_2O)_6]^{2+}$ ions are converted into red–brown $[Co(NH_3)_6]^{2+}$ ions on adding ammonia:

$$[Co(H_2O)_6]^{2+}(aq) + 6NH_3(aq) \rightleftharpoons [Co(NH_3)_6]^{2+}(aq) + 6H_2O(l)$$
(pink)   (red–brown)

Addition of OH^- ions to a solution of cobalt(II) chloride precipitates blue cobalt(II) hydroxide, which forms dark green cobalt(II) oxide on heating. Addition of S^{2-} ions precipitates black cobalt(II) sulphide, and CO_3^{2-} ions give a pink hydrated carbonate, $CoCO_3.6H_2O$.

29 Complexes of cobalt

It is very difficult to oxidise Co^{2+}(aq) ions to Co^{3+}(aq), but cobalt(II) complexes with ligands other than H_2O can be easily oxidised. This is due to the effect of the ligand on the electrode potential values (p. 337). The values below show that the $[Co(H_2O)_6]^{2+}$ ion is a very weak reducing agent, whereas $[Co(CN)_6]^{4-}$ is a strong one.

$[Co(CN)_6]^{3-}(aq)/[Co(CN)_6]^{4-}(aq)$ $E^\ominus = -0.80\,V$
$[Co(NH_3)_6]^{3+}(aq)/[Co(NH_3)_6]^{2+}(aq)$ $E^\ominus = +0.10\,V$
$[Co(H_2O)_6]^{3+}(aq)/[Co(H_2O)_6]^{2+}(aq)$ $E^\ominus = +1.82\,V$

Addition of ammonia to an aqueous solution of a cobalt(II) salt first precipitates the hydroxide, but this reacts with excess NH_3 to form $[Co(NH_3)_6]^{2+}$ ions. On passing air through the solution, particularly in the presence of a catalyst of activated charcoal, the $[Co(NH_3)_6]^{2+}$ ions are oxidised to yellow $[Co(NH_3)_6]^{3+}$ ions. The oxidation can also be brought about by other oxidising agents, such as hydrogen peroxide. In summary

$$[Co(H_2O)_6]^{2+} \xrightarrow[\text{oxidation}]{NH_3+} [Co(NH_3)_6]^{3+}$$
$$\text{(pink)} \qquad\qquad\qquad \text{(yellow)}$$

The conversion of $[Co(H_2O)_6]^{2+}$ into $[Co(CN)_6]^{3-}$, by adding CN^- ions, is even easier.

Similar ligand exchange and oxidation occurs when sodium nitrite is added to an acidified solution of a cobalt(II) salt. The product, sodium hexanitrocobaltate(III), $Na_3[Co(NO_2)_6]$, can be used as a test for K^+ ions, with which it forms an insoluble complex, $K_3[Co(NO_2)_6]$.

NICKEL AND ITS COMPOUNDS

30 The metal

Nickel occurs as pentlandite. This is a sulphide ore, containing mainly nickel(II), iron(II) and copper(II) sulphides, but even a rich ore contains only about 2.5 per cent of nickel, so the separation of nickel from the valuable by-products is a lengthy process. The final stage is the reduction of nickel(II) oxide by carbon into crude nickel, which is refined by electrolysis. Alternatively, in the Mond process, the crude nickel is converted into tetracarbonyl nickel(0), $[Ni(CO)_4]$, by reaction with carbon monoxide at about 50 °C. The carbonyl is then decomposed at about 180 °C,

$$Ni(s) + 4CO(g) \underset{150\,°C}{\overset{50\,°C}{\rightleftharpoons}} [Ni(CO)_4](l)$$

and the carbon monoxide can be recirculated.

Nickel is a hard metal which is quite lustrous. It is relatively unreactive, being unaffected by moist air at room temperature. It will form nickel(II) oxide on heating in air or oxygen, reacts slowly with most strong acids, reacts with steam when red-hot, and combines with halogens and sulphur on heating.

Nickel is used in cladding steel, and in making many alloys, e.g. stainless and heat-resisting steels (nichrome), cupronickel alloys (monel and constantan), alloys with low coefficient of expansion (invar), and nickel–silvers (as used in EPNS).

It is also used in electroplating, often as an undercoat for chromium, and as a catalyst in the hydrogenation of fats and in making synthesis gas. Finely divided nickel is required, and this is obtained by reduction of the oxides or by removal of aluminium from an aluminium–nickel alloy (Raney nickel).

31 Simple compounds of nickel

Nickel(III) compounds are rare and are of doubtful composition. The nickel(II) compounds are generally green, and they contain the $[Ni(H_2O)_6]^{2+}$ ion when hydrated or in solution.

Nickel(II) sulphate is typical. It is made by reaction between nickel(II) oxide, hydroxide or carbonate and dilute sulphuric acid. As a hydrate $NiSO_4.7H_2O$, it is green, and it is a vitriol, isomorphous with iron(II) and cobalt(II) sulphates. It forms double sulphates, e.g. $(NH_4)_2SO_4.NiSO_4.6H_2O$, which are used as electrolytes in nickel plating.

Addition of S^{2-}, CO_3^{2-} or OH^- ions to a solution of nickel(II) sulphate precipitates, respectively, black nickel(II) sulphide, green basic nickel(II) carbonate and light green nickel(II) hydroxide.

Nickel(II) hydroxide forms the positive plate of a nickel–cadmium battery, the negative plate being cadmium hydroxide, and the electrolyte potassium hydroxide solution. On charging, the nickel(II) hydroxide is converted into hydrated nickel(III) oxide, and the cadmium hydroxide into cadmium. The reverse changes take place on discharge:

$$2Ni(OH)_2 + Cd(OH)_2 \xrightleftharpoons[\text{discharge}]{\text{charge}} Ni_2O_3.H_2O + Cd + 2H_2O$$

32 Complexes of nickel

Nickel complexes usually have nickel in an oxidation state of $+2$. Typical examples are listed below.

$[Ni(H_2O)_6]^{2+}$	octahedral	green
$[Ni(NH_3)_6]^{2+}$	octahedral	deep blue
$[Ni(CN)_4]^{2-}$	square	orange
$[NiCl_4]^{2-}$	tetrahedral	blue
$[Ni(CO)_4]$	tetrahedral	colourless

The hexaammine, hexacyano and hexachloro complexes are made from solutions of nickel(II) salts by adding excess NH_3, CN^- ions or Cl^- ions respectively. Tetracarbonylnickel(0) is a neutral complex, in which nickel has an oxidation state of zero.

The complex of Ni^{2+} with butanedione dioxime (dimethylglyoxime), which is red and insoluble, provides a convenient qualitative and quantitative test for Ni^{2+} ions (p. 337).

COPPER AND ITS COMPOUNDS

33 Extraction, properties and uses of copper

Copper is extracted from sulphide ores such as copper pyrites, $CuFeS_2$. A series of operations eventually produces an impure form of the metal, which is made into anodes in an electrolytic cell containing pure copper cathodes and acidified copper(II) sulphate solution. On electrolysis, the copper passes from the anode to the cathode. Valuable impurities, e.g. selenium, tellurium, silver and gold, can be extracted from the impurities in the anode, which collect as an anode slime.

Copper is a tough, malleable and ductile metal with a characteristic colour. It reacts slowly with air at room temperature, being tarnished by the formation of oxide and sulphide films. At higher temperatures it reacts with oxygen to form mixtures of copper(I) and copper(II) oxides. Copper is not attacked by water or steam, but, on exposure to moist air for long periods, it becomes coated with a green film of basic sulphate and basic carbonate, known as *verdigris*. The metal does not react with non-oxidising acids, except in the presence of air, but does react with oxidising acids, the product depending on the conditions used. Hot concentrated sulphuric acid gives sulphur dioxide; cold 50 per cent nitric acid gives nitrogen monoxide; and hot concentrated nitric acid gives nitrogen dioxide:

$$Cu(s) + 2H_2SO_4(aq) \longrightarrow CuSO_4(aq) + SO_2(g) + 2H_2O(l)$$
$$3Cu(s) + 8HNO_3(aq) \longrightarrow 3Cu(NO_3)_2(aq) + 2NO(g) + 4H_2O(l)$$
$$Cu(s) + 4HNO_3(aq) \longrightarrow Cu(NO_3)_2(aq) + 2NO_2(g) + 2H_2O(l)$$

Copper can be rolled into sheets, hammered into foil, and drawn into wire and piping. As it also has high electrical and thermal conductivities, it is widely used in the electrical industry, and in making boilers and boiler tubes. Because it is resistant to corrosion, it is also used in roofing and other building operations.

Copper is also an important component of many alloys, e.g. brasses (20–40% Zn, 60–80% Cu), bronzes (90–95% Cu, 5–10% Sn) and cupro-nickels (75% Cu, 25% Ni).

34 Copper(I) compounds

Simple covalent compounds of copper(II) are generally less thermally stable than the corresponding copper(I) compounds. Where compounds with weakly electronegative groups are concerned, e.g. copper(II) iodide and copper(II) cyanide, the decomposition into the corresponding copper(I) compound is spontaneous at room temperature:

$$2CuI_2(s) \longrightarrow 2CuI(s) + I_2(s)$$
$$2Cu(CN)_2(s) \longrightarrow 2CuCN(s) + C_2N_2(g)$$

With more electronegative groups, the decomposition requires a higher temperature, but many copper(II) compounds change into the corresponding copper(I) compounds on strong heating, e.g.

$$2CuCl_2(s) \longrightarrow 2CuCl(s) + Cl_2(g)$$
$$4CuO(s) \longrightarrow 2Cu_2O(s) + O_2(g)$$
$$2CuS(s) \longrightarrow Cu_2S(s) + S(s)$$

For simple ionic compounds the situation is different because Cu^+ ions disproportionate in aqueous solution (p. 93 and Fig. 47, p. 92) and the equilibrium

$$2Cu^+(aq) \rightleftharpoons Cu^{2+}(aq) + Cu(s)$$

lies well to the right. As a result, only very low concentrations of Cu^+ ions can exist in aqueous solutions. That is why the only common copper(I) compounds are the insoluble ones; water soluble ones must be made in non-aqueous solutions.

a Copper(I) oxide, Cu_2O This dark red, insoluble solid is made by partially reducing alkaline copper(II) solutions. It is necessary to prevent the initial formation of a precipitate of copper(II) hydroxide, and this is done by using Fehling's solution (p. 336) Addition of a suitable reducing agent, e.g. ethanal or reducing sugars, forms copper(I) oxide; the reaction is used as a qualitative and quantitative test for reducing sugars.

Copper(I) oxide is basic and covalent. It reacts with acids to form copper(I) salts, but these immediately decompose into the copper(II) salts unless they are insoluble, e.g.

dark red
$$Cu_2O(s) + 2HCl(aq) \longrightarrow 2CuCl(s) + H_2O(l)$$
$$\text{(insoluble)}$$

$$Cu_2O(s) + H_2SO_4(aq) \longrightarrow Cu_2SO_4(aq) + H_2O(l)$$
$$\downarrow$$
$$CuSO_4(aq) + Cu(s)$$

b Copper(I) sulphate, Cu_2SO_4 Dimethyl sulphate reacts with copper(I) oxide to form copper(I) sulphate,

$$Cu_2O(s) + (CH_3)_2SO_4(l) \longrightarrow Cu_2SO_4(s) + (CH_3)_2O(l)$$

but this decomposes in water, giving copper(II) sulphate and copper.

c Copper(I) chloride, CuCl Boiling copper(II) chloride solution with excess copper and concentrated hydrochloric acid, and pouring the mixture into air-free water precipitates copper(I) chloride:

$$CuCl_2(aq) + Cu(s) \longrightarrow 2CuCl(s)$$

Alternatively, sulphur dioxide and other reducing agents can be used to reduce copper(II) chloride.

Copper(I) chloride is a white, insoluble solid with a zinc blende structure; density measurements show that it has a formula of Cu_2Cl_2 in the vapour state. It turns green on exposure to air, due to the formation of some copper(II) chloride and some basic copper(II) chloride. It is used as a catalyst.

35 Copper(II) compounds

a Copper(II) oxide, CuO   This is a black, insoluble covalent solid, and is made by heating the metal in air or oxygen, or by heating the nitrate or carbonate.

It is a basic oxide, giving copper(II) salts with acids, and it is readily reduced to copper by heating with carbon, hydrogen or many organic chemicals. It melts at about 1150 °C and decomposes to copper(I) oxide and oxygen at higher temperatures.

b Copper(II) hydroxide, $Cu(OH)_2$ Addition of ammonia to a boiling solution of copper(II) sulphate produces first a green, and then a blue, precipitate. If the blue precipitate is warmed with sodium hydroxide solution, copper(II) hydroxide is formed. Mixing sodium hydroxide and copper(II) sulphate solutions gives basic copper(II) sulphate, $CuSO_4.3Cu(OH)_2$.

Copper(II) hydroxide is a pale blue solid, which is insoluble in water but, because it is slightly amphoteric, is soluble in acids and in concentrated alkaline solutions. It also dissolves in aqueous ammonia to form $[Cu(NH_3)_4(H_2O)_2]^{2+}$ ions.

c Copper(II) carbonate   This is only known as a basic salt. It occurs naturally as malachite, $CuCO_3.Cu(OH)_2$, and as azurite, $Cu(OH)_2.2CuCO_3$, and is precipitated on mixing sodium carbonate and copper(II) salt solutions.

The carbonate is a blue–green, insoluble solid, which decomposes into copper(II) oxide on heating.

d Copper(II) sulphide, CuS   This is made by gently heating powdered copper with excess sulphur, or by passing hydrogen sulphide into an acidic solution of a copper(II) salt.

It is a black, insoluble solid, which is slowly oxidised in moist air to the sulphate, and converted into copper(I) sulphide and sulphur on strong heating.

e Copper(II) chloride, $CuCl_2$ Hot copper reacts with dry chlorine to form anhydrous copper(II) chloride, which is dark brown; and a green hydrate, $CuCl_2.2H_2O$, is made by reaction between hydrochloric acid and copper(II) oxide. The hydrate can be converted into the anhydrous salt by heating in a stream of hydrogen chloride. Stronger heating of the copper(II) chloride converts it into copper(I) chloride and chlorine.

A dilute solution of copper(II) chloride in water is blue, due to the presence of the $[Cu(H_2O)_6]^{2+}$ ion. If excess concentrated hydrochloric acid is added, yellow $[CuCl_4]^{2-}$ ions are formed:

$$[Cu(H_2O)_6]^{2+}(aq) + 4Cl^-(aq) \longrightarrow [CuCl_4]^{2-}(aq) + 6H_2O(l)$$
(blue)   (yellow)

With a lower concentration of Cl^- ions, both the coloured complexes will be present, so the solution will look green.

f Copper(II) sulphate, $CuSO_4.5H_2O$ (blue vitriol) This is prepared in the laboratory by adding dilute sulphuric acid to copper(II) oxide, hydroxide or carbonate. Industrially, it is obtained by reacting scrap copper with hot dilute sulphuric acid and blowing air through the mixture. The air oxidises the copper and the resulting oxide reacts with the acid:

$$2Cu(s) + 2H_2SO_4(aq) + O_2(g) \longrightarrow 2CuSO_4(aq) + 2H_2O(l)$$

The pentahydrate is blue, but it loses four molecules of water of crystallisation at about 100°C, and the white, anhydrous salt is formed above 300°C. The anhydrous salt readily picks up water to re-form the coloured hydrate, and this provides a convenient test to detect the presence of water. The structure of the pentahydrate involves hydrogen bonds and is described on p. 54.

Copper(II) sulphate decomposes on heating:

$$CuSO_4(s) \longrightarrow CuO(s) + SO_3(g)$$

The sulphate is the most important copper(II) compound. It is used in electroplating, as a timber preservative, in treating water, and in dyeing. Most of it is used, however, as an agricultural fungicide, e.g. *Bordeaux mixture* ($CuSO_4$ and $Ca(OH)_2$) and *Burgundy mixture* ($CuSO_4$ and Na_2CO_3).

36 Complexes of copper

a Copper(I) complexes Copper(I) chloride will form complexes with NH_3 and with CN^- ligands, i.e.

$$[Cu(NH_3)_2]^+ \qquad [Cu(CN)_4]^{3-}$$

Unlike $Cu^+(aq)$, both these complexes are stable in solution, and this is another example of the stabilising influence of NH_3 and CN^- ligands (p. 337).

Solutions that contain the diamminecopper(I) ion will absorb carbon monoxide, and will react with ethyne to precipitate red, explosive copper(I) dicarbide, Cu_2C_2.

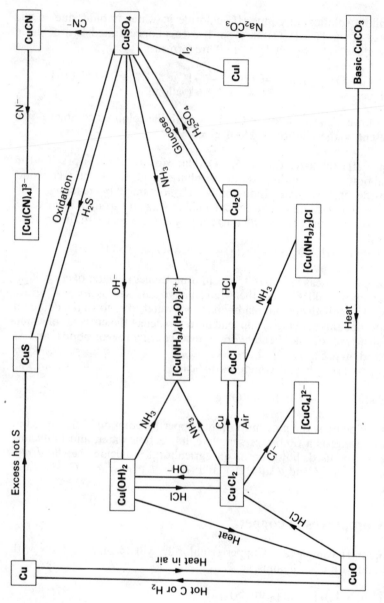

Chart 17 Copper and its simple compounds.

b Copper(II) complexes The ammines are the commonest copper(II) complexes. Addition of NH_3 to solutions of copper(II) salts gives a mixture of $[Cu(NH_3)_{6-x}(H_2O)_x]^{2+}$ ions, with x having a value of 5 when the concentration of NH_3 is low, and a value of 1 when it is high. To get $[Cu(NH_3)_6]^{2+}$ ions, it is necessary to treat a copper(II) salt with liquid ammonia.

Tetraamminecopper(II) sulphate-1-water, $[Cu(NH_3)_4]SO_4.H_2O$, is made by adding excess ammonia solution to copper(II) sulphate solution. Addition of ethanol precipitates the tetraammine as a deep purple solid. Its aqueous solution has the unusual property of dissolving cellulose, and is used in the 'cuprammonium' process for manufacturing rayon.

$[CuCl_4]^{2-}$ and other chloro-complexes are also well known, and Fehling's solution contains a copper(II) complex (p. 336).

ZINC AND ITS COMPOUNDS

37 The extraction of zinc

Zinc blende, ZnS, is the main source of zinc, but the extraction of the metal is complicated by the fact that the ore generally contains considerable amounts of lead sulphide, and lesser amounts of cadmium sulphide and other impurities.

The ore is roasted, and this converts the zinc sulphide into the oxide, which is then reduced to the metal by heating with coke. The zinc is led off as a vapour, which is then cooled. Redistillation is used to purify the zinc, and 99.99 per cent pure metal is available commercially.

Where there are cheap supplies of electricity, zinc is also manufactured by the electrolysis of zinc sulphate solution, the solution being made from zinc oxide and sulphuric acid.

38 Properties and uses of zinc

Zinc is a bluish-white, moderately hard, brittle metal. It is chemically reactive, displacing hydrogen from dilute acids and from alkalis,

$$Zn(s) + 2H^+(aq) \longrightarrow Zn^{2+}(aq) + H_2(g)$$
$$Zn(s) + 2OH^-(aq) + 2H_2O(l) \longrightarrow [Zn(OH)_4]^{2-}(aq) + H_2(g)$$

and reacting with steam (to form an oxide), with halogens (to form halides) and with sulphur (to form a sulphide). In moist air it becomes coated with a layer of basic carbonate, $ZnCO_3.Zn(OH)_2$, which protects it.

Zinc is used as a protective coating for steel. In galvanisation, the steel is cleaned by dipping it in sulphuric acid and then coated by dipping it in molten zinc or by electroplating. *Sherardised* iron is made by heating iron or steel articles with zinc dust in rotating drums. Zinc coating may also be applied by spraying.

The main zinc alloy is brass (p. 384) but other zinc alloys, containing small amounts of aluminium and copper, have low melting temperatures and are useful in making die-cast articles. Zinc is also used in making dry batteries, lithographic plates, foil for packing, and as a reducing agent.

39 Compounds of zinc

The only characteristic that zinc has in common with the transition metals is its ability to form complexes. Otherwise, it is very different from them, in so far as it exerts a fixed oxidation state of $+2$, its compounds are colourless unless they contain a coloured anion, and the Zn^{2+} ion is not paramagnetic.

a Zinc oxide, ZnO   This is made by roasting zinc ores or by burning zinc metal. It is a white, insoluble powder which turns yellow on heating.

Zinc oxide is amphoteric, and it will react with acid and alkalis:

$$ZnO(s) + 2H^+(aq) \longrightarrow Zn^{2+}(aq) + H_2O(l)$$
$$ZnO(s) + 2OH^-(aq) + H_2O(l) \longrightarrow [Zn(OH)_4]^{2-}(aq)$$

It is the most important zinc compound, being used as a white pigment or filler. Zinc sulphide is also white, so that reaction of atmospheric hydrogen sulphide with the pigment does not cause discoloration, as it does with lead pigments. Zinc oxide is also used medicinally in zinc ointments, and in lotions and powders.

b Zinc hydroxide, Zn(OH)₂   This is precipitated on adding a little sodium hydroxide solution to a solution of a zinc salt. It is a white, insoluble solid, and is amphoteric, like the oxide:

$$Zn(OH)_2(s) + 2H^+(aq) \longrightarrow Zn^{2+}(aq) + 2H_2O(l)$$
$$Zn(OH)_2(s) + 2OH^-(aq) \longrightarrow [Zn(OH)_4]^{2-}(aq)$$

It is also soluble in excess ammonia solution, forming an ammine:

$$Zn(OH)_2(s) + 4NH_3(aq) \longrightarrow [Zn(NH_3)_4]^{2+}(aq) + 2OH^-(aq)$$

c Zinc carbonate, ZnCO₃ (calamine)   This occurs naturally or can be made by reaction between solutions of sodium hydrogencarbonate and zinc salts:

$$Zn^{2+}(aq) + 2HCO_3^-(aq) \longrightarrow ZnCO_3(s) + H_2O(l) + CO_2(g)$$

If sodium carbonate is used, basic zinc carbonate is precipitated.

Zinc carbonate is a white, insoluble, typical carbonate. It is used, as a suspension of the basic carbonate and tinged with iron(III) oxide, in calamine lotion for the treatment of sore skin.

d Zinc chloride, $ZnCl_2$ Anhydrous zinc chloride is made from dry chlorine or hydrogen chloride and zinc. In the hydrated form, it is made by treating the metal, oxide, hydroxide or carbonate with dilute hydrochloric acid.

Anhydrous zinc chloride is very deliquescent, and it is soluble in many organic solvents. As it is also a poor conductor in aqueous solutions, and has a low melting temperature, its structure displays considerable covalent character.

Zinc chloride is used as a wood preservative and as a catalyst or dehydrating agent in organic reactions.

e Zinc sulphate, $ZnSO_4$ (white vitriol) The metal, oxide, hydroxide and carbonate react with dilute sulphuric acid to form zinc sulphate, which crystallises as the heptahydrate at room temperature. At higher temperatures other hydrates form, and at 450 °C, the anhydrous salt is formed. On stronger heating, it decomposes to give zinc oxide and sulphur(VI) oxide.

f Zinc sulphide, ZnS (zinc blende) This occurs naturally or can be made by direct combination of heated zinc and sulphur, or by precipitation from zinc salt solutions with hydrogen sulphide or ammonium sulphide. Zinc sulphide is used in paints and, when impure, it is luminous.

g Complexes of zinc Solutions of zinc salts contain the colourless, octahedral $[Zn(H_2O)_6]^{2+}$ ion, but tetrahedral complexes are more common. Typical examples are

$$[Zn(NH_3)_4]^{2+} \qquad [Zn(OH)_4]^{2-} \qquad [Zn(CN)_4]^{2-}$$

They are formed by adding excess NH_3, OH^- ions or CN^- ions, respectively, to solutions of zinc(II) salts. They are all colourless.

Questions on Chapters 25 and 26

1 What is meant by a transition element? Summarise, with examples, the distinctive properties of transition elements and show how they are related to the electronic structure of these elements. (L)

2 To what extent should (a) copper and (b) zinc be regarded as transition elements?

3 'The general trend across a transition series is for a decrease in the ease of oxidation of a lower to a higher oxidation state'. Illustrate and discuss.

4 'For each transition metal, the lowest oxidation states are basic, the

oxidation state of $+4$ is usually amphoteric, whilst the higher oxidation states, if they exist, are acidic.' Illustrate and discuss.

5 Give an account of the uses of the metals in the first transition series.

6 Outline how you would prepare in the laboratory a crystalline compound of a metal in which the oxidation state (or valency) of the metal is (a) three, (b) six and (c) seven. You must start each preparation either from the metal itself or from a common compound in an oxidation state (or valency) other than that asked for. A different metal is to be used for each of the cases (a), (b) and (c). What would happen if hydrogen sulphide was passed through an acid solution of each of the compounds you have prepared? (L)

7 Explain the meaning of the terms transition metal and complex ion, illustrating your answers by reference to *one* of the following elements: chromium, manganese, iron or copper. Discuss *two* examples of the use of complex ion formation in qualitative analysis. (OC)

8 Write a comparative account of the chemistry of any three transition elements of your own choice, selected from the series titanium to copper inclusive. In your essay, provide illustrative examples of principal oxidation states and the formation of coloured ions, and describe the preparation and structure of two complexes, of each of the metals chosen. (SUJB)

9 (a) Describe briefly how any one transition element is extracted from one of its ores. (b) How would you prepare from any one transition metal a compound in which the metal forms part of a complex ion? (c) How would you prepare from any one transition metal a compound in which the metal forms part of an anion? (OC)

10 Explain (a) why the Cu^{2+} ion is colourless whilst the $[Cu(H_2O)_6]^{2+}$ ion is blue and that of $[Cu(NH_3)_4(H_2O)_2]^{2+}$ is deeper blue, (b) why the $[Co(H_2O)_6]^{2-}$ ion is pink whilst that of $[CoCl_4]^{2-}$ is blue.

11 Discuss the chemical characteristics of two salts whose anions contain a metal. Give an account of their preparation or manufacture, and of their uses in analysis. (OC)

12 For either chromium or manganese or nickel or copper, describe briefly (a) how you would prepare a compound of the metal in its lowest valency state, (b) how you would prepare a salt of the metal in its highest oxidation state, (c) one use of the salt described in (b). (OC)

13 Describe how you would carry out the following conversions: (a) sulphur to sodium thiosulphate(VI), (b) copper to copper(I) oxide, (c) chromium (III) oxide to potassium dichromate(VI). (SUJB)

14 Compare and contrast the chemistry of (a) iron and cobalt, (b) cobalt and nickel, (c) nickel and copper, (d) chromium and manganese.

15 Give an account of the reactions of compounds of (a) iron and (b) silver which can be used for (i) the qualitative and (ii) the quantitative determination of *other* elements or radicals. (OC)

16 Describe briefly the extraction of titanium from its ores. Why is the process which you describe used in preference to other methods of extraction? Give two important uses of titanium and/or its compounds. (OC)

17 Discuss the electronic basis for the classification of iron and copper as transition elements. To what extent is the relative position of these two elements in the first transition series useful as a guide to their chemistry? (C)

18 Describe briefly how zinc may be extracted from one common ore. With zinc as starting material, how would you prepare a sample of (a) anhydrous zinc(II) chloride, (b) zinc(II) carbonate, (c) brass? How, and under what conditions, does zinc(II) chloride react with (i) water, (ii) ammonia, (iii) sodium hydroxide, (iv) hydrogen sulphide? (S)

19 Write an account of the properties of the first row transition metals by considering, with examples wherever possible, (a) the physical properties which are characteristic of transition metals, (b) the common oxidation states in aqueous solution for each element, (c) the formation of, and bonding in, complexes, (d) the reactions of complexes, restricted to substitution reactions of hydrated ions. (JMB)

20 (a) Explain what is meant by each of the following, illustrating your answer with appropriate examples from the chemistry of the elements in the period scandium to zinc: (i) oxidation state or number, (ii) transition element, (iii) disproportionation, (iv) complex ion, (v) coordination numbers in complex ions. (b) Discuss the effect of change of ligand on the oxidation–reduction behaviour of cobalt ions in aqueous solution. (JMB)

21 Outline the physico-chemical principles underlying the operation of the blast furnace, commenting on the Ellingham diagrams for carbon and iron. Give an account of the uses of iron, emphasising the advantages and disadvantages in each case. Give the structure of iron(III) chloride. Describe, giving essential experimental details, a method for obtaining a pure sample of anhydrous iron(III) chloride from iron. (L)

22 Give concise explanations for the following: (a) Manganese can exist in oxidation states up to a maximum of $+7$, whereas scandium has no oxidation state higher than $+3$. (b) In aqueous solution, the scandium(III) ion is colourless whereas the iron(II) ion is coloured. (c) The chromium(II) ion is a better reducing agent than the manganese(II) ion. (d) The electron configuration of chromium is not $Ar\,3d^4 4s^2$ as might be expected, but $Ar\,3d^5 4s^1$. (e) The compound $Pt(NH_3)_2Cl_2$ and the ion $[Cr(NH_3)_4Cl_2]^+$ each exist in two isomeric forms. (L)

23 For each of the elements, sodium, sulphur and chromium, (a) describe a convenient method for preparing its chloride or one of its chlorides (you are expected to give a different method for each), (b) discuss the effect of water on these chlorides, and (c) give an account of the acid–base nature of the oxides of these elements. The three elements named are in different blocks of the periodic table. To what extent may the properties which you have just discussed be correlated with the block they are in? (L)

24 (a) You are provided with a large number of test-tubes and with nine unlabelled bottles and it is known that these contain approximately $2\,mol\,dm^{-3}$ solutions of copper nitrate, copper sulphate, aluminium sulphate, sodium hydroxide, sodium carbonate, barium chloride, sulphuric acid, potassium iodate, potassium iodide. Using no other apparatus or chemicals, devise a scheme whereby you could positively identify the contents of all the bottles. (b) Give examples from the chemistry of copper to illustrate and explain the following concepts: variable oxidation states, non-stoicheiometry, formation of complex ions and the dependence of their colour on the surrounding ligands. (W)

25 (a) Discuss the assertion that the division of the elements in the periodic table into metals and non-metals is an arbitrary one. (b) Describe the amphoteric behaviour of aluminium and zinc and discuss the suggestion that amphoteric behaviour is a general property of the metals, other than those in Groups 1 and 2 of the periodic table. (W)

26 The standard electrode potential of the Fe^{3+}(aq), Fe^{2+}(aq)| Pt electrode is $+0.77$ volt. Explain the meaning of this statement and describe how the value can be determined in the laboratory. Include essential experimental detail.

Trends in the chemical properties of the elements in the periodic table can, in some cases, be illustrated by standard electrode potential values. Use the values given in this question to illustrate *three* such trends and as far as possible show that the trends you describe can also be related to the changes in the atomic properties of the elements involved. [Standard electrode potentials (volts): $Cr^{3+}_{(aq)}$, $Cr^{2+}_{(aq)}$| Pt, -0.41; $Mg^{2+}_{(aq)}$| Mg, -2.38; $Ba^{2+}_{(aq)}$| Ba, -2.89; Cl_2| $Cl^-_{(aq)}$| Pt, $+1.36$.] (W)

27 (a) Explain why the lattice energy of zinc sulphide is different from that based on a purely ionic model. (b) Give the reactions for the production of zinc from its sulphide and mention *briefly* the reaction conditions. If the enthalpy changes of formation (kJ mol^{-1}) for ZnO(s) and CO(g) are -348.1 and -110.5 respectively, deduce the enthalpy change for (i) the reduction of zinc oxide with carbon and (ii) for the decomposition of zinc oxide to zinc and oxygen. Give the effect of temperature on the reactions and comment on why reaction (i) and not (ii) is used in the extraction of zinc. (c) Discuss similarities and differences in the chemistry of zinc and copper. Include in your answer the respective electronic configurations, the valencies in their compounds, the characteristics of their ions in solution including reactions of the ions with aqueous ammonia, potassium iodide, hydrochloric acid and potassium cyanide. (W)

28 (a) Describe and discuss the Haber process for the manufacture of ammonia. Mention the sources of the raw materials. (b) Describe and explain what happens when an aqueous solution of ammonia is added slowly, until in excess, to separate aqueous solutions of the following: (i) copper(II) sulphate, (ii) cobalt(II) chloride, (iii) silver nitrate. What use is made in the laboratory of the final product in the reaction in (iii)? (JMB)

29 Complex ions are important in chemistry. In *each* of the following examples, give the oxidation state (number) of the central atom (metal). (a) By describing *two* typical reactions of *each* complex ion, illustrate the role of (i) $[Al(H_2O)_6]^{3+}$ in the chemistry of aluminium and (ii) $[Fe(H_2O)_6]^{3+}$ in the chemistry of iron. (b) Give *one* example of the use of a complex hydride, e.g. $[BH_4]^-$, as a reducing agent. (c) Describe the shape of $[FeCl_4]^-$ and of $[Ni(CN)_4]^{2-}$. (d) Explain how the following ions are important for the identification of the respective species: (i) $[Ag(NH_3)_2]^+$ for chloride ions, (ii) $[Fe(CN)_6]^{3-}$ for ions of iron. (W)

30 (a) State *five* properties which are typical of the transition elements and illustrate each with a specific example. (b) For *each* of the oxidation numbers $+2$, $+3$, $+4$, $+6$ and $+7$, write the formula of *one* compound or ion in which the transition element *either* chromium *or* manganese has that oxidation number. (c) For *each* of the five examples chosen in (b), name a reagent which will change the oxidation number and write the redox equation for the reaction. (AEB)

Appendix I
The detection of simple cations

1 Aluminium, Al^{3+} (a) Aqueous solution plus NaOH(aq) gives a white precipitate, which is soluble in excess NaOH.
(b) Aqueous solution plus dilute HCl plus ammonium ethanoate solution plus aluminon gives a red colour or precipitate on standing.

2 Ammonium, NH_4^+ Solid or solution warmed with NaOH(aq) gives NH_3 gas.

3 Barium, Ba^{2+} Aqueous solution plus K_2CrO_4(aq) gives a yellow precipitate.

4 Calcium, Ca^{2+} Aqueous solution plus ammonium ethanedioate solution gives a white precipitate.

5 Chromium(III), Cr^{3+} Boil aqueous solution with NaOH(aq) and H_2O_2(aq). Acidify with ethanoic acid and add $Pb(NO_3)_2$(aq). A yellow precipitate forms.

6 Cobalt, Co^{2+} Heat borax on wire in bunsen flame to get a colourless bead. Reheating with a little solid gives a blue bead.

7 Copper(II), Cu^{2+} Aqueous solution plus $K_4[Fe(CN)_6]$(aq) gives a chocolate-brown precipitate.

8 Iron(II), Fe^{2+} Aqueous solution plus $K_3[Fe(CN)_6]$(aq) gives a blue colour or precipitate.

9 Iron(III), Fe^{3+} Aqueous solution plus NH_4SCN(aq) gives a blood-red colour.

10 Lead, Pb^{2+} Aqueous solution plus K_2CrO_4(aq) or KI(aq) gives a yellow precipitate.

11 Magnesium, Mg^{2+} To a slightly acidic solution, add two to three drops of magneson and then NaOH(aq) until alkaline. A blue precipitate forms.

12 Manganese(II), Mn^{2+} Aqueous solution plus NaOH(aq) gives a white precipitate, which slowly turns brown or black.

13 Mercury(II), Hg^{2+} Aqueous solution plus KI(aq) gives reddish precipitate, which is soluble in excess KI.

14 Nickel(II), Ni^{2+}   Make solution just alkaline with NH_3(aq) and add butanedione dioxime (dimethyl glyoxime). A red precipitate forms.

15 Potassium, K^+   Aqueous solution plus $Na_3[Co(NO_2)_6]$(aq) gives a yellow precipitate.

16 Silver, Ag^+   Aqueous solution plus dilute HCl gives a white precipitate, which is soluble in excess NH_3(aq).

17 Sodium, Na^+   Aqueous solution plus zinc uranyl ethanoate solution gives a yellow precipitate.

18 Zinc, Zn^{2+}   To an aqueous solution, add solid NH_4Cl and then NH_3(aq) until alkaline. Pass H_2S(g) through. A white precipitate forms.

Appendix II
The detection of simple anions

1 Bromide, Br^- (a) Heat solid with concentrated H_2SO_4. Fumes of HBr and reddish Br_2 vapour formed.

(b) Aqueous solution plus 1,1,1-trichloroethane plus chlorine water gives a red–brown colour in the organic layer.

2 Carbonate, CO_3^{2-} Solid or solution warmed with dilute HCl gives CO_2 gas.

3 Chloride, Cl^- (a) Heat solid with solid $K_2Cr_2O_7$ and concentrated H_2SO_4. Red fumes of CrO_2Cl_2 evolved, which give a yellow precipitate of $PbCrO_4$ on passing into aqueous $Pb(NO_3)_2$.

(b) Aqueous solution plus dilute HNO_3 plus $AgNO_3(aq)$ gives a white precipitate of AgCl, which is soluble in $NH_3(aq)$.

4 Chromate(VI), CrO_4^{2-} (a) Aqueous solution plus aqueous $Pb(NO_3)_2$ gives a yellow precipitate of $PbCrO_4$.

(b) Aqueous solution plus $H_2O_2(aq)$ plus dilute H_2SO_4 plus ethoxyethane gives a deep blue colour of $CrO(O_2)_2$ in the ethereal layer.

5 Dichromate(VI), $Cr_2O_7^{2-}$ Aqueous solution plus $H_2O_2(aq)$ plus dilute H_2SO_4 plus ethoxyethane gives a deep blue colour in the ethereal layer.

6 Iodide, I^- (a) Heat solid with concentrated H_2SO_4. Fumes of HI and purple I_2 vapour evolved.

(b) Aqueous solution plus 1,1,1-trichloroethane plus chlorine water gives a purple colour in organic layer.

7 Nitrate, NO_3^- Aqueous solution plus $FeSO_4(aq)$ plus concentrated H_2SO_4, to form a lower layer. A brown ring, containing $[Fe(NO)(H_2O)_5]^{2+}$ ions, forms at the junction.

8 Nitrite, NO_2^- Aqueous solution plus $FeSO_4(aq)$ gives a brown colour containing $[Fe(NO)(H_2O)_5]^{2+}$ ions.

9 Phosphate, PO_4^{3-} Aqueous solution plus concentrated HNO_3 plus ammonium molybdate(VI) solution. A yellow colour or precipitate forms on boiling.

10 Sulphate, SO_4^{2-} Aqueous solution plus dilute HCl plus $BaCl_2(aq)$ gives a white precipitate of $BaSO_4$.

11 Sulphide, S^{2-} (a) Warm solid with dilute HCl. H_2S gas evolved.

(b) Aqueous solution plus NaOH(aq) plus solution of disodium pentacyanonitrosylferrate(III) (p. 380), gives a violet colour.

12 Sulphite, SO_3^{2-} Solid or solution warmed with dilute HCl gives SO_2 gas.

13 Thiosulphate, $S_2O_3^{2-}$ (a) Solid or solution warmed with dilute HCl gives SO_2 gas plus yellow precipitate of sulphur.

(b) Aqueous solution will decolorise dilute solution of I_2 in aqueous KI.

Appendix III
Standard electrode potentials at 25 °C

oxidising agent	reducing agent	$E^\ominus$/V
$Li^+(aq)+e^-$	$Li(s)$	-3.045
$K^+(aq)+e^-$	$K(s)$	-2.929
$Ba^{2+}(aq)+2e^-$	$Ba(s)$	-2.90
$Ca^{2+}(aq)+2e^-$	$Ca(s)$	-2.87
$Na^+(aq)+e^-$	$Na(s)$	-2.714
$Mg^{2+}(aq)+2e^-$	$Mg(s)$	-2.37
$Al^{3+}(aq)+3e^-$	$Al(s)$	-1.67
$Mn^{2+}(aq)+2e^-$	$Mn(s)$	-1.18
$Zn^{2+}(aq)+2e^-$	$Zn(s)$	-0.763
$Cr^{3+}(aq)+3e^-$	$Cr(s)$	-0.74
$Fe^{2+}(aq)+2e^-$	$Fe(s)$	-0.44
$Cr^{3+}(aq)+e^-$	$Cr^{2+}(aq)$	-0.41
$Co^{2+}(aq)+2e^-$	$Co(s)$	-0.28
$Ni^{2+}(aq)+2e^-$	$Ni(s)$	-0.25
$Sn^{2+}(aq)+2e^-$	$Sn(s)$	-0.14
$Pb^{2+}(aq)+2e^-$	$Pb(s)$	-0.13
$H^+(aq)+e^-$	$\frac{1}{2}H_2(g)$	0.00
$Sn^{4+}(aq)+2e^-$	$Sn^{2+}(aq)$	$+0.15$
$Cu^{2+}(aq)+e^-$	$Cu^+(aq)$	$+0.153$
$Cu^{2+}(aq)+2e^-$	$Cu(s)$	$+0.3$
$[Fe(CN)_6]^{3-}(aq)+e^-$	$[Fe(CN)_6]^{4-}(aq)$	$+0.36$
$Cu^+(aq)+e^-$	$Cu(s)$	$+0.52$
$\frac{1}{2}I_2(\text{in KI aq})+e^-$	$I^-(aq)$	$+0.54$
$O_2(g)+2H^+(aq)+2e^-$	$H_2O_2(aq)$	$+0.68$
$Fe^{3+}(aq)+e^-$	$Fe^{2+}(aq)$	$+0.77$
$Ag^+(aq)+e^-$	$Ag(s)$	$+0.80$
$Hg^{2+}(aq)+e^-$	$\frac{1}{2}Hg_2^{2+}(aq)$	$+0.92$
$\frac{1}{2}Br_2(aq)+e^-$	$Br^-(aq)$	$+1.07$
$\frac{1}{2}O_2(g)+2H^+(aq)+2e^-$	$H_2O(l)$	$+1.23$
$\frac{1}{2}Cr_2O_7^{2-}(aq)+7H^+(aq)+3e^-$	$Cr^{3+}(aq)+\frac{7}{2}H_2O(l)$	$+1.33$
$\frac{1}{2}Cl_2(aq)+e^-$	$Cl^-(aq)$	$+1.36$
$MnO_4^-(aq)+8H^+(aq)+5e^-$	$Mn^{2+}(aq)+4H_2O(l)$	$+1.52$
$MnO_4^-(aq)+4H^+(aq)+3e^-$	$MnO_2(s)+2H_2O(l)$	$+1.6$
$Pb^{4+}(aq)+2e^-$	$Pb^{2+}(aq)$	$+1.70$
$\frac{1}{2}H_2O_2(aq)+H^+(aq)+e^-$	$H_2O(l)$	$+1.77$
$Co^{3+}(aq)+e^-$	$Co^{2+}(aq)$	$+1.82$
$\frac{1}{2}S_2O_8^{2-}(aq)+e^-$	$SO_4^{2-}(aq)$	$+2.01$
$\frac{1}{2}F_2(aq)+e^-$	$F^-(aq)$	$+2.87$

increasing strength as oxidising agent (left column, downward)

increasing strength as reducing agent (middle column, upward)

Appendix IV
Some thermodynamic data at 25 °C

	$\Delta H^{\ominus}/\text{kJ mol}^{-1}$	$\Delta G^{\ominus}/\text{kJ mol}^{-1}$	$S^{\ominus}/\text{J mol}^{-1}\,\text{K}^{-1}$
$Ag^+(aq)$	105.9	77.1	73.9
$AgCl(s)$	-127	-109.6	96.1
$Al^{3+}(g)$	5484		
$Al^{3+}(aq)$	-531	-481	-313.3
$AlCl_3(s)$	-695	-637	170
$Al_2O_3(s)$	-1669	-1576	51
$Ba^{2+}(g)$	1661		
$Ba^{2+}(aq)$	-538.6	-560.7	12.55
$BaCl_2(s)$	-860	-811	130
$BaO(s)$	-558	-528	70.3
$BeCl_2(s)$	-512		
$BeO(s)$	-611	-582	14
$Br_2(g)$	31.96	3.15	245.4
$Br(g)$	112.1	82.42	175
$Br^-(g)$	-234		
$Br^-(aq)$	-120.9	-102.9	82.4
$C(diamond)$	1.883	2.90	2.38
$Ca^{2+}(g)$	1926		
$Ca^{2+}(aq)$	-543.3	-553	-55.23
$CaCl_2(s)$	-795	-750	114
$CaO(s)$	-635	-604	40
$CCl_4(l)$	-139	-68.6	214
$CH_4(g)$	-74.89	-50.63	186.2
$Cl(g)$	120.9	105.0	165.3
$Cl^-(g)$	-246.4		
$Cl^-(aq)$	-167.4	-131.4	56.48
$CO(g)$	-110.5	-137.3	197.9
$CO_2(g)$	-393.5	-394.4	213.6
$CO_3^{2-}(aq)$	-676.1	-527	-56.9
$CS_2(l)$	87.9	63.6	151
$Cs^+(g)$	461		
$Cs^+(aq)$	-248	-282	133.1
$CsCl(s)$	-447	-419	
$Cs_2O(s)$	-318		
$Cu^{2+}(aq)$	64.4	64.8	-98.74
$F(g)$	790	616	158.7
$F^-(g)$	-270.7		
$F^-(aq)$	-332.6	-278.8	-13.8
$H(g)$	218	203.3	114.6
$H^+(g)$	1535		

	$\Delta H^{\ominus}/\text{kJ mol}^{-1}$	$\Delta G^{\ominus}/\text{kJ mol}^{-1}$	$S^{\ominus}/\text{J mol}^{-1}\,\text{K}^{-1}$
$H^-(g)$	142.7		
$H^+(aq)$	0	0	0
$HBr(g)$	-36.2	-53.2	198.6
$HCl(g)$	-92.47	-95.4	186.6
$HF(g)$	-271.1	-273.2	173.6
$HI(g)$	26.5	2.1	206.5
$H_2O(l)$	-285.8	-237.2	70
$H_2O(g)$	-241.8	-228.4	188.7
$H_2O_2(l)$	-187.4	-120.4	109.6
$I_2(g)$	62.34	19.25	260.5
$I(g)$	106.7	61.09	180.7
$I^-(g)$	-197		
$I^-(aq)$	-55.2	-51.5	111.3
$K^+(g)$	514.6		
$K^+(aq)$	-251.2	-282.4	102.5
$KCl(s)$	-436	-408	82.7
$K_2O(s)$	-362		
$Li^+(g)$	687		
$Li^+(aq)$	-278	-294	14.2
$LiCl(s)$	-402	-377	
$Li_2O(s)$	-596	-560	39.2
$Mg^{2+}(g)$	2348		
$Mg^{2+}(aq)$	-461.8	-456.1	-118
$MgCl_2(s)$	-642	-592	89.5
$MgO(s)$	-602	-570	27
$Na^+(g)$	610		
$Na^+(aq)$	-239.7	-262	60.25
$NaCl(s)$	-411	-384	72.4
$Na_2O(s)$	-416	-377	72.8
$NH_3(g)$	-46.2	-16.65	192.3
$NH_4^+(aq)$	-132.6	-79.5	113.4
$NH_4Cl(s)$	-315.1	-204.6	94.56
$NO(g)$	90.4	86.6	210.5
$NO_2(g)$	33.18	51.3	240
$O_3(g)$	142.7	163.2	238.8
$OH^-(aq)$	-230	-157	-11
$Rb^+(g)$	491		
$Rb^+(aq)$	-246	-282	124.3
$RbCl(s)$	-433	-404	119
$Rb_2O(s)$	30		
$SiO_2(s)$	-912	-856	41.84
$SO_2(g)$	-296.9	-300.4	248.1
$SO_3(g)$	-395.7	-371.1	256.6
$Sr^{2+}(g)$	1790		
$Sr^{2+}(aq)$	-546	-556	-39.3

Some thermodynamic data at 25°C

	$\Delta H^{\ominus}/\text{kJ mol}^{-1}$	$\Delta G^{\ominus}/\text{kJ mol}^{-1}$	$S^{\ominus}/\text{J mol}^{-1}\,\text{K}^{-1}$
$SrCl_2(s)$	-828	-781	120
$SrO(s)$	-590	-560	54.4
$Zn^{2+}(g)$	2783		
$Zn^{2+}(aq)$	-153.1	-147.3	-108.8

Appendix V
Data for the elements

	element	atomic number	relative atomic mass	melting temperature/K	boiling temperature/K	density/g cm^{-3}	ionisation energies/ kJ mol^{-1} 1st	2nd	3rd	metallic radius/nm
	H	1	1.007 97	14.0	20.7	0.071	1311			
noble gases	He	2	4.002 6		4.2	0.126	2372	5250		
	Ne	10	20.179	24.5	27.2	1.21	2080	3964	6150	
	Ar	18	39.948	84	87	1.40	1520	2665	3947	
	Kr	36	83.80	117	121	2.16	1351	2370	3565	
	Xe	54	131.30	161	166	3.52	1169	2050	3100	
	Rn	86	(222)	202	211	4.4	1037			
Group 1A	Li	3	6.939	453	1550	0.53	520	7297	11810	0.156
	Na	11	22.989 8	371	1165	0.97	496	4563	6912	0.191
	K	19	39.102	336	1047	0.86	420	3069	4439	0.238
	Rb	37	85.47	312	961	1.53	403	2650	3849	0.253
	Cs	55	132.705	302	960	1.87	376	2420	3376	0.270
	Fr	87	(223)	303	920					
Group 2A	Be	4	9.012 2	1556		1.85	899	1758	14850	0.112
	Mg	12	24.305	923	1380	1.74	738	1450	7730	0.160
	Ca	20	40.08	1123	1760	1.55	590	1146	4941	0.196
	Sr	38	87.62	1043	1657	2.6	549	1064	4226	0.215
	Ba	56	137.34	998	1910	3.5	502	965		0.224
	Ra	88	(226)	973		5.0	509	977		
Group 3B	B	5	10.811			2.34	800	2427	3658	
	Al	13	26.981 5	933	2730	2.70	578	1816	2745	0.142
	Ga	31	69.72	303	2676	5.91	579	1979	2962	0.135
	In	49	114.82	430	2320	7.31	558	1820	2705	0.167
	Tl	81	204.37	577	1740	11.85	589	1970	2880	0.171
Group 4B	C	6	12.011 15			2.26	1086	2352	4619	
	Si	14	28.086	1680	2628	2.33	786	1577	3228	
	Ge	32	72.59	1210	3100	5.32	760	1537	3301	0.139
	Sn	50	118.69	505	2540	7.30	708	1411	2941	0.158
	Pb	82	207.19	601	2020	11.4	715	1452	3080	0.174
Group 5B	N	7	14.006 7	63	77	0.81	1403	2858	4578	
	P	15	30.973 8	317	554	1.82	1012	1896	2908	
	As	33	74.921 6			5.72	947	1798	2735	
	Sb	51	121.75	903		6.62	834	1590	2443	0.161
	Bi	83	208.80	545	1830	9.8	703	1611	2452	0.182
Group 6B	O	8	15.994	54	90	1.14	1314	3388	5297	
	S	16	32.064	392	718	2.07	999	2258	3381	
	Se	34	78.96	490	958	4.79	941	2075	3096	
	Te	52	127.60	723	1260	6.24	869	1799	2996	
	Po	84	(210)	527	1235	9.32	813			
Group 7B	F	9	18.998 4	53	85		1681	3381 *	6046	
	Cl	17	35.453	172	239	1.56	1255	2287	3849	
	Br	35	79.904	266	332	3.12	1142	2084	3464	
	I	53	126.904 4	387	456	4.94	1010	1845	3138	
	At	85	(210)	575	650		916			
1st transition series	Sc	21	44.956	1670	3000	3.0	631	1235	2389	0.160
	Ti	22	47.9	1950	3550	4.51	656	1309	2650	0.146
	V	23	50.942	2160	3650	6.1	650	1414	2828	0.135
	Cr	24	51.996	2176	2915	7.19	653	1592	3056	0.128
	Mn	25	54.938	1577	2368	7.42	717	1509	3257	0.136
	Fe	26	55.847	1808	3160	7.86	762	1561	2956	0.127
	Co	27	58.933	1760	3150	8.90	758	1644	3231	0.125
	Ni	28	58.71	1728	3100	8.90	737	1752	3489	0.124
	Cu	29	63.54	1357	2868	8.94	745	1958	3545	0.128
	Zn	30	65.37	693	1180	7.13	906	1734	3831	0.137

Data for the elements

ionic radii of M^{x+} cations/nm	ionic radii of M^{x-} anions/nm	standard electrode potential for M^{x+}/M or M/M^{x-} couples/V		Pauling electronegativity	$\Delta H_{a,m}$ (p. 8)/ kJ mol^{-1}	$S^{\ominus}$/ J mol^{-1} K^{-1}	
	0.154$^-$	0.00$^+$		2.1	218	65	
						126	noble gases
						147	
						155	
						164	
						170	
						176	
0.060$^+$		-3.04^+		1.0	161	28	Group 1A
0.095$^+$		-2.71^+		0.9	109	51	
0.133$^+$		-2.92^+		0.8	90	64	
0.148$^+$		-2.93^+		0.8	86	77	
0.169$^+$		-2.95^+		0.7	79	83	
0.176$^+$				0.7	73	94	
0.031^{2+}		-1.85^{2+}		1.5	324	9	Group 2A
0.065^{2+}		-2.37^{2+}		1.2	148	33	
0.099^{2+}		-2.87^{2+}		1.0	178	41	
0.113^{2+}		-2.89^{2+}		1.0	164	52	
0.135^{2+}		-2.90^{2+}		0.9	180	63	
0.143^{2+}				0.9	159	71	
0.020^{3+}		-1.66^{3+}		2	573	6	Group 3B
0.050^{3+}		-0.53^{3+}		1.5	326	28	
0.062^{3+}	0.113$^+$	-0.34^{3+}	-0.25^+	1.6	286	41	
0.093^{3+}	0.132$^+$	-0.72^{3+}	-0.34^+	1.7	243	58	
0.115^{3+}	0.144$^+$			1.8	182	64	
0.015^{4+}		0.260^{4-}		2.5	715	6	Group 4B
0.041^{4+}		0.271^{4-}		1.8	440	19	
0.053^{4+}		0.272^{4-}		1.8	377	31	
0.071^{4+}	0.112^{2+}	0.294^{4-}	-0.136^{2+}	1.8	303	52	
0.084^{4+}	0.120^{2+}	0.01^{4+}	-0.126^{2+}	1.8	196	65	
0.011^{5+}		0.171^{3-}		3.0	473	96	Group 5B
0.034^{5+}		0.212^{3-}		2.1	316	41	
0.047^{5+}		0.222^{3-}		2.0	302	35	
0.062^{5+}		0.245^{3-}		1.9	262	46	
0.074^{3+}	0.120^{3+}			1.9	207	57	
0.009^{6+}		0.140^{2-}		3.5	249	103	Group 6B
0.029^{6+}		0.184^{2-}		2.5	279	32	
0.042^{6+}		0.198^{2-}		2.4	227	42	
0.056^{6+}		0.221^{2-}		2.1	197	50	
				2.0	144	63	
0.007^{7+}		0.136$^-$	2.87$^-$	4.0	79	101	Group 7B
0.026^{7+}		0.181$^-$	1.36$^-$	3.0	121	112	
0.039^{7+}		0.195$^-$	1.07$^-$	2.8	112	76	
0.050^{7+}		0.216$^-$	0.54$^-$	2.5	107	58	
				2.2	92	61	
0.081^{3+}	0.090^{2+}	-2.1^{3+}	-1.63^{2+}	1.3	343	35	1st transition series
0.076^{3+}	0.088^{2+}	-1.2^{3+}	-1.2^{2+}	1.5	471	31	
0.074^{3+}	0.084^{2+}	-0.86^{3+}	-0.91^{2+}	1.6	514	29	
0.069^{3+}	0.080^{2+}	-0.74^{3+}	-1.18^{2+}	1.6	398	24	
0.066^{3+}	0.076^{2+}	-0.28^{3+}	-0.44^{2+}	1.5	281	32	
0.064^{3+}	0.074^{2+}	-0.04^{3+}	-0.28^{2+}	1.8	418	27	
0.063^{3+}	0.072^{2+}	0.4^{3+}	-0.25^{2+}	1.8	425	30	
0.062^{3+}	0.096$^+$			1.8	430	30	
0.069^{2+}		0.34^{2+}	0.52$^+$	1.9	338	33	
0.074^{2+}		-0.763^{2+}		1.6	131	42	

series	element	atomic number	relative atomic mass	melting temperature/K	boiling temperature/K	density/g cm^{-3}	ionisation energies/kJ mol^{-1} 1st	2nd	3rd	metallic radius/nm
2nd transition series	Y	39	88.905	1770	3500	4.40	616	1180	1979	0.181
	Zr	40	91.22	2125	4650	6.53	674	1268	2217	0.160
	Nb	41	92.906	2760	5200	8.55	664	1381	2416	0.147
	Mo	42	95.94	2895	5100	10.2	685	1557	2618	0.140
	Tc	43	(99)	2600	5000	11.5	703	1473	2850	0.135
	Ru	44	101.07	2770	4380	12.2	711	1617	2746	0.133
	Rh	45	102.905	2230	4230	12.4	720	1744	2996	0.134
	Pd	46	106.4	1828	3200	12.0	804	1874	3177	0.137
	Ag	47	107.87	1235	2485	10.5	731	2072	3368	0.144
	Cd	48	112.4	594	1038	8.65	876	1630	3615 *	0.152
3rd transition series	La	57	138.91	1193	3640	6.17	541	1103	1849	0.187
	Hf	72	178.49	2495	5500	13.3	760	1440	2250	0.158
	Ta	73	180.948	3270	5700	16.6	760	1560	2155	0.147
	W	74	183.85	3660	5800	19.3	770	1677	2322	0.141
	Re	75	186.2	3450	5900	21.0	759	1598	2510	0.137
	Os	76	190.2	2970	4700	22.6	840	1640	2410	0.135
	Ir	77	192.2	2720	4620	22.5	900	1640	2602	0.135
	Pt	78	195.09	2062	4100	21.4	870	1791	2745	0.138
	Au	79	196.967	1337	3000	19.3	889	1978	2895	0.144
	Hg	80	200.59	243	630	13.6	1007	1809	3310	0.155
lanthanoids	Ce	58	140.12	1070	3740	6.67	540	1008	1925	0.182
	Pr	59	140.907	1210	3400	6.77	527	1017	2238	0.182
	Nd	60	144.24	1300	3300	7.0	527	1033		0.181
	Pm	61	(147)	1300	3000		540			
	Sm	62	150.35	1350	2200	7.54	540	1079		
	Eu	63	151.96	1100	1710	5.26	540	1084		0.204
	Gd	64	157.25	1580	3000	7.89	547	1155		0.178
	Tb	65	158.924	1630	3100	8.27	594	1113		0.177
	Dy	66	162.50	1680	2900	8.54	577	1125		0.175
	Ho	67	164.93	1730	2900	8.80	573	1138		0.176
	Er	68	167.26	1770	3200	9.05	582	1151		0.173
	Tm	69	168.934	1820	2000	9.33	586	1163		0.174
	Yb	70	173.04	1100	1700	6.98	561	1167		0.193
	Lu	71	174.97	1930	3600	9.84	598	1381		0.173
actinoids	Ac	89	(227)	1320	3470	10	499			
	Th	90	232.038	2000	4500	11.7	587			0.180
	Pa	91	(231)	1500	4300	15.4	568			0.163
	U	92	238.03	1405	4090	19.0	584			0.154
	Np	93	(237)	1410		19.8	597			0.150
	Pu	94	(242)	913	3500		585			0.164
	Am	95	(243)			11.7	578			
	Cm	96	(247)							
	Bk	97	(247)							
	Cf	98	(249)							
	Es	99	(254)							
	Fm	100	(253)							
	Md	101	(256)							
	No	102	(254)							
	Lr	103	(257)							

ionic radii of M^{x+} cations/nm	ionic radii of M^{x-} anions/nm	standard electrode potential for M^{x+}/M or M/M^{x-} couples/V		Pauling electronegativity	$\Delta H_{a,m}$ (p. 8)/ kJ mol^{-1}	$S^{\ominus}$/ J mol^{-1} K^{-1}	
0.093^{3+}		-2.37^{3+}	-1.53^{4+}	1.2	427	44	2nd transition series
				1.4	611	39	
		-1.1^{3+}		1.6	726	37	
		-0.2^{3+}		1.8	658	29	
		0.4^{2+}		1.9	677	34	
		0.45^{2+}	0.6^{2+}	2.2	643	29	
		0.8^{3+}	0.99^{3+}	2.2	556	32	
		1.3^{4+}	0.8^{+}	2.2	378	38	
		1.4^{2+}		1.9	284	43	
0.097^{2+}	0.126^{+}	-0.4^{2+}		1.7	112	52	
0.115^{3+}		-2.52^{3+}		1.1	423	57	3rd transition series
		-1.7^{4+}		1.3	619	46	
				1.5	782	42	
		-0.11^{3+}		1.7	849	33	
		$+0.3^{3+}$		1.9	770	37	
		$+0.9^{2+}$		2.2	791	33	
		$+1.2^{3+}$		2.2	665	36	
		$+1.2^{3+}$		2.2	565	42	
0.137^{+}		$+1.4^{3+}$	$+1.7^{+}$	2.4	366	47	
0.110^{2+}		$+0.85^{2+}$	$+0.79^{+}$	1.9	61	76	
0.111^{3+}		-2.34^{3+}		1.1			lanthanoids
0.109^{3+}				1.1			
0.108^{3+}		-2.25^{3+}		1.1			
0.106^{3+}				1.1			
0.104^{3+}				1.1			
0.103^{3+}				1.1			
0.102^{3+}				1.2			
0.100^{3+}				1.2			
0.099^{3+}				1.2			
0.097^{3+}				1.2			
0.096^{3+}				1.2			
0.095^{3+}				1.2			
0.094^{3+}				1.2			
0.093^{3+}				1.2			
0.118^{3+}				1.1	385	63	actinoids
0.114^{3+}		-1.90^{4+}		1.3	576	53	
0.112^{3+}				1.5	527		
0.111^{3+}		-1.8^{3+}		1.7	490	50	
0.109^{3+}		-1.9^{3+}		1.3			
0.107^{3+}				1.3			
0.106^{3+}				1.3			

Logarithms

	0	1	2	3	4	5	6	7	8	9	differences								
											1	2	3	4	5	6	7	8	9
10	0000	0043	0086	0128	0170	0212	0253	0294	0334	0374	4	8	12	17	21	25	29	33	37
11	0414	0453	0492	0531	0569	0607	0645	0682	0719	0755	4	8	11	15	19	23	26	30	34
12	0792	0828	0864	0899	0934	0969	1004	1038	1072	1106	3	7	10	14	17	21	24	28	31
13	1139	1173	1206	1239	1271	1303	1335	1367	1399	1430	3	6	10	13	16	19	23	26	29
14	1461	1492	1523	1553	1584	1614	1644	1673	1703	1732	3	6	9	12	15	18	21	24	27
15	1761	1790	1818	1847	1875	1903	1931	1959	1987	2014	3	6	8	11	14	17	20	22	25
16	2041	2068	2095	2122	2148	2175	2201	2227	2253	2279	3	5	8	11	13	16	18	21	24
17	2304	2330	2355	2380	2405	2430	2455	2480	2504	2529	2	5	7	10	12	15	17	20	22
18	2553	2577	2601	2625	2648	2672	2695	2718	2742	2765	2	5	7	9	12	14	16	19	21
19	2788	2810	2833	2856	2878	2900	2923	2945	2967	2989	2	4	7	9	11	13	16	18	20
20	3010	3032	3054	3075	3096	3118	3139	3160	3181	3201	2	4	6	8	11	13	15	17	19
21	3222	3243	3263	3284	3304	3324	3345	3365	3385	3404	2	4	6	8	10	12	14	16	18
22	3424	3444	3464	3483	3502	3522	3541	3560	3579	3598	2	4	6	8	10	12	14	15	17
23	3617	3636	3655	3674	3692	3711	3729	3747	3766	3784	2	4	6	7	9	11	13	15	17
24	3802	3820	3838	3856	3874	3892	3909	3927	3945	3962	2	4	5	7	9	11	12	14	16
25	3979	3997	4014	4031	4048	4065	4082	4099	4116	4133	2	3	5	7	9	10	12	14	15
26	4150	4166	4183	4200	4216	4232	4249	4265	4281	4298	2	3	5	7	8	10	11	13	15
27	4314	4330	4346	4362	4378	4393	4409	4425	4440	4456	2	3	5	6	8	9	11	13	14
28	4472	4487	4502	4518	4533	4548	4564	4579	4594	4609	2	3	5	6	8	9	11	12	14
29	4624	4639	4654	4669	4683	4698	4713	4728	4742	4757	1	3	4	6	7	9	10	12	13
30	4771	4786	4800	4814	4829	4843	4857	4871	4886	4900	1	3	4	6	7	9	10	11	13
31	4914	4928	4942	4955	4969	4983	4997	5011	5024	5038	1	3	4	6	7	8	10	11	12
32	5051	5065	5079	5092	5105	5119	5132	5145	5159	5172	1	3	4	5	7	8	9	11	12
33	5185	5198	5211	5224	5237	5250	5263	5276	5289	5302	1	3	4	5	6	8	9	10	12
34	5315	5328	5340	5353	5366	5378	5391	5403	5416	5428	1	3	4	5	6	8	9	10	11
35	5441	5453	5465	5478	5490	5502	5514	5527	5539	5551	1	2	4	5	6	7	9	10	11
36	5563	5575	5587	5599	5611	5623	5635	5647	5658	5670	1	2	4	5	6	7	8	10	11
37	5682	5694	5705	5717	5729	5740	5752	5763	5775	5786	1	2	3	5	6	7	8	9	10
38	5798	5809	5821	5832	5843	5855	5866	5877	5888	5899	1	2	3	5	6	7	8	9	10
39	5911	5922	5933	5944	5955	5966	5977	5988	5999	6010	1	2	3	4	5	7	8	9	10
40	6021	6031	6042	6053	6064	6075	6085	6096	6107	6117	1	2	3	4	5	6	8	9	10
41	6128	6138	6149	6160	6170	6180	6191	6201	6212	6222	1	2	3	4	5	6	7	8	9
42	6232	6243	6253	6263	6274	6284	6294	6304	6314	6325	1	2	3	4	5	6	7	8	9
43	6335	6345	6355	6365	6375	6385	6395	6405	6415	6425	1	2	3	4	5	6	7	8	9
44	6435	6444	6454	6464	6474	6484	6493	6503	6513	6522	1	2	3	4	5	6	7	8	9
45	6532	6542	6551	6561	6571	6580	6590	6599	6609	6618	1	2	3	4	5	6	7	8	9
46	6628	6637	6646	6656	6665	6675	6684	6693	6702	6712	1	2	3	4	5	6	7	8	9
47	6721	6730	6739	6749	6758	6767	6776	6785	6794	6803	1	2	3	4	5	5	6	7	8
48	6812	6821	6830	6839	6848	6857	6866	6875	6884	6893	1	2	3	4	4	5	6	7	8
49	6902	6911	6920	6928	6937	6946	6955	6964	6972	6981	1	2	3	4	4	5	6	7	8
50	6990	6998	7007	7016	7024	7033	7042	7050	7059	7067	1	2	3	3	4	5	6	7	8
51	7076	7084	7093	7101	7110	7118	7126	7135	7143	7152	1	2	3	3	4	5	6	7	8
52	7160	7168	7177	7185	7193	7202	7210	7218	7226	7235	1	2	2	3	4	5	6	7	8
53	7243	7251	7259	7267	7275	7284	7292	7300	7308	7316	1	2	2	3	4	5	6	7	7
54	7324	7332	7340	7348	7356	7364	7372	7380	7388	7396	1	2	2	3	4	5	6	6	7

Logarithms

	0	1	2	3	4	5	6	7	8	9	differences								
											1	2	3	4	5	6	7	8	9
55	7404	7412	7419	7427	7435	7443	7451	7459	7466	7474	1	2	2	3	4	5	5	6	7
56	7482	7490	7497	7505	7513	7520	7528	7536	7543	7551	1	2	2	3	4	5	5	6	7
57	7559	7566	7574	7582	7589	7597	7604	7612	7619	7627	1	2	2	3	4	5	5	6	7
58	7634	7642	7649	7657	7664	7672	7679	7686	7694	7701	1	1	2	3	4	4	5	6	7
59	7709	7716	7723	7731	7738	7745	7752	7760	7767	7774	1	1	2	3	4	4	5	6	7
60	7782	7789	7796	7803	7810	7818	7825	7832	7839	7846	1	1	2	3	4	4	5	6	6
61	7853	7860	7868	7875	7882	7889	7896	7903	7910	7917	1	1	2	3	4	4	5	6	6
62	7924	7931	7938	7945	7952	7959	7966	7973	7980	7987	1	1	2	3	3	4	5	6	6
63	7993	8000	8007	8014	8021	8028	8035	8041	8048	8055	1	1	2	3	3	4	5	5	6
64	8062	8069	8075	8082	8089	8096	8102	8109	8116	8122	1	1	2	3	3	4	5	5	6
65	8129	8136	8142	8149	8156	8162	8169	8176	8182	8189	1	1	2	3	3	4	5	5	6
66	8195	8202	8209	8215	8222	8228	8235	8241	8248	8254	1	1	2	3	3	4	5	5	6
67	8261	8267	8274	8280	8287	8293	8299	8306	8312	8319	1	1	2	3	3	4	5	5	6
68	8325	8331	8338	8344	8351	8357	8363	8370	8376	8382	1	1	2	3	3	4	4	5	6
69	8388	8395	8401	8407	8414	8420	8426	8432	8439	8445	1	1	2	2	3	4	4	5	6
70	8451	8457	8463	8470	8476	8482	8488	8494	8500	8506	1	1	2	2	3	4	4	5	6
71	8513	8519	8525	8531	8537	8543	8549	8555	8561	8567	1	1	2	2	3	4	4	5	5
72	8573	8579	8585	8591	8597	8603	8609	8615	8621	8627	1	1	2	2	3	4	4	5	5
73	8633	8639	8645	8651	8657	8663	8669	8675	8681	8686	1	1	2	2	3	4	4	5	5
74	8692	8698	8704	8710	8716	8722	8727	8733	8739	8745	1	1	2	2	3	4	4	5	5
75	8751	8756	8762	8768	8774	8779	8785	8791	8797	8802	1	1	2	2	3	3	4	5	5
76	8808	8814	8820	8825	8831	8837	8842	8848	8854	8859	1	1	2	2	3	3	4	5	5
77	8865	8871	8876	8882	8887	8893	8899	8904	8910	8915	1	1	2	2	3	3	4	4	5
78	8921	8927	8932	8938	8943	8949	8954	8960	8965	8971	1	1	2	2	3	3	4	4	5
79	8976	8982	8987	8993	8998	9004	9009	9015	9020	9025	1	1	2	2	3	3	4	4	5
80	9031	9036	9042	9047	9053	9058	9063	9069	9074	9079	1	1	2	2	3	3	4	4	5
81	9085	9090	9096	9101	9106	9112	9117	9122	9128	9133	1	1	2	2	3	3	4	4	5
82	9138	9143	9149	9154	9159	9165	9170	9175	9180	9186	1	1	2	2	3	3	4	4	5
83	9191	9196	9201	9206	9212	9217	9222	9227	9232	9238	1	1	2	2	3	3	4	4	5
84	9243	9248	9253	9258	9263	9269	9274	9279	9284	9289	1	1	2	2	3	3	4	4	5
85	9294	9299	9304	9309	9315	9320	9325	9330	9335	9340	1	1	2	2	3	3	4	4	5
86	9345	9350	9355	9360	9365	9370	9375	9380	9385	9390	1	1	2	2	3	3	4	4	5
87	9395	9400	9405	9410	9415	9420	9425	9430	9435	9440	0	1	1	2	2	3	3	4	4
88	9445	9450	9455	9460	9465	9469	9474	9479	9484	9489	0	1	1	2	2	3	3	4	4
89	9494	9499	9504	9509	9513	9518	9523	9528	9533	9538	0	1	1	2	2	3	3	4	4
90	9542	9547	9552	9557	9562	9566	9571	9576	9581	9586	0	1	1	2	2	3	3	4	4
91	9590	9595	9600	9605	9609	9614	9619	9624	9628	9633	0	1	1	2	2	3	3	4	4
92	9638	9643	9647	9652	9657	9661	9666	9671	9675	9680	0	1	1	2	2	3	3	4	4
93	9685	9689	9694	9699	9703	9708	9713	9717	9722	9727	0	1	1	2	2	3	3	4	4
94	9731	9736	9741	9745	9750	9754	9759	9763	9768	9773	0	1	1	2	2	3	3	4	4
95	9777	9782	9786	9791	9795	9800	9805	9809	9814	9818	0	1	1	2	2	3	3	4	4
96	9823	9827	9832	9836	9841	9845	9850	9854	9859	9863	0	1	1	2	2	3	3	4	4
97	9868	9872	9877	9881	9886	9890	9894	9899	9903	9908	0	1	1	2	2	3	3	4	4
98	9912	9917	9921	9926	9930	9934	9939	9943	9948	9952	0	1	1	2	2	3	3	4	4
99	9956	9961	9965	9969	9974	9978	9983	9987	9991	9996	0	1	1	2	2	3	3	3	4

Antilogarithms

	0	1	2	3	4	5	6	7	8	9	differences 1	2	3	4	5	6	7	8	9
.00	1000	1002	1005	1007	1009	1012	1014	1016	1019	1021	0	0	1	1	1	1	2	2	2
.01	1023	1026	1028	1030	1033	1035	1038	1040	1042	1045	0	0	1	1	1	1	2	2	2
.02	1047	1050	1052	1054	1057	1059	1062	1064	1067	1069	0	0	1	1	1	1	2	2	2
.03	1072	1074	1076	1079	1081	1084	1086	1089	1091	1094	0	0	1	1	1	1	2	2	2
.04	1096	1099	1102	1104	1107	1109	1112	1114	1117	1119	0	1	1	1	1	2	2	2	2
.05	1122	1125	1127	1130	1132	1135	1138	1140	1143	1146	0	1	1	1	1	2	2	2	2
.06	1148	1151	1153	1156	1159	1161	1164	1167	1169	1172	0	1	1	1	1	2	2	2	2
.07	1175	1178	1180	1183	1186	1189	1191	1194	1197	1199	0	1	1	1	1	2	2	2	2
.08	1202	1205	1208	1211	1213	1216	1219	1222	1225	1227	0	1	1	1	1	2	2	2	2
.09	1230	1233	1236	1239	1242	1245	1247	1250	1253	1256	0	1	1	1	1	2	2	2	3
.10	1259	1262	1265	1268	1271	1274	1276	1279	1282	1285	0	1	1	1	1	2	2	2	3
.11	1288	1291	1294	1297	1300	1303	1306	1309	1312	1315	0	1	1	1	2	2	2	2	3
.12	1318	1321	1324	1327	1330	1334	1337	1340	1343	1346	0	1	1	1	2	2	2	2	3
.13	1349	1352	1355	1358	1361	1365	1368	1371	1374	1377	0	1	1	1	2	2	2	2	3
.14	1380	1384	1387	1390	1393	1396	1400	1403	1406	1409	0	1	1	1	2	2	2	3	3
.15	1413	1416	1419	1422	1426	1429	1432	1435	1439	1442	0	1	1	1	2	2	2	3	3
.16	1445	1449	1452	1455	1459	1462	1466	1469	1472	1476	0	1	1	1	2	2	2	3	3
.17	1479	1483	1486	1489	1493	1496	1500	1503	1507	1510	0	1	1	1	2	2	2	3	3
.18	1514	1517	1521	1524	1528	1531	1535	1538	1542	1545	0	1	1	1	2	2	2	3	3
.19	1549	1552	1556	1560	1563	1567	1570	1574	1578	1581	0	1	1	1	2	2	3	3	3
.20	1585	1589	1592	1596	1600	1603	1607	1611	1614	1618	0	1	1	1	2	2	3	3	3
.21	1622	1626	1629	1633	1637	1641	1644	1648	1652	1656	0	1	1	2	2	2	3	3	3
.22	1660	1663	1667	1671	1675	1679	1683	1687	1690	1694	0	1	1	2	2	2	3	3	3
.23	1698	1702	1706	1710	1714	1718	1722	1726	1730	1734	0	1	1	2	2	2	3	3	3
.24	1738	1742	1746	1750	1754	1758	1762	1766	1770	1774	0	1	1	2	2	2	3	3	4
.25	1778	1782	1786	1791	1795	1799	1803	1807	1811	1816	0	1	1	2	2	2	3	3	4
.26	1820	1824	1828	1832	1837	1841	1845	1849	1854	1858	0	1	1	2	2	3	3	3	4
.27	1862	1866	1871	1875	1879	1884	1888	1892	1897	1901	0	1	1	2	2	3	3	3	4
.28	1905	1910	1914	1919	1923	1928	1932	1936	1941	1945	0	1	1	2	2	3	3	4	4
.29	1950	1954	1959	1963	1968	1972	1977	1982	1986	1991	0	1	1	2	2	3	3	4	4
.30	1995	2000	2004	2009	2014	2018	2023	2028	2032	2037	0	1	1	2	2	3	3	4	4
.31	2042	2046	2051	2056	2061	2065	2070	2075	2080	2084	0	1	1	2	2	3	3	4	4
.32	2089	2094	2099	2104	2109	2113	2118	2123	2128	2133	0	1	1	2	2	3	3	4	4
.33	2138	2143	2148	2153	2158	2163	2168	2173	2178	2183	0	1	1	2	2	3	3	4	4
.34	2188	2193	2198	2203	2208	2213	2218	2223	2228	2234	1	1	2	2	3	3	4	4	5
.35	2239	2244	2249	2254	2259	2265	2270	2275	2280	2286	1	1	2	2	3	3	4	4	5
.36	2291	2296	2301	2307	2312	2317	2323	2328	2333	2339	1	1	2	2	3	3	4	4	5
.37	2344	2350	2355	2360	2366	2371	2377	2382	2388	2393	1	1	2	2	3	3	4	4	5
.38	2399	2404	2410	2415	2421	2427	2432	2438	2443	2449	1	1	2	2	3	3	4	4	5
.39	2455	2460	2466	2472	2477	2483	2489	2495	2500	2506	1	1	2	2	3	3	4	5	5
.40	2512	2518	2523	2529	2535	2541	2547	2553	2559	2564	1	1	2	2	3	3	4	5	5
.41	2570	2576	2582	2588	2594	2600	2606	2612	2618	2624	1	1	2	2	3	4	4	5	5
.42	2630	2636	2642	2649	2655	2661	2667	2673	2679	2685	1	1	2	2	3	4	4	5	5
.43	2692	2698	2704	2710	2716	2723	2729	2735	2742	2748	1	1	2	3	3	4	4	5	6
.44	2754	2761	2767	2773	2780	2786	2793	2799	2805	2812	1	1	2	3	3	4	4	5	6
.45	2818	2825	2831	2838	2844	2851	2858	2864	2871	2877	1	1	2	3	3	4	5	5	6
.46	2884	2891	2897	2904	2911	2917	2924	2931	2938	2944	1	1	2	3	3	4	5	5	6
.47	2951	2958	2965	2972	2979	2985	2992	2999	3006	3013	1	1	2	3	3	4	5	5	6
.48	3020	3027	3034	3041	3048	3055	3062	3069	3076	3083	1	1	2	3	4	4	5	6	6
.49	3090	3097	3105	3112	3119	3126	3133	3141	3148	3155	1	1	2	3	4	4	5	6	6

Antilogarithms

	0	1	2	3	4	5	6	7	8	9	differences								
											1	2	3	4	5	6	7	8	9
.50	3162	3170	3177	3184	3192	3199	3206	3214	3221	3228	1	1	2	3	4	4	5	6	7
.51	3236	3243	3251	3258	3266	3273	3281	3289	3296	3304	1	2	2	3	4	5	5	6	7
.52	3311	3319	3327	3334	3342	3350	3357	3365	3373	3381	1	2	2	3	4	5	5	6	7
.53	3388	3396	3404	3412	3420	3428	3436	3443	3451	3459	1	2	2	3	4	5	6	6	7
.54	3467	3475	3483	3491	3499	3508	3516	3524	3532	3540	1	2	2	3	4	5	6	6	7
.55	3548	3556	3565	3573	3581	3589	3597	3606	3614	3622	1	2	2	3	4	5	6	7	7
.56	3631	3639	3648	3656	3664	3673	3681	3690	3698	3707	1	2	3	3	4	5	6	7	8
.57	3715	3724	3733	3741	3750	3758	3767	3776	3784	3793	1	2	3	3	4	5	6	7	8
.58	3802	3811	3819	3828	3837	3846	3855	3864	3873	3882	1	2	3	4	4	5	6	7	8
.59	3890	3899	3908	3917	3926	3936	3945	3954	3963	3972	1	2	3	4	5	5	6	7	8
.60	3981	3990	3999	4009	4018	4027	4036	4046	4055	4064	1	2	3	4	5	6	6	7	8
.61	4074	4083	4093	4102	4111	4121	4130	4140	4150	4159	1	2	3	4	5	6	7	8	9
.62	4169	4178	4188	4198	4207	4217	4227	4236	4246	4256	1	2	3	4	5	6	7	8	9
.63	4266	4276	4285	4295	4305	4315	4325	4335	4345	4355	1	2	3	4	5	6	7	8	9
.64	4365	4375	4385	4395	4406	4416	4426	4436	4446	4457	1	2	3	4	5	6	7	8	9
.65	4467	4477	4487	4498	4508	4519	4529	4539	4550	4560	1	2	3	4	5	6	7	8	9
.66	4571	4581	4592	4603	4613	4624	4634	4645	4656	4667	1	2	3	4	5	6	7	9	10
.67	4677	4688	4699	4710	4721	4732	4742	4753	4764	4775	1	2	3	4	5	7	8	9	10
.68	4786	4797	4808	4819	4831	4842	4853	4864	4875	4887	1	2	3	4	6	7	8	9	10
.69	4898	4909	4920	4932	4943	4955	4966	4977	4989	5000	1	2	3	5	6	7	8	9	10
.70	5012	5023	5035	5047	5058	5070	5082	5093	5105	5117	1	2	4	5	6	7	8	9	11
.71	5129	5140	5152	5164	5176	5188	5200	5212	5224	5236	1	2	4	5	6	7	8	10	11
.72	5248	5260	5272	5284	5297	5309	5321	5333	5346	5358	1	2	4	5	6	7	9	10	11
.73	5370	5383	5395	5408	5420	5433	5445	5458	5470	5483	1	3	4	5	6	8	9	10	11
.74	5495	5508	5521	5534	5546	5559	5572	5585	5598	5610	1	3	4	5	6	8	9	10	12
.75	5623	5636	5649	5662	5675	5689	5702	5715	5728	5741	1	3	4	5	7	8	9	10	12
.76	5754	5768	5781	5794	5808	5821	5834	5848	5861	5875	1	3	4	5	7	8	9	11	12
.77	5888	5902	5916	5929	5943	5957	5970	5984	5998	6012	1	3	4	5	7	8	10	11	12
.78	6026	6039	6053	6067	6081	6095	6109	6124	6138	6152	1	3	4	6	7	8	10	11	13
.79	6166	6180	6194	6209	6223	6237	6252	6266	6281	6295	1	3	4	6	7	9	10	11	13
.80	6310	6324	6339	6353	6368	6383	6397	6412	6427	6442	1	3	4	6	7	9	10	12	13
.81	6457	6471	6486	6501	6516	6531	6546	6561	6577	6592	2	3	5	6	8	9	11	12	14
.82	6607	6622	6637	6653	6668	6683	6699	6714	6730	6745	2	3	5	6	8	9	11	12	14
.83	6761	6776	6792	6808	6823	6839	6855	6871	6887	6902	2	3	5	6	8	9	11	13	14
.84	6918	6934	6950	6966	6982	6998	7015	7031	7047	7063	2	3	5	6	8	10	11	13	15
.85	7079	7096	7112	7129	7145	7161	7178	7194	7211	7228	2	3	5	7	8	10	12	13	15
.86	7244	7261	7278	7295	7311	7328	7345	7362	7379	7396	2	3	5	7	8	10	12	13	15
.87	7413	7430	7447	7464	7482	7499	7516	7534	7551	7568	2	3	5	7	9	10	12	14	16
.88	7586	7603	7621	7638	7656	7674	7691	7709	7727	7745	2	4	5	7	9	11	12	14	16
.89	7762	7780	7798	7816	7834	7852	7870	7889	7907	7925	2	4	5	7	9	11	13	14	16
.90	7943	7962	7980	7998	8017	8035	8054	8072	8091	8110	2	4	6	7	9	11	13	15	17
.91	8128	8147	8166	8185	8204	8222	8241	8260	8279	8299	2	4	6	8	9	11	13	15	17
.92	8318	8337	8356	8375	8395	8414	8433	8453	8472	8492	2	4	6	8	10	12	14	15	17
.93	8511	8531	8551	8570	8590	8610	8630	8650	8670	8690	2	4	6	8	10	12	14	16	18
.94	8710	8730	8750	8770	8790	8810	8831	8851	8872	8892	2	4	6	8	10	12	14	16	18
.95	8913	8933	8954	8974	8995	9016	9036	9057	9078	9099	2	4	6	8	10	12	15	17	19
.96	9120	9141	9162	9183	9204	9226	9247	9268	9290	9311	2	4	6	8	11	13	15	17	19
.97	9333	9354	9376	9397	9419	9441	9462	9484	9506	9528	2	4	7	9	11	13	15	17	20
.98	9550	9572	9594	9616	9638	9661	9683	9705	9727	9750	2	4	7	9	11	13	16	18	20
.99	9772	9795	9817	9840	9863	9886	9908	9931	9954	9977	2	5	7	9	11	14	16	18	20

Answers to questions

Chapter 4 (p. 58)   **8** (a) $1.18 \times 10^{-23}\,cm^3$ (b) $1.28 \times 10^{-8}\,cm$

Chapter 5 (p. 68)   **9** $-390\,kJ\,mol^{-1}$
11 $-65.5\,kJ\,mol^{-1}$
15 (a) $-2824\,kJ\,mol^{-1}$ (b) $-1489\,kJ\,mol^{-1}$
     (c) $-1564\,kJ\,mol^{-1}$ (d) $-627\,kJ\,mol^{-1}$

Chapter 6 (p. 78)   **4** (a) 1.55×10^{15} (b) 10.54×10^4 (c) 3.16×10^{41}
5 2.97
6 (a) $-80\,kJ\,mol^{-1}$ (b) 10^{14}
8 $-80\,kJ\,mol^{-1}$; 10^{14}
10 1.47×10^{-5}
11 $-9.83\,kJ\,mol^{-1}$

Chapter 8 (p. 117)   **9** $700\,°C$
11 $704\,°C$

Chapter 11 (p. 152) **15** $11.2\,dm^3$

Chapter 12 (p. 163)   **1** $743\,°C$
4 $851\,kJ\,mol^{-1}$

Chapter 16 (p. 211) **12** $1074.5\,kJ\,mol^{-1}$; $1080\,kJ\,mol^{-1}$

Chapter 17 (p. 220) **11** $1521\,°C$

Chapter 18 (p. 228)   **1** 96 per cent

Chapter 19 (p. 247) **15** $-642.6\,kJ\,mol^{-1}$

Chapter 20 (p. 261) **18** $-104.15\,kJ\,mol^{-1}$

Chapter 21 (p. 280) **15** $301.6\,kJ\,mol^{-1}$
16 $463.4\,kJ\,mol^{-1}$; $140.2\,kJ\,mol^{-1}$

Index